THEORY OF RADIATION PROCESSES IN METAL SOLID SOLUTIONS

THEORY OF RADIATION PROCESSES IN METAL SOLID SOLUTIONS

Yu. V. Trushin

Nova Science Publishers Inc.

Art Director: Maria Ester Hawrys
Assistant Director: Elenor Kallberg
Graphics: Denise Dieterich and Kerri Pfister
Manuscript Coordinator: Roseann Pena
Book Production:Tammy Sauter, and Benjamin Fung
Circulation: Irene Kwartiroff and Annette Hellinger

Library of Congress Cataloging-in-Publication Data

Theory of radiation processes in metal solid solutions/
Trushin, Yu. V.
p. cm.
Includes bibliographical references and index.
ISBN 1-56072-260-6 :
1.Metals. --Effect of radiation on. 2. Metals--Defects.
I. Title.
TA460.T785 1995 95-42713
620. 1'628--dc20 CIP

6080 Jericho Turnpike, Suite 207
Commack, New York 11725
Tele. 516-499-3103 Fax 516-499-3146
E Mail Novasci1@aol.com

Printed in the United States of America

To commemoration of my teacher

Profesor A.N. Orlov

CONTENTS

INTRODUCTION

Development of modern production processes, improvement of available power equipment and design of thermonuclear apparatus require that structural materials would work their resource being exposed to diverse outside factors such as a fast-particle irradiation.

The last decades investigations in the field of radiation physics of solids and radiation metal science rapidly progress. Since the radiation experiments are very expensive, development of the theory as a tool to predict changes in properties and modify the materials was provided with a power impetus. Light has been thrown upon some questions of radiation process theory in a number of monographs and collected works (for example in Refs. 1-23).

The importance of materials with specially introduced impurities is hard to overestimate. In particular, the structural materials which are manufactured and used by modern energetics and production engineering, are alloys doped by various impurities. Especial importance is attached to the materials with the structure of solid solutions which are decomposed in the working process, their behavior being changed by an intensive and high-energy irradiation. To investigate and use the regularities of such changes is one of the problems of modern radiation physics of solids and radiation metal science.

Influence of the impurity atoms introduced into a matrix crystal manifests itself at all stages of radiation damage, ranging from the dynamic-stage elementary act to the physical processes which lead to occurring macro changes in the materials irradiated (swelling, hardening, creep, etc.). There is no consistent physical theory yet, and the material of this book, where, basing on the available original achievements, some concepts are developed, and a theory of radiation processes in biatomic materials is constructed, may be envisaged as an introduction to the theory of such processes for the materials used in modern technics.

In the solid solutions at various concentrations as well as various stages of their decomposition, the radiation processes proceed as influenced both by the individual impurities and the formation of secondary-phase precipitates. In its turn, the mechanical properties of solid-solution alloys, if changed under irradiation, result from the foresaid features and the defect structure thus forming.

By the present time we observe extensive development in the field of radiation physics of metal materials (for example Refs. 24-77). However, most theoretical results have been obtained for pure crystals. The most developments, with few exceptions, fail to give an experimentally authentic physical interpretation of the radiation processes and the regularities of changing the material properties. As a result, there is need for time-consuming theoretical investigations of radiation defect kinetics to be performed in alloys and, in the first place, solid solutions. Individual theoretical considerations of specific compositions were run to clarify, for instance, the importance of phase instability [78-82], the effect of irradiation upon phase transitions [83-87], the effect of impurities [88], and so on.

However, there are different factors to affect the radiation processes at different stages of solid-solution decomposition or at different impurity concentrations (diluted or concentrated solid solutions) [7,8,14,89-100]. In this connection, the book give a theoretical consideration to the impurity contribution to fundamental physical processes occurring in the irradiated metal solid solutions.

While the solid-solution alloys were used as structural materials basically important to nuclear engineering from the very beginning of its manufacturing, the physical (kinetic) investigations of such systems under hard irradiation are only at initial stages of their progressing. The material used in this book permits to discuss a sequence of radiation processes calculated for the solid solutions being at their various decomposition stages.

The first two chapters are of a great importance. They gather the facts essential to he radiation-process theory to be built up. Chapter 1 discusses physically fundamental mechanisms of radiation damage in pure metals. Chapter 2 which is based on the experimentally known dependencies of impurity crystals and solid solutions and also on the theoretical considerations of pure metals, formulates a self-consistent system of equations which describes basic mechanisms of the structure evolution in solid solutions under irradiation.

Chapter 3 is devoted to describing cascades of moving atoms, and accompanying processes in biatomic materials. Chapter 4 considers point-defect kinetics in solid solutions under irradiation. Attention is drawn to the necessity to account for the peculiarities of an inhomogeneous defect distribution close to sinks existing in the materials and for the coherent precipitates contributing to the decomposition process.

As fast as the solid-solution decomposition proceeds, the secondary-phase precipitates are setting apart and a boundary between the precipitates and matrix is forming. Chapter 5 describes the calculations of point-defect concentration profiles and discusses mechanisms of this stage of radiation-induced depletion of solid solution.

Important characteristics of structural sinks of the materials (precipitate, dislocations, small-range dislocation loops) are their adsorbing abilities. The Chapter 6 exemplified the calculations of the point-defect calculations of the point-defect rates and efficiency of such sinks.

Isolation of the secondary-phase precipitates exposed to irradiation is going to evolve with the mechanisms of solid-solution depletion. Their sizes are time-dependent. Therefore, the Chapter 7 deals with the calculation of radius-dependent spherical precipitates.

The solid-solution decomposition, the peculiarities of forming their structure under irradiation affect the macroscopic properties of such a material. Chapter 8 follows how the solid-solution decomposition is related to the radiation swelling being one of such properties. Finally, the Conclusion discusses a possibility to approach to the description of radiation changes in the irradiated materials in a single manner.

It is natural that the author intended to cover as large as possible number of presently realized developments. However, a book size and a selected topical framework not always permitted to mention many interesting results. The author wanted an independent and logical exposition to make the book not only helpful to specialists but students and researchers working in adjacent branches of physics. What result has been obtained, this is to be judged by a reader.

At the very beginning of my work on the book I have got an inestimable support and encouragement from my teacher Prof. A.N.Orlov to whose commemoration this book is dedicated.

I am very grateful to many physicists whose stimulating discussions have helped me to complete this book. Primarily, I should like to express my gratitude to my colleagues from A.F.Ioffe Physical-Technical Institute of Academy of Sciences and the other institutes. It is impossible to mention all but I am especially grateful to: Profs. V.Kirsanov (Polytechnic Institute, Tver'), A.Suvorov (Institute of Theoretical and Experimental Physics, Moscow), S.Saralidze (Institute of Physics, Tbilisi), A.Parshin (St.Petersburg Technical University), A.Bakaj, J.Nekhludov, V.Slesov (Physical-Technical Institute, Kharkov), H.Weber (Atominstitut der sterreichischen Universitten, Wien), H.Wollenberger, J.Biersack (Hahn-Meitner Institut, Berlin), K.Ehrlich (Kernforschungszentrum, Karlsruhe), W.Pompe (Zentralinstitut der Festkrperphysik und Metallkunde, Dresden), D.Karpusov (Institute of Electronics, Sofia), Drs. H.Wiedersich

(Argonne National Laboratory), C.Abromeit, V.Naundorf, N.Wanderka (Hahn-Meitner Institut, Berlin), M.Jenkins (University of Oxford), L.Mansur (Oak Ridge National Laboratory), K.Anderko (Kernforschungszentrum, Karlsruhe), F.Dworschak, H.Trinkaus, P.Jung (Institut fr Festkrperforschung, Jlich).

I especially appreciate A.Plisko's help in preparing a Russian version of the manuscript and M.Bryzhina's assistance in translating and preparing the book in English. Finally, I am greatly thankful to my wife and daughter for their support during all the period of working on the book which has happened to be more longer than expected.

Leningrad - Dresden - Wien - Berlin -
St.Petersburg, 1991

Yu.V.Trushin

CHAPTER 1

RADIATION DAMAGE OF METALS

1.1. General Concepts of Radiation Damage in Materials

Radiation damage of materials and their all technical impact are presented by a wide range of physical processes including elastic and inelastic interactions of bombarding radiation particles with the atoms (molecules) of a solid, cascades of atomic collisions, diffusion of atoms presenting in every way the processes of defect structure evolution, diverse material deformations, and so on, which causes their properties to change (Fig. 1.1).

The radiation damage of materials has pioneered in attributing considerable interest as the first atomic reactors were in the course of their advent. Swelling of fissioned materials, deformation of graphite blocks set the originators of atomic reactors making effects and means to establish the reasons for such phenomena. E. Wiegner was one of the first who suggested [97] that the high-energy fission neutrons should displace the atoms of crystalline solids from their equilibrium positions and that it is these displacements that will lead to serious technological effects.

An emphasis in the theoretical research of radiation processes in metals was made on the investigation of pure (without impurities) materials. A wide class of alloys and steels was experimentally examined, but kinetic regularities of the defect structure formed have been known awfully scarcely.

Based on the investigation available (see, e.g., Refs. 1-96), a general physical pattern of radiation damage in metals can be concerned as follows (Fig. 1.1).

Interacting with the metal, the irradiating particles displace the atoms from the crystal lattice sites and transfer them a portion of their energy. If this energy exceeds the threshold displacement energy ε_d which is characteristic of the given substance, then the atom proves to be knocked on from the lattice site. Such atoms are called primarily knocked-on atoms (PKA). In further collisions of PKA with the lattice atoms, there appear secondary particles giving rise, in their turn, to tertiary ones, and so on. So a cascade of moving atoms forms. In the region of irradiated solid solution which the cascade has passed through (a cascade zone), there occurs a large amount (up to some thousand) of Frenkel pairs (vacancies and interstitials) and takes place a local heating of the material. As a result of passing the cascade there occurs, around the sites of irradiating particle collision with the atoms of the material, a complexly composed region consisting of the depleted and concentrated zones. Taken together, they form a so-called embryo damage of the material. A modern experimental procedure of studying the defects in crystals, and primary autoionic microscopy, permits to observe not only aggregates of structure defects, but some individual interstitials and vacancies (see, e.g., Ref. 36).

Using the present-day theoretical and experimental data and computations as the base, the pattern of radiation cascade development since the instant of occurring ε_0-energy PKA to that of forming characteristic structures in this region of a solid, can be presented as a diagram (Fig. 1.2) (see Refs. 11, 98-100). Some elements of the diagram presented have been theoretically and partially experimentally understood. For example, some models have been constructed and autoionic microscopic studies of the depleted and concentrated zones [36, 47, 100-103] and thermal spikes [104] performed, the conditions of forming and propagating the channelons, dynamic crowdions and focusons understood [23, 104-113], the defect distributions in the cascade region, average number of displaced atoms, average dimensions of the cascade region obtained [36, 42, 47, 98, 99, 102, 120-130], the questions of impurity influence on the development of moving atom cascades [36, 131-140] partially investigated.

At high PKA energies (10^4-10^5eV), the cascade region forming in the crystal, can be splitted into the subregions, that is, subcascades [103, 124, 141-143]. This, naturally, even more complicates the structure of the material irradiated. However, the process of splitting into subcascades essentially depends on the PKA deceleration by the crystal electrons and on the crystal temperature [144].

Thus, the region of embryo damage remaining in the crystal which the moving-atom cascade has passed, is a complicated structural segregate whose further evolution leads to the formation of more stable defect structures in the material irradiated.

If the irradiation fluencies are large, the probability for the cascade to pass through the already damaged crystal areas increases. Each after-cascade defect distribution in dimensions and in space which is formed in the time of the order of 10^{-13}-10^{-14}s tends, in the course of time, to equilibrate with the surrounding crystal lattice. But on irradiation by the high-density particle fluxes, a new cascade of the atom - atom collisions can pass through this unsteady defect state, its evolution standing in the way. Superimposed on all this is the influence of the lattice thermal oscillations, mechanical stresses and other external effects.

Simultaneously with the point defect generation, the radiation energy is put to ionization and excitation of the electrons of a regular crystal lattice, as well as to an increase in the thermal energy of conduction electrons. There takes place a complex rearrangement in the electron subsystem of the crystal, and as a result some additional mechanisms of point defect formation are coming into force.

Furthermore, a strong excitation of the electron subsystem can lead to the initiation of a sharp local temperature increase and, naturally, to the effects analogous to the local heating. The future of the point defects essentially depends on some external factors: temperature, applied stresses. As a result, the vacancies and interstitials initiating in the crystal, begin to migrate over it. In this case, they either outgo to the sinks (dislocations, grain boundaries, secondary-phase precipitates) or unite into clusters, thus forming voids and dislocation - vacancy and interstitial loops [2, 13, 22, 25, 28-31, 33, 37, 39, 40, 145, 146]. Opposite defects recombine on their rendezvous, not only in the form of single defects unbounded in clusters, but find their way to the opposite defects connected into clusters. With the computerized modeling (see, for example, Refs. 148-150) one has managed to derive the volume and shape of the spontaneous recombination zone for both a Frenkel pair and a dislocation. Usually the irradiation gives rise to a sufficiently large amount of unsteady Frenkel pairs, i.e. such vacancies and interstitials which are at the distance less than the size of the spontaneous recombination zone. Now such unsteady pairs disappear in the time of the order of 10^{-11}- 10^{-12}s. Their concentration is two or more orders as high as that of stable Frenkel pairs. Creation and later "collapsing" of the unsteady pairs lead to the longitudinal waves occurring in the crystal and damping out with the distance - first extension and then compression waves being able to additionally transfer the available point defects. There appears a so called "radiation buffeting" of the crystal [44,151,152]. The radiation buffeting can be attributed to the subthreshold phenomena when the energy of irradiating particles is not enough to form some steady defects, but nevertheless the radiation effects do take place.

After the dynamic stage, further thermal rearrangements based on the diffusion processes markedly change the dimensions, shape and distribution of the radiation defects, which is expressed in changing their macroscopic properties under irradiation.

Pass on to the definition of the investigations at various stages of developing the radiation crystal damage.

1.2. Dynamic Processes in Crystals

By dynamic processes are meant in the radiation physics of materials some cascades of moving atoms and orientation effects following the cascade development.

1.2.1. Basic Concepts

A general state of investigations in studying the cascade processes in metals has been reviewed in some survey papers [11,93,98-100,153]. We shall enlarge on general regularities of cascade description in pure metals with the aim to have in future a possibility to consider the cascades in impure crystals.

A primarily knocked-on atom or another atom moving with the kinetic energy ε in the direction Ω interacts with the crystal lattice atoms with the probability determined by the energy transfer cross section $\sigma(\varepsilon,\varepsilon'')$. In this case the atom changes the direction of motion to Ω' and the energy to ε'. It becomes a scattered atom (*s*) and transfers to atom displaced from a lattice site (*d*) the energy ε'' and a momentum in the direction Ω''. If the energy of moving atom is less than ε_d, its progressive motion is stopped and it comes to rest forming an interstitial configuration (e.g. dumbbell one) with one of the lattice atoms. When all the moving atoms will have been stopped, the first stage of cascade development , the atomic collision stage comes to an end. It may last the time equal to a fraction of picosecond.

If $\varepsilon''>\varepsilon_d$, the lattice atom is displaced from its site and can, in its turn, knock on some secondary atoms. Remained in place of the knocked-on atoms are vacancies. If $\varepsilon''<\varepsilon_d$, excited are only oscillations of the atom being caught in, i.e. the energy ε'' transforms to heat. The moving atoms also loose their energy by the excitation of the electron subsystem of the crystal. However, at the energies lower than 10-100 keV, which are of principal interest for the cascade formation, these losses are unessential.

The atomic motion in a crystal can be oriented (channelons, focusons, dynamic crowdions) or nonoriented one (see Fig.1.2). The scattered atom can move nonorientedly with the probability $P^s(\varepsilon,\Omega)$ or be channeled with the probability $P^s_C(\varepsilon,\Omega)$ so that

$$P^s(\varepsilon,\Omega) + P^s_C(\varepsilon,\Omega) = 1. \tag{1.1}$$

The displaced atom can move nonorientedly with the probability $P^d(\varepsilon,\Omega)$, be channeled ($P^d_C(\varepsilon,\Omega)$), transferred as a focuson ($P^d_F(\varepsilon,\Omega)$) or a dynamic crowdion ($P^d_D(\varepsilon,\Omega)$):

$$P^d(\varepsilon,\Omega) + P^d_C(\varepsilon,\Omega) + P^d_F(\varepsilon,\Omega) + P^d_D(\varepsilon,\Omega) = 1. \tag{1.2}$$

The atoms moving nonorientedly produce, for the most part, the radiation point defects. The function of the atoms moving orientedly is to remove part of atoms and energy from the cascade process, inasmuch as part of the energy is not lost to the displacements (knocking-on) the atoms from the lattice sites.

The analytical theory of cascade processes (see,for example, Refs. 120,

121, 154-168) is based on the investigation of the kinetic equation describing the vacancy and interstitial formation and their further evolution. For this purpose, one considers a cascade ensemble formed by the PKA with the same energy ε_0 at the origin of the coordinates ($\mathbf{r}$=0), but outgoing in different directions Ω_0. A distribution time-dependent function is inserted for the k-type moving atoms, $\Phi_k(\mathbf{r},\varepsilon,\Omega,t)$, of the coordinates $\mathbf{r}$, energy ε, directions Ω. It is related to the density of moving atom flux $\Psi_k(\mathbf{r},\varepsilon,\Omega,t)$ by the relations

$$\Phi_k(\mathbf{r},\varepsilon,\Omega,t) = v\Psi_k(\mathbf{r},\varepsilon,\Omega,t), \tag{1.3}$$

where v is the atomic velocity equal to $\sqrt{2\varepsilon/m}$, and m is the atomic mass.

The function $\Phi_k(\mathbf{r},\varepsilon,\Omega,t)$ satisfies the kinetic Boltzmann equation. The moving atoms are obtained as the result of two processes: scattering on various centers of j type which are distributed with the concentration $C_j(\mathbf{r},t)$, and displacement from the lattice sites. The scattering centers are the lattice atoms (in solid solutions they are atoms of various types), the interstitials (for example in the dumbbell configuration) and other lattice defects: vacancies (j=v), bivacancies (j=2v), and so on. Consequently, on the right side of the equation for the distribution function there must be two integral terms describing the scattering and displacement processes, the orientation atomic movements having to be taken into account by the relations (1.1) and (1.2). Then the equation for the function $\Phi_k(\mathbf{r},\varepsilon,\Omega,t)$ can be written in the form

$$\frac{1}{v}\,\frac{\partial\Phi_k(\mathbf{r},\varepsilon,\Omega,t)}{\partial t}+\Omega\nabla\Phi_k(\mathbf{r},\varepsilon,\Omega,t)+\sum_j C_j(\mathbf{r},t)\Phi_k(\mathbf{r},\varepsilon,\Omega,t)\sigma_{jk}(\varepsilon)=$$

$$\sum_j C_j(\mathbf{r},t)\int_{\varepsilon}^{\varepsilon_0} d\varepsilon_1 \int_{4\pi} d\Omega_1 \Phi_k(\mathbf{r},\varepsilon_1,\Omega_1,t)\sigma_{jk}(\varepsilon_1,\varepsilon_1-\varepsilon)K_s(\varepsilon_1,\Omega_1 \to \varepsilon,\Omega)+$$

$$\sum_j C_j(\mathbf{r},t)\int_{\varepsilon}^{\varepsilon_0} d\varepsilon_1 \int_{4\pi} d\Omega_1 \Phi_k(\mathbf{r},\varepsilon_1,\Omega_1,t)\sigma_{jk}(\varepsilon_1,\varepsilon)K_d(\varepsilon_1,\Omega_1;\varepsilon,\Omega), \tag{1.4}$$

where k is the type of moving atom (a matrix or impurity atom); j is the type of an object at rest (scatterer) including defects; $\sigma_{jk}(\varepsilon_1,\varepsilon)$ is a differential cross section of transferring the energy ε to the resting center of j type from the incoming atom of k type with the energy ε_1, and

$$\sigma_{jk}(\varepsilon)=\int_0^{\varepsilon}\sigma_{jk}(\varepsilon,\varepsilon'')d\varepsilon'' \tag{1.5}$$

is the total cross section of the interaction of the moving ε-energy atom with a center of j type. The functions K_s and K_d have the following form (cf. Ref. 170, p. 266).

$$K_s(\varepsilon_1,\Omega_1 \to \varepsilon,\Omega)=P_s(\varepsilon_1,\Omega_1 \to \varepsilon,\Omega)P^s(\varepsilon,\Omega),$$

$$K_d(\varepsilon_1,\Omega_1;\varepsilon,\Omega)=P_d(\varepsilon_1,\Omega_1;\varepsilon,\Omega)P^d(\varepsilon,\Omega), \tag{1.6}$$

where

$$P_s(\varepsilon_1,\Omega_1 \to \varepsilon,\Omega)=\frac{1}{2\pi}\int_0^{\varepsilon_1} d\varepsilon' \int_{4\pi} d\Omega' p_s(\varepsilon',\Omega' \to \varepsilon,\Omega)\delta\left[\Omega\Omega' - \sqrt{\varepsilon'/\varepsilon}\right],$$

$$P_d(\varepsilon_1,\Omega_1;\varepsilon,\Omega)=\frac{1}{2\pi}\int_0^{\varepsilon_1} d\varepsilon'' \int_{4\pi} d\Omega'' p_d(\varepsilon'',\Omega'';\varepsilon,\Omega)\delta\left[\Omega\Omega'' - \sqrt{\varepsilon''/\varepsilon}\right]$$

are the probabilities of scattering and displacing, respectively, and

$p_s(\varepsilon',\Omega' \rightarrow \varepsilon,\Omega)$ and $p_d(\varepsilon'',\Omega'';\varepsilon,\Omega)$ are the probabilities of taking place the scattering and displacing processes whose result is the fact that the parameters of motion will be ε and Ω.

A knowledge of the function $\Phi_k(\mathbf{r},\varepsilon,\Omega,t)$ will make it possible to find the cascade function $\nu(\varepsilon_0)$ and other cascade characteristics. To solve equation (1.4) analytically is a complicated problem. Therefore one is needed to insert additional simplifying assumptions which will be discussed in the course of our presentation

The most popular characteristic of the moving-atom cascade is the cascade function $\nu(\varepsilon_0)$ being an average (over the above mentioned ensemble) number of the displaced atoms in the cascade. It is an integral cascade characteristic, that is why it does not carry information on defect dislocation in the cascade region. The cascade function can be expressed through the distribution function of the moving atoms $\Phi_k(\mathbf{r},\varepsilon,\Omega,t)$. Since each particle with the energy on the interval from ε_d to $2\varepsilon_d$ creates no more than one Frenkel pair, the number of displaced atoms in the cascade is that of the atoms which move with the energies in this prethreshold region, i.e. those which come to rest in the next act of interaction with the rested atoms. In the stationary case $\partial\Phi_k(\mathbf{r},\varepsilon,\Omega,t)/\partial t = \partial\Psi_k(\mathbf{r},\varepsilon,\Omega,t)/\partial t = 0$, the cascade function $\nu(\varepsilon_0)$ can be expressed through the flux density of moving atoms as follows:

$$\nu(\varepsilon_0) = \int_{\varepsilon_d}^{2\varepsilon_d} d\varepsilon \int d\mathbf{r} \int d\Omega \Psi_k(\mathbf{r},\varepsilon,\Omega) = \qquad (1.7)$$

$$\int_{\varepsilon_d}^{2\varepsilon_d} d\varepsilon \int d\mathbf{r} \int d\Omega \frac{\Phi_k(\mathbf{r},\varepsilon,\Omega)}{\nu(\varepsilon)} = \int_{\varepsilon_d}^{2\varepsilon_d} \nu^{-1}(\varepsilon)\Phi_k(\varepsilon)d\varepsilon.$$

Consequently, to calculate the cascade function $\nu(\varepsilon_0)$ is enough to know not all the function $\Phi_k(\mathbf{r},\varepsilon,\Omega)$, but only its integral $\Phi_k(\varepsilon)$ over the directions of motion and the cascade volume.

The cascade function was calculated for various models using different simplifications. The most popular model is the Kinchin-Pease one [120] proposing that the atoms behave as hard spheres, all collisions are elastic, and the cascade is a sequence of collisions of identical atoms when each of them obtains the same energy in a collision. As a result one has for the cascade function $\nu(\varepsilon_0)$

$$\nu_{KP} = \begin{cases} 1, & 0 < \varepsilon_0 < 2\varepsilon_d, \\ \varepsilon_0/2\varepsilon_d, & 2\varepsilon_d < \varepsilon_0 < \varepsilon_i, \\ \varepsilon_i/2\varepsilon_d, & \varepsilon_0 > \varepsilon_i, \end{cases} \tag{1.8}$$

where ε_i is the energy of moving atom higher than this all energy is lost to electron excitation and lower to the displacements. The value ε_i depends on the electron structure of the solid. For metals, for example, $\varepsilon_i \approx A$ keV, where A is the atomic weight of the moving atom.

Sneider and Neufeld [121] assuming that the energy loss in each collision is ε_d and that both atoms continue to move after the collision, no matter how small the energy may be, derived that

$$\nu_{SN}(\varepsilon_0)=0.56(1 + \varepsilon_0/\varepsilon_d). \tag{1.9}$$

In virtue of the first assumption, $\nu_{SN} < \nu_{KP}$ must take place, however, the second assumption leads to increasing the cascade function, because of this there takes place a partial compensation of these contributions and the functions ν_{SN}, and ν_{KP} differ not so critically. One can adduce some other instances of the cascade function calculations which yield similar results (see Ref. 9).

The role of orientation effects, and channeling in particular, was evaluated in Refs. 109, 154. In this case, the form of the function depends on the probabilities of channeling the scattered (*s*) and displaced (*d*) atoms, P^s_C and P^d_C, respectively:

$$\nu_{OR}(\varepsilon_0) = \left[\varepsilon_0/2\varepsilon_d\right]^{1-P^s_C-P^d_C}, \quad \varepsilon_0 > \varepsilon_d.$$

A consideration of the focusing phenomenon merely reduces to replacing $2\varepsilon_d$ by the value of the focusing energy ε_F with the resulting indication of the energy boundaries for $\nu(\varepsilon_0)$ in the form

$$\left[\frac{\varepsilon_C}{\varepsilon_F}\right]^{1-P^s_C-P^d_C} < \nu(\varepsilon_0) \le \left[\frac{\varepsilon_C}{2\varepsilon_d}\right]^{1-P^s_C-P^d_C}, \quad \varepsilon_F > 2\varepsilon_d,$$

where ε_C is the critical channeling energy, and ε_F is the focusing energy average per a set of closely packed directions.

Another simple expression for the cascade function uses a notion of "damage energy" $E_D(\varepsilon_0)$ that is all the energy extended to form the moving atoms. According to Ref. 171,

$$\nu(\varepsilon_0) = 0.8E_D(\varepsilon_0)/2\varepsilon_d.$$

In a general case, as indicated above, the determination of a cascade function amounts to focusing an expression for the flux of moving atoms $\Phi_k(\varepsilon)$ which is the integral (1.7). To find a solution to equation (1.4) for $\Phi_k(\mathbf{r},\varepsilon,\Omega,t)$, it is necessary to make some simplifying assumptions [100,156,162]:

1. Considered is a stationary problem, $\partial\Phi_k(\mathbf{r},\varepsilon,\Omega,t)/\partial t=0$. This means that in an ensemble given a sufficiently large number of the cascades arising from a PKA with the energy ε_0 to $\varepsilon_0+d\varepsilon_0$ at a point $\mathbf{r}=0$, there occurs, on the average, the same PKA number over the time interval dt at any t, that is equal to Ψ_0.

2. Cascades in an ensemble do not overlap.

3. The defect concentration in the source crystal is low, the scattering on those can be ignored.

4. The atomic interaction is assumed paired.

To solve equation (1.4) with the above-listed suppositions, the method of expanding the sought-for function $\Phi_k(\mathbf{r},\varepsilon,\Omega)$ into infinite series is used over the systems of orthogonal polynomials [157-162, 170, 172]. The momentum of the function $\Phi_{00}^{00}(\varepsilon)$ is connected with the density of moving atoms of the energy ε by the relation $2\pi\Phi_{00}^{00}(\varepsilon)=\Phi_k(\varepsilon)$. The interaction cross sections $\sigma_{jk}(\varepsilon_1,\varepsilon'')$ appearing in (1.5) are taken in the form of Lindhard cross section [173]

$$\sigma_{jk}(\varepsilon_1,\varepsilon'')=F_{jk}\varepsilon_1^{-\zeta}\varepsilon''^{-(1+\zeta)}, \tag{1.10}$$

where

$$F_{jk}=F\left(\frac{m_k Z_k^2}{mZ^2}\right)^{\zeta}\left[\frac{2}{1+(Z_k/Z)^{2/3}}\right]^{1-\zeta},$$

$$F=\pi\zeta^{\zeta+1}(3-\zeta)^{\zeta}(Ze)^{4\zeta}(0.8853a_0/Z^{1/3})^{2(1-\zeta)}2^{-1}, \tag{1.11}$$

m_k, Z_k and m, Z are the mass and the charge of the outcoming (k) and resting (j) atoms, respectively; $0 < \zeta < 1$ is a parameter of the Lindhard cross section which weakly depends on the energy ε_1; e is the electron charge; and a_0 is the Bohr radius.

The probabilities in expression (1.6) have the form [157, 170]

$$p_s(\varepsilon',\Omega' \to \varepsilon,\Omega) = \delta(\varepsilon'-\varepsilon)\delta(\Omega'-\Omega),$$

$$p_d(\varepsilon'',\Omega'';\varepsilon,\Omega) = \delta(\varepsilon''-\varepsilon)\delta(\Omega''-\Omega).$$

The solution to the equation resulting for $\phi_k(\varepsilon)$ in an isotropic material and its substitution into (1.7) yield [11, 98, 99, 164]

$$\nu_0(\varepsilon_0)=\chi_0\left(2^{\kappa_0-0.5}-1\right)(\kappa_0-0.5)^{-1}\left(\frac{\varepsilon_0}{2\varepsilon_d}\right)^{\kappa_0+0.5}, \tag{1.12}$$

where

$$\kappa_0=0.5\left[1-\zeta+\sqrt{4\beta_0-(1-\zeta)^2}\right],\quad \chi_0=(\kappa_0+\zeta-1)(2\kappa_0+\zeta-1)^{-1},$$

$$\beta_0=2(2\zeta-1)(2\zeta+1)^{-1}.$$

Expressions for the cascade function are found analogously with allowance for the orientation effects (for example dynamic crowdions) [11, 88, 93, 100, 153]. So, if the probability of the dynamic crowdion formation

P^d_D does not depend on the energy ε on some interval of its changing, then we have instead of (1.2)

$$P^d = 1 - P^d_D,$$

from where

$$\nu_D(\varepsilon_0) = \nu_0(\varepsilon_0)H_D, \qquad (1.13)$$

is obtained, where

$$H_D=(2\varepsilon_d/\varepsilon_F)^{\delta_0 P^d_D}\left[1+\delta_0 P^d_D(\kappa_0-0.5)^{-1}\right]\left(2^{\kappa_0-0.5-\delta_0 P^d_D}-1\right)\left(2^{\kappa_0-0.5}-1\right)^{-1},$$

$$\delta_0 = P^d_D\left[4\beta_0-(1-\zeta)^2\right]^{-1/2}.$$

Correlation of the cascade functions for copper calculated by the given expressions, is presented in Fig. 1.3. It is seen that the consideration of the structure factors lowers the number of displaced atoms in the cascade, as compared to the value given by the Kinchin-Pease formula (1.8)

The expressions considered here for the cascade functions characterize the defect cascade structure in pure (without impurities) materials to the moment of completing the process of atomic collisions in the cascade. For their correlation with the experimental data, one needs to take into account a subsequent evolution of the cascade region in the course of which the structure is formed which is observed in the experiments.

Considerable information on kinetics of developing the atomic-collision cascades, their structure and parameters are provided by numerous computations [42, 125, 127-129, 144, 148, 174-176].

The elucidation of the atomic mechanisms of radiation defect formation in passing the neutrons through the crystal, was the first problem solved by the method of computerized modeling [107]. Now with computers the cascades resulting from PKA with the energy up to 100 keV are investigated which contain thousands of defects. Some albums of cascades in various materials have been prepared for various PKA energies (for example Refs. 177, 178). Studied are the details of propagating the chains of focused collisions for various crystallographic directions [144, 179], the atomic motion between the atomic rows of the lattice [109, 115, 116]; discovered are the phenomena of cascade breakdown into the subcascades [102, 126, 180], destructive channeling [181] in which the channeled atom knocks the atoms from the channel wall by side impacts. Investigated is the influence of thermal motion on the cascade propagation (for example in Ref. 178), had are the defect distributions in the cascade region and many their details (see, for example, Refs. 19-21, 150, 175, 176, 182). Computerized modeling validates that it is in developing the cascades of atomic collisions and their short-term annealing that all the atomic rearrangements considered take place, furnishing insights into their subtle details.

The predictions of the computerized cascade modeling find a direct experimental corroboration in the experiments with a field ionic microscope, which permits to regain the structure of the cascade regions in an irradiated needle point. It is, however, to be noted that studied in the experimental microscopic investigation of the cascades [36, 38, 103, 183-185] is not the cascade itself (as a process), but a cascade region, i.e. a defect distribution in the crystal which is formed as a result of passing a cascade of moving atoms. One should keep in mind that the indicated defect region only indirectly accounts for the cascade structure, inasmuch as a distinct contribution to its formation has been made by the processes of recombination, clusterization, and so on, in both the time of developing the cascade itself and after it. By and large the defect region of a single atomic-collision cascade is a so-called denuded zone [186] (the region of locally elevated vacancy concentration) with a mantle on the periphery consisting of the interstitials.

It is clear that the concentration of denuded zones depends on both the material temperature (in the process after the irradiation) and its composition. Furthermore, one can observe an interstitial "shell" in pure metals only in the very low-temperature experiments (depending on the nature of the material irradiated), since the interstitials have an essentially higher mobility than the vacancies [187].

To calculate the spatial defect distribution in the cascade region it is necessary to take into account its complex evolution. The cascade of moving atoms is, after its passing, the vacancies and interstitials forming a nonrelaxed cascade region to the moment of finishing the atomic-collision stage. In further time its structure relaxes, and the fields of local stresses are formed around the vacancies and interstitials, in this case those of the opposite defects which proves to be in the limits of Frenkel-pair instability zone, recombine instantaneously. The time of proceeding this process is of the order of thermal oscillation period and amounts about 0.5 ns. In this time the kinetic energy of the oscillating atoms does not manage to be markedly redistributed, and the cascade region reserves the high effective temperature, which can exceed the melting temperature. The following stage is cooling-down the cascade. It lasts some picoseconds. Finally, the last stage, annealing, comes. Now the diffusion relaxation processes proceed which include the recombination of opposite defects and the cluster formation from like defects.

In the course of cascade evolution, the local supersaturations of j-type defects $C_j(\mathbf{r},t)$ are determined by their generation rates $g_j(\mathbf{r},t)$ and instantaneous recombination ones $g_{jR}(\mathbf{r},t)$, as well as the annealing processes $Q_j(\mathbf{r},t)$. All the processes enumerated can be described by the balance equations

$$\frac{\partial C_j(\mathbf{r},t)}{\partial t} = g_j(\mathbf{r},t) - g_{jR}(\mathbf{r},t) - Q_j(\mathbf{r},t). \tag{1.14}$$

The local equation of particle number conservation has the form

$$\rho_{\mathrm{nuc}} = \sum_j \hat{C}_j(\mathbf{r},t), \tag{1.15}$$

where ρ_{nuc} is the nuclear density. The generation rates of vacancies and interstitial atoms can be as follows [160-164]:

$$g_v(\mathbf{r},t) = \rho_{\mathrm{nuc}} \sum_k \int_{\varepsilon}^{\varepsilon_0} d\varepsilon_1 \int_{4\pi} d\Omega_1 \Phi_k(\mathbf{r},\varepsilon_1,\Omega_1,t) \int_0^{\varepsilon_1} d\varepsilon'' \sigma_{1k}(\varepsilon_1,\varepsilon'') \times$$

$$\left[p(\varepsilon_1 - \varepsilon'') p(\varepsilon'') + p(\varepsilon_1 - \varepsilon'') q(\varepsilon'') P_D^d(\varepsilon'') \right] - \tag{1.16}$$

$$C_v(\mathbf{r},t)\sum_k \int_{\varepsilon}^{\varepsilon_0} d\varepsilon_1 \int_{4\pi} d\Omega_1 \Phi_k(\mathbf{r},\varepsilon_1,\Omega_1,t) \int_0^{\varepsilon} d\varepsilon'' \sigma_{vk}(\varepsilon_1,\varepsilon'') q(\varepsilon_1-\varepsilon'') -$$

$$C_v(\mathbf{r},t)\sum_k \left[\mu_{vD} g_{Dk}(\mathbf{r},t)+\mu_{vC} g_{Ck}(\mathbf{r},t)\right] - g_{vR}(\mathbf{r},t);$$

$$g_i(\mathbf{r},t)=\rho_{\text{nuc}}\sum_k \int_{\varepsilon}^{\varepsilon_0} d\varepsilon_1 \int_{4\pi} d\Omega_1 \Phi_k(\mathbf{r},\varepsilon_1,\Omega_1,t) \int_0^{\varepsilon_1} d\varepsilon'' \sigma_{1k}(\varepsilon_1,\varepsilon'')\times$$

$$\left[q(\varepsilon_1-\varepsilon'')q(\varepsilon'')-q(\varepsilon_1-\varepsilon'')q(\varepsilon'')P_D^d(\varepsilon'')\right]-$$

$$C_i(\mathbf{r},t)\sum_k \int_{\varepsilon}^{\varepsilon_0} d\varepsilon_1 \int_{4\pi} d\Omega_1 \Phi_k(\mathbf{r},\varepsilon_1,\Omega_1,t) \int_0^{\varepsilon_1} d\varepsilon'' 2\sigma_{ik}(\varepsilon_1,\varepsilon'') p(\varepsilon_1-\varepsilon'') p(\varepsilon'')+ \tag{1.17}$$

$$\rho_{\text{nuc}}\sum_k \left[\mu_{iD} g_{Dk}(\mathbf{r},t) + \mu_{iC} g_{Ck}(\mathbf{r},t)\right] - g_{iR}(\mathbf{r},t),$$

where $p(\varepsilon)$ is the probability for the atom obtaining the energy ε to go out of the interaction point as nonorientally moving one; $q(\varepsilon)=1-p(\varepsilon)$; $\mu_{j\varphi} g_{\varphi k}(\mathbf{r},t)$ is the number of j-type particles formed in unit time at the cost of stopping the orientedly moving atom of k type ($\varphi=D$ relates to dynamic crowdions, $\varphi=C$ to channelons); the rate of instantaneous j-defect recombination has the form

$$g_{jR}(\mathbf{r},t)=C_j(\mathbf{r},t) \int_{\mathbf{r}-\mathbf{r}_{0R}}^{\mathbf{r}+\mathbf{r}_{0R}} d\mathbf{r}'\, g_n(\mathbf{r}',t), \tag{1.18}$$

where r_{0R} is the size of the recombination zone, n is the type ($n\neq j$) of an opposite defect with that the j-type defect recombines, $\check{\varepsilon}$ is a minimum energy of the moving atom.

The equations (1.4), (1.14)-(1.18) form the system describing kinetics of defect evolution in the cascade region. Known in the literature are various solutions to this system, e.g. Refs. 157, 161, 163, 166, 169, which permit, with a specific set of simplifications, to get information on the cascade-region defect distribution, its sizes and configuration and further compare it to the data of nature [36, 38, 183, 184] and numerical experiments [174-176,179], as well as to the computations of such systems (e.g. Refs. 125, 127-129).

1.2.2. Focusing Atom-Atom Collisions

The cascades of moving atoms in the crystals with lowering the energy (in an ε_0 energy exchange by PKA) are terminated by focusing atomic collisions along the close-packed crystallographic directions [9,11,104-114]. The first calculations of the focusing effects for the hard-sphere model were carried out in Refs. 106, 108 with the Born-Meier potential

$$V_{\mathrm{BM}}(r)=A\exp(-r/a_{\mathrm{BM}}),$$

where A and a_{BM} are the potential constants. If a set of the atoms with the interaction radius R_n, each moving successively with the energies ε_n at the angles ϑ_n to the chain direction, is considered, one can find a recurrence relation between the angles and energies as follows [106]:

$$\sin\vartheta_{n+1} = \sin\vartheta_n\left[\alpha_0\cos\vartheta_n - \sqrt{1-\alpha_0^2\sin^2\vartheta_n}\right], \tag{1.19}$$

$$\varepsilon_{n+1} = \varepsilon_n(1-\alpha_0\sin^2\vartheta_n), \tag{1.20}$$

where $\alpha_0=a/R_0$, and $R_0=a_{\mathrm{BM}}\ln(2A/\varepsilon)$ is the distance of a maximum approaching. As an low-angle approximation we have instead of (1.19)

$$\vartheta_{n+1} \approx \vartheta_n(\alpha_0-1). \tag{1.19a}$$

The value

$$\Lambda \equiv \frac{\vartheta_{n+1}}{\vartheta_n} = \alpha_0 - 1 = \frac{a}{R_0} - 1 \tag{1.21}$$

is called the focusing parameter.

In the scope of the simple model described, one can evaluate the critical focusing energy $\varepsilon_F^{<hkl>}$ for the crystallographic direction <*hkl*> lower than that the value R_0 is long enough the focusing to take place in the given direction (see Refs. 11, 106, 110),

$$\varepsilon_F^{<hkl>} = 2A\exp\left[-a^{<hkl>}/2a_{\mathrm{BM}}\right], \tag{1.22}$$

where $a^{<hkl>}$ is the interatomic spacing along the direction <*hkl*>. The critical angle $\vartheta_F^{<hkl>}$ of focusing the pulses along the direction <*hkl*>, i.e. the angle lower than that the focusing is realized (i.e. $\Lambda<1$), is determined from the condition that

$$a^{<hkl>} = 2R_0\cos\vartheta_F\ . \tag{1.23}$$

Sequences of the focused collisions can be considered as the motion of a peculiar kind of quasiparticle which, depending on Λ, can be of various kinds [106]. This classification can be performed, with the use of just listed characteristics [104]:

(1) Focusing, $\Lambda<1$, $\vartheta_{n+1}<\vartheta_n$. A focused pulse transfer is carried out when $\varepsilon<\varepsilon_F$ and $\vartheta<\vartheta_F$.

(2) Defocusing, $\Lambda>1$, $\vartheta_{n+1}>\vartheta_n$. The correlation of impact parameters ceases in such process after a definite leg of path length, and the motion becomes chaotic. The defocused series of collisions with a substantial correlation are also possible at the energies $\varepsilon>\varepsilon_F$. They occur at $\varepsilon<\varepsilon_F$ for the angles $\vartheta>\vartheta_F$ and at $\varepsilon<\varepsilon_F$ for $\vartheta<\vartheta_F$. The maximum angle ϑ_s in whose ranges a sequence of atomic collisions will take place, can be determined from the condition

$$\sin\vartheta_s = \frac{R_0}{a} = (\Lambda+1)^{-1}$$

or for small angles,

$$\vartheta_s \simeq (\Lambda+1)^{-1}.$$

(3) Energy transfer. If the initial energy is small, all atoms are back to their initial positions, and it is only the energy that transmitted in the sequence of collisions. In this case a vacancy - implanted atom pair does not form.

(4) Mass and energy transfer. At large initial energy the collision succession consists of a number of substituting collisions at the insertion of which remains a vacancy. After some rundown an excess atom is set free from the chain and goes to rest as an interstitial.

The combination of these parameters permits the collision sequences to be classed into the following groups:

(a) focusons (1) and (3), i.e. for the angles, the condition $\Lambda < 1$ is fulfilled and an energy transfer is only takes place;

(b) dynamic crowdions (1) and (4), (2) and (4), i.e. upon different angle conditions there occurs, in parallel with the energy transfer, a mass transfer. In certain situations there can occur a mutual focuson - dynamic crowdion transformation.

The calculations by the hard-sphere model reflect the significance of the phenomenon qualitatively realistically, showing that the focused collisions are able to transfer impulses for considerable distances in the chains.

The hard-sphere approximation quantitatively overestimates the focusing mechanism and gives only a qualitative pattern. Here one does take into account skewed central collisions in which the particles slide at the great distance one from another, as well as the nearest neighboring atoms whose distance to the colliding particles is more than the field where the potential selected is valid. That is why to obtain some comprehensive data on this pattern, it is necessary to account for: (a) a real interatomic potential, (b) the influence of the neighboring rows. Such calculations were made (see, for example, Refs. 108, 188). The allowance for the potential smoothness makes it possible to realize some focused substitutions.

Analytical calculation possibilities are bounded for the moment by the small-angle case, which leads to some overestimation of the focusing energy and other process parameters.

The study of the atomic collision focusing has been shown that the atomic collision chains arise at an essentially higher energy than is given by the evaluations in the simple focusing theory. This process is determined by the atoms of neighboring rows which, during passing the collision chain, obtain a return clear of the chain. The influence of the neighboring atoms tells on the development of the focusing process, which brings into being an effect of indirect or lens focusing. The results of calculations taking account of the lens focusing are listed in Refs. 23, 110.

The upper energy limit for the indirect focusing is more than that for the simple one to amount $\sim 10^3$eV. However, great energy losses per one interatomic spacing make the radius of this process almost the same as in the direct focusing. When the energy of primary collision is higher than 10^3eV, the lens focusing process can be transformed into channeling.

The existence of focused collision chains as dynamic crowdions determines an experimentally supported fact (e.g. in Refs. 47, 101, 112-114) of increasing the size of a single cascade defect region due to forming an interstitial mantle around the denuded zone. Furthermore, a numerical experiment also shows that such oriented forms of motion do exist in the crystals [102, 104, 107, 111, 144, 175, 179, 180].

1.3. Radiation-Induced Nucleation of Point Defect Clusters

After a cascade of moving atoms and any attendant dynamic processes had been developed, the dynamic stage of radiation damage of the materials completes, and the diffusion stage begins (see Fig. 1.1, Level II). The first step of this stage is a radiation-induced nucleation of point-defect clusters of both intrinsic ones (dislocation loops, voids) and those with participation of impurities (secondary-phase precipitates, heterogeneous nucleation). We shall look shortly at this set of processes, which will not be considered further, in the major part of the book, but will be in fact a starting state to study the processes at the diffusion stage.

In describing kinetics of mobile-point-defect formation, three characteristic steps of this evolution are recognized [189, 194]: nucleation, further growth and coalescence. A major driving force of the cluster nucleation processes is a tend of the material to diminish its free energy [189]. A typical curve of the ΔF_q free energy dependence of q point-defect cluster on its size R_q has the form given in Fig. 1.4. The value R_q is the critical dimension of a point-defect cluster with which the free energy of the cluster reaches its maximum value $\Delta F_q = \Delta F_q$.

It is seen from Fig.1.4 that when $R_q < R_{qC}$, the free energy of the material increases with increasing the nucleus. Fine clusters possess a relatively great surface energy and are energy-gained [189]. However, if the rate of the nucleus dissociation is less than that of its growth due to the connection of the point defects diffused in the material volume, the nucleus sizes can reach the critical value $R_q = R_{qC}$. A further nucleus growth becomes energy-gained, which correspond to the beginning of the second step in evolving the mobile-point-defect cluster.

Numerous experimental data (e.g. Refs. 195-197) and the results of computerized modeling (see, e.g., the review in Ref. 42) show that the nucleus of vacancy voids and interstitial dislocation loops are formed abundantly at the initial stage of irradiation of metals and alloys by fast particles.

The fluctuation nucleation of voids in the materials under irradiation was first theoretically considered in Refs. 24, 25. The vacancy-void fluxes were calculated in the phase space of intrinsic-point-defect cluster dimensions, as well as the rate of void nucleation, were calculated on the basis of the homogeneous nucleation theory [198]. A further development of the theory of fluctuation nucleation and growth of voids were performed in Refs. 26, 28, 199-202. In Ref. 202, the theory of the fluctuation nucleation of vacancy voids in metals was generalized [24, 25] with allowance for a thermal mobility of bivacancies. In Ref. 199, a homogeneous nucleation and growth of vacancy voids and interstitial dislocation loops in pure crystals were investigated basing on the kinetic equations for the functions of intrinsic point defect distribution in sizes. It was shown that the theoretical dependence of the function of point-defect-cluster distribution $f_q(R_q)$ in sizes R_q on the irradiation dose and temperature is qualitatively proper.

It is to be noted that the formation of vacancy voids and interstitial loops depend on external attacks (irradiation temperature, mechanical loads, etc.) and the form of the initial defect structure of the crystal (see, e.g., Refs. 203-206). Thus, in Ref. 203 the effect of edge dislocation elastic fields on the radiation-induced nucleation and growth of interstitial loops and vacancy voids in metals was studied. The probability of forming a steady point-defect cluster and the rate of its growth were shown to essentially depend on the position of the nucleus relative to the dislocation.

1.4. Diffusion Stage of Radiation Damage

Realized after the dynamic stage in material irradiation are diffusion rearrangements in the defect structure accompanied by the nucleation processes shortly discussed in Section 1.3.

Nowadays the theoretical investigations of the structural inhomogeneity kinetics in metals and alloys are carried out, as a rule, in the approximation of homogeneous mobile point defect distribution over the material volume [19, 40, 207-211]. A more proper approach (see, e.g., Refs. 27, 212) defining more precisely the already available theoretical representations about the evolution of defect material structure, is based on calculating fluxes of mobile point defects toward the structural inhomogeneities in the crystal. Therefore, in studying the effect of irradiation on the materials, one needs to find a solution to the inhomogeneous balance equations which describe the evolution of their defect structure. As a rule, the diffusion of mobile point defects is a nonstationary and inhomogeneous process. That is why considered must be the space-time point defect kinetics which is described by a system of nonlinear partial differential equations for the defect supersaturations $C_j(\mathbf{r},t)$.

In Ref. 213 the kinetics of intrinsic point defects has been investigated for pure crystals when the vacancies and interstitials are distributed over the material volume homogeneously. In this case the equation for the defect concentrations $C_{j0}(\mathbf{r},t)$ can be presented as follows (for $j=i,v$):

$$\frac{dC_{j0}(t)}{dt} = g - \mu D_i C_{i0}(t) C_{v0}(t) - D_j C_{j0}(t) k_j^2, \tag{1.24}$$

where g is the defect generation rate,

$$D_j = a^2 \nu_0 \exp\left[-\varepsilon_j^m / kT\right] \tag{1.25}$$

is the diffusion coefficient of the j defect type, ε_j^m is their migration energy, T is the absolute temperature, k is the Boltzmann constant, ν_0 is a frequency of the order of Debye one, $\mu=4\pi r_R$ is the recombination coefficient of vacancies and interstitials, r_R is the radius of the recombination zone,

$$k_j^2 = \sum_{q=1}^{n_j} S_q^j(t, k_j^2) \tag{1.26}$$

is the sum of q-type sink strengths S_q^j, n_j is the number of sink types for j defects with k_j^{-1} being the length of diffusion path of j defects before their adsorption by any nearest sink (see, e.g., Refs. 214, 215).

The nonstationary balance equations of (1.24)-type were considered in many works (see,e.g., Refs. 207, 216-218). However, the sink strength sum (k_j^2) was there considered to be constant and independent of the time t. Taking account of the t-dependence of k_j^2 has permitted the authors of Ref. 213 to analytically study and compute the behavior of the intrinsic point defects concentrations in an initial period of irradiation and to evaluate the characteristic time of interstitial loop nucleation, their concentrations and sizes.

Furthermore, in Ref. 213 a feature of the irradiated intrinsic point defect kinetics was found which is important for a subsequent discussion. It consists in the fact that after the lapse of the critical time t_*, after the switching-on the irradiation, the vacancy and interstitial concentrations reach their stationary values C_j^+ (see Fig. 1.5). In Ref. 213 a simple analytical expression has been obtained for t_*

$$t_* \simeq (D_v k_v^2)^{-1}, \tag{1.27}$$

from where one can see that the nonstationary period is governed by the diffusion of less mobile (relative to the interstitials) vacancies. On the strength of (1.27), it is shown in Ref. 213 that only a small amount of interstitial loops has had time to nucleate in the material in the initial irradiation time ($t < t_*$), and hence, when analyzing the evolution of intrinsic point defect clusters, one can neglect the period of nonstationary distribution for the vacancies and interstitials ($t < t_*$).

The most simple solutions to the systems of equations of the form

(1.27) have been obtained in the stationary case

$$\frac{dC_{j0}(t)}{dt} = 0. \tag{1.18}$$

There is a good reason to write equation (1.24) with allowance for C_{j0} being in fact related to j-defect precipitates with respect to the thermodynamic concentration C^e_j, i.e. $C^+_{j0} = \hat{C}^+_{j0} - C^e_j$, where $\hat{C}^+_{j0}$ is a true absolute stationary concentration of j defects in the pure case (0), i.e.

$$g - \mu D_i(\hat{C}^+_{v0} - C^e_v)(\hat{C}^+_{i0} - C^e_i) - D_j(\hat{C}^+_{j0} - C^e_j)k^2_j = 0. \tag{1.24a}$$

The solution to the algebraic equation (1.24a) has the form, analogously to Ref. 27,

$$\hat{C}^+_{j0} = C_0 H_j\left[\sqrt{1 + \eta_0^2} - \eta_0\right] + C^e_j, \tag{1.29}$$

where

$$C_0 = \sqrt{g/\mu D_i}; \quad \eta_0 = \sqrt{D_v k^2_v k^2_i/4\mu g}; \quad H_i = \sqrt{D_v k^2_v/D_i k^2_i} \equiv H^{-1}_v.$$

When the generation rates are large and the sinks are weak ($\eta_0 \ll 1$), one has (see, e.g., Ref. 27)

$$\hat{C}^+_{j0} \simeq C_0 H_j(1-\eta_0) + C^e_j \simeq \frac{g}{D_j k^2_j} + C^e_j. \tag{1.29a}$$

1.5. Diffusion Fluxes of Point Defects toward the Sinks

In theoretical studies of structure inhomogeneity kinetics in the pure crystals it is essential to know the diffusion fluxes $\mathbf{J}^j_q(\mathbf{r},t)$ of the intrinsic point defects through the surfaces $\Lambda_q(R_q)$ bounding the sinks q. A reasonable solution of such problem calls for the calculation of the concentration profiles for the vacancies and interstitials nearby the sinks q. For pure crystals, the system of balance equations has the form

$$\frac{\partial C_{j0}(\mathbf{r},t)}{\partial t} = g - \mathrm{div}\mathbf{J}^j_{q0}(\mathbf{r},t) - \mu D_i C_{i0} C_{v0} - D_j C_{j0} k_j^2, \tag{1.30}$$

where

$$\mathbf{J}^j_{q0}(\mathbf{r},t) = - D_j \left[\nabla C_{j0}(\mathbf{r},t) + \frac{C_{j0}(\mathbf{r},t)}{kT} \nabla E^j_q(\mathbf{r},t) \right] \tag{1.31}$$

is the flux density of mobile point defects (j=v,i) toward the sink of q type, and $E^j_q(\mathbf{r},t)$ is the elastic energy of j-type defect interaction with the sink q [190, 191].

The solution of the system (1.30) with fixed initial and boundary condition still remains an intricate problem, since (1.30) are nonlinear second-order differential equations in partial derivatives with variable coefficients. Usually the concentration profiles of intrinsic point defects are calculated either on the assumption of a homogeneous distribution of vacancies and interstitials over the material volume (see, e.g., Refs. 13, 207), or on the assumption of their stationary inhomogeneous distribution [212, 218, 219, 222, 227-232].

As the diffusion fluxes of intrinsic point defects toward a specific q sink (see (1.31)) are found, it is essentially to know a spatial distribution of vacancies and interstitials in the vicinity of this sink. Such a problem is usually solved within the approximation of a stationary inhomogeneous distribution of intrinsic point defects over the material volume:

$$\frac{\partial C_j(\mathbf{r},t)}{\partial t} = 0. \qquad (1.28a)$$

The verity of the given approach is attested, for example, to the results of Ref. 213.

Consider, in turn, solutions to the balance equations (1.30) under the condition (1.28a) in the vicinity of dislocations, interstitial loops and vacancy voids, because the structure of these sinks determines the geometry of the problem, as well as the matter transfer in the material.

1.5.1. Diffusion Fluxes of Intrinsic Point Defects toward Various Sinks

Being nonsaturable sinks for the intrinsic point defects, the dislocations are always present in the material. The elastic interaction of the dislocation with the vacancies and interstitials changes their equilibrium concentrations at close range of the dislocation, which, in its turn, leads to asymmetry in the diffusion fluxes of intrinsic point defects toward the sink of a given type.

When the concentration profiles of vacancies and interstitials close by an edge dislocation (q=D) are calculated, it will normally (see,e.g., Refs. 212, 219-228) suffice to take into account only the elastic point defect interaction with the given sink type.[1] Now the energy $E^j_D(\mathbf{r})$ is

---

[1] In Ref. 222 the concentration profiles of intrinsic point defects have been found in the elastic field of an edge dislocation with account for a modular point-defect

representable in the polar coordinates (r,ϑ_j) in the form [223, 224]

$$E_D^j(r,t) = \frac{a|\Delta\Omega_j|G}{2\pi} \cdot \frac{1-2\nu}{1-\nu} \cdot \frac{\sin\vartheta_j}{r} \equiv A_D^j \frac{\sin\vartheta_j}{r}. \tag{1.32}$$

Here G is the shear modulus of the material, ν is the Poissoncoefficient; $\Delta\Omega_j$ is a local change in the volume related to the point defect of j type; and $\vartheta_n = \vartheta_j \pm \pi$; $n,j = v,i$; $n \neq j$, as far as $\Delta\Omega_v$ and $\Delta\Omega_i$ have opposite signs [224]. The elastic interaction of intrinsic point defects with an edge dislocation was first taken into account in Refs. 264, 282. So, for example, determined in Ref. 264 were the concentration profiles of vacancies and interstitials, as well as the efficiencies $K_D^j(r_*)$ of their adsorption by a unit length of the edge dislocation:

$$C_j(r,\vartheta_j) = C_j^e \exp\left[-E_D^j(r,\vartheta_j)/kT\right] + V\left(\frac{R_0^j}{2r}\right)(\hat{C}_j^+ - C_j^e)\exp\left[-E_D^j(r,\vartheta_j)/2kT\right], \tag{1.33}$$

$$K_D^j(r_*) \simeq \alpha_{DO}^j D_j(\hat{C}_j^+ - C_j^e). \tag{1.34}$$

Here we have

$$\alpha_{DO}^j = \frac{2\pi}{\ln(2L_D/R_0^j)}; \quad R_0^j \equiv \frac{A_D^j}{kT}; \tag{1.35}$$

---

interaction with the given sink type. As a rule, this interaction is weak.

$$V\left(\frac{R_0^j}{2r}\right) = \frac{K_0\left(\frac{R_0^j}{2r}\right)\cdot I_0\left(\frac{R_0^j}{2r_*}\right)-K_0\left(\frac{R_0^j}{2r_*}\right)\cdot I_0\left(\frac{R_0^j}{2r}\right)}{K_0\left(\frac{R_0^j}{2L_D}\right)\cdot I_0\left(\frac{R_0^j}{2r_*}\right)-K_0\left(\frac{R_0^j}{2r_*}\right)\cdot I_0\left(\frac{R_0^j}{2L_D}\right)};$$

$\hat{C}_j^+$ is an average stationary concentration of intrinsic point defects far from the dislocation; r_* is the radius of the edge dislocation core; $I_0(R_0^j/2r)$ and $K_0(R_0^j/2r)$ are the Bessel functions of a purely imagine argument (see,e.g., Ref. 225).

In Refs. 219, 226, the method developed in Ref. 212 to calculate the efficiencies $K_D^j(r_*)$ in the field of applied stress σ. It is well to bear in mind that the expression (1.35) is inconsistent with the experimental data on the dependence of the preference factor $\alpha_{D0}^i/\alpha_{D0}^v$ on the temperature and dose of irradiation. To remedy this contradiction it was suggested in Refs. 227, 228 that the intrinsic point defect adsorption takes place not at any point of the dislocation line but only at the kinks. However, the computations (see, e.g., Ref. 149) show that the energies of bounding the vacancies and interstitials with the edge dislocations are too high the kinks on the dislocation line to be essential in irradiation. In Refs. 229-231 the calculation was performed, within the effective-medium approximation, for the concentration profiles of intrinsic point defects and the adsorption efficiencies in case when the dislocation is a continuous sink for the point defects. In Ref. 229 it was assumed complementary that the dislocation is surrounded by the potential barrier inhibiting the point defect precipitation on it. Cumbersome expressions were found for α^j_{D0}. The authors of Refs. 229, 230 have shown that the preference factor $\alpha^i_{D0}/\alpha^v_{D0}$ increases with the temperature and irradiation dose. This agrees with the experimental data on the stainless steel irradiated in a high-voltage electron microscope at the temperatures higher than 400°C (see Ref. 230).

In Refs. 212, 219-222, 227-232, the volume generation of intrinsic point defects close by an edge dislocation was not taken into consideration. Such a consideration of g was originally carried out in Ref. 233. However, given in Refs. 212 and 233 were different boundary conditions, and the limiting transition ($g \longrightarrow 0$) from the results of Ref. 233 to the relation (1.33) has not met with success.

The concentration profiles of movable point defects close by screw dislocations are routinely computerized by numerical methods (see, e.g., Refs. 42, 234, 235), because the point defect interaction with a screw dislocation is determined by the atomic structure of dislocation core rather than the elastic forces.

It is a complex problem to solve analytically the balance equations (1.30) close by a dislocation loop (q=L). Such problems are commonly investigated in two limiting cases: large-sized ($R_L \geq L_L$) and small-sized ($R_L \ll L_L$) loops, where R_L is the loop radius, and $2L_L$ is an average separation between the loops.

If the sizes of dislocation loops are more, or of the order of, the average separation between them ($R_L \geq L_L$), then each small section of such a loop can be thought as a straight line having the elastic properties of an edge dislocation. Analogous problems were solved by many authors (see, e.g., Refs. 206, 236-239).

Unlike the edge dislocation, the energy of interacting $E_L^j(\mathbf{r})$ of a small-sized dislocation loop with an intrinsic point defect depends on both the *j*-defect type and the structure of the loop itself [223, 224, 240]. This adds complexity to the calculation of the diffusion fluxes of vacancies and interstitials toward the given sink type. Such problems are routinely solved by numerical methods (see, e.g., Refs. 42, 241, 242).

In the limiting case $R_L \ll L_L$, by the dislocation loop surface must be meant that of enclosing R_L-radius toroid [243], while the boundary of the region of loop influence on the fluxes of vacancies and interstitials is a sphere of the radius L_L. It was shown in Ref. 22 that in the case of the prismatic dislocation loop growth, the sole difference between the intrinsic point defect fluxes toward a void and a dislocation loop is determined by the renormalization of the concentration of vacancies and interstitials on the outside surface $\Lambda_L(L_L)$. According to the results of Ref. 22, the growth of an equivalent spherical R_L-radius cluster will be further considered rather than that of a dislocation loop, the energy of the former being representable in the form

$$E_L^j(r)=\frac{A_L^j}{r^n}, \tag{1.36}$$

where A_L^j is a constant independent of the elastic properties of the material, the loop size and the intrinsic point-defect type, and n is a parameter of the theory.

So the kinetics of intrinsic point defects nearly a large-sized dislocation loop has been studied rather extensively. In the limiting case of small loops, the analytical solution to the balance equations (1.30) has not been obtained. If assumed reasonably [22], it can be found within the approximation (1.36), however, up to the present this problem has remained unsettled.

As distinguished from the above considered cases of dislocations and loops, the interaction energy E^j_V of intrinsic point defects with a spherical void is constant over the material volume (see, e.g., Refs. 244, 245). In this case the balance equations (1.30) are essentially simplified and can be represented by

$$\frac{\partial C_j(r,t)}{\partial t} = g + D_j \Delta C_j(r) - \mu D_i C_i C_v - D_j C_j k_j^2 = 0, \qquad (1.37)$$

the elastic field effect on the concentration profiles of vacancies and interstitials being exerted through the boundary conditions.

There are many works devoted to the study of vacancy void kinetics [13, 26, 199, 214, 239, 246-253]. However, as a rule, they, when solving the balance equations (1.30), either use the assumption of a nonstationary and homogeneous distribution of the intrinsic point defects over the material volume (see, e.g., Ref. 13), or ignore the effect of the recombination term in (1.37) on the point defect concentration profiles and on their diffusion fluxes toward a void (see, e.g., Ref. 214). In Ref. 253 has been offered an approximate method of solving the stationary diffusion equations (1.37) taking account of the recombination term, the dependence being found of the void radius $R_V(t)$ on the time t.

1.5.2. Point Defect Adsorption by Various Sinks

With a knowledge of the concentration profiles of the movable point defects close by the structure inhomogeneity of the material, one can find, by (1.31), the densities of their fluxes $\mathbf{J}^j_q(R_q)$ toward each specific q-type sink. The rates $I^j_q(R_q)$ of point defect adsorption by q sinks are related to the flux densities as follows (see, e.g., Refs. 190, 191):

$$I^j_q(R_q) = 4\pi R_q^2 \left|\mathbf{J}^j_q(r)\right|_{r=R_q} \tag{1.38}$$

in case of spherical sinks, and

$$I^j_q(R_q) = 2\pi R_q^2 \left|\mathbf{J}^j_q(r)\right|_{r=R_q} \tag{1.39}$$

in case of plane sinks (of the type of small-sized dislocation loops:$R_L \ll L_L$). The efficiencies of point defect adsorption by an edge dislocation have the form [212]

$$K^j_D(r_*) = r_* \int_0^{2\pi} \left|\mathbf{J}^j_D(r_*,\vartheta_j)\right| d\vartheta_j \ . \tag{1.40}$$

The analytical calculations of $I^j_q(R_q)$ and $K^j_D(r_*)$ were carried out as the stationary approximation for the edge dislocations, vacancy voids and interstitial loops in pure metals (see, e.g., Refs. 206, 212, 214, 219, 222, 226-231, 233). In this case the corresponding rates $I^j_q(R_q)$ and the efficiencies $K^j_D(r_*)$ of the intrinsic point defect adsorption by isolated q sinks have been found by basing on the solutions of (1.30)-type balance equations by (1.38)-(1.40). The effect of various sink types on one other was first taken into account in Refs. 214, 254-259. However, as noted above, they neglected the effect of elastic fields of structure inhomogeneities on the point defect diffusion mobility, which leads to a perceptible change in the values $I^j_q(R_q)$ and $K^j_D(r_*)$.

1.6. Coalescence of Point Defect Clusters

When the structural inhomogeneities in the material (voids, interstitial loops, precipitates and so on) reach specified critical dimensions, their growth substantively speeds up through "evaporation" of finer clusters of the same q type. There is beginning a coalescence stage. This phenomenon was first theoretically studied in Refs. 192-194, where it was shown that the asymptotic behavior of the distribution function $f_q(R_q)$ of the material structure inhomogeneities in the sizes R_q does not depend on its original form. The irradiation effect on the defect structure evolution at the coalescence stage was considered in a number of further publications (see, e.g., Refs. 260-262), where an asymptotic form of $f_q(R_q)$ was found for various structure inhomogeneities in the material.

The consideration of similar class problems does not enter into the scope of the present book.

CHAPTER 2

RADIATION DAMAGE OF ALLOYS

2.1. General Concepts of the Solid Solution Decomposition under Irradiation

For the solid solution materials which are in equilibrium in the absence of irradiation, it is necessary to take into account that when exposed to radiation, they may evolve into nonequilibrium materials where decomposition is in progress. A distinguishing feature of such radiation-induced decomposition is that the phase diagram is changed in the irradiation process, and the radiation-induced rearrangements are reversed when this process had come to an end. If exposed to irradiation is the nonequilibrium solid solution in which the decomposition process is either progressing with a given rate or frozen due to the low temperature, this process is referred to as a radiation-accelerated one. Analogous definition may be given to the radiation-induced and radiation-accelerated dissolution in case of nonequilibrium precipitates.

As opposed to the vacancy voids and loops, the precipitates are composed of several kinds of atoms and their lattice symmetry is usually different from that of the matrix lattice. Therefore, their morphology and formation kinetics demonstrate a greater variety. Coherent and incoherent precipitates may be distinguished. Being various, their shape can be approximated by ellipsoids of various axial ratio. The decomposition processes may be divided, by their kinetics, into the nucleation and growth processes for a new composition-fixed phase and those of spinodal decomposition when the fluctuation amplitude of the solid-solution concentration progressively increases until a new-phase composition is reached. The nucleation can proceed homogeneously over the matrix volume (homogeneous decomposition) or at some lattice defects, in particular at the grain or subgrain boundaries. In this case, the new phase is growing into the grain volume out of the boundary usually having the shape of a lamellar structure (cellular and eutectoidal decomposition). In the cellular decomposition alternating are the lamellas of the matrix and the new phase, in the eutectoidal decomposition those of two new phases (for example in perlite α-Fe+Fe_3C formation from austenite, i.e. the solid solution of C in γ-Fe).

Finally, if in the homogeneous decomposition the impurity concentration in the solid solution has been reduced down to the equilibrium one, the atomic rearrangements do not die away. The fact is that the small-sized precipitates possess a relatively high surface energy, which becomes lower in their enlargement (coalescence) when the small-sized precipitates are dissolving, and the coarse ones are growing at their expense. In binary alloys, and especially in many-component ones, there can co-exist various kinds of precipitates which undergo mutual transformations at their successive decomposition stages. All the processes listed are influenced by the fast-particle irradiation. By the recent time, this influence is only investigated in some particular cases.

If kinetics of precipitation growth and their influence on the redistribution of diffusion point-defect fluxes is studied theoretically, this requires (analogously to pure crystal case) to estimate the concentration profiles of near-precipitate point defects of various types. In this case, two limiting cases may be distinguished: individual diffusion of foreign atoms toward the precipitates [184,185] and their diffusion through the complexes [15,200].

2.2. The Effect of Irradiation on the Alloy Phase Diagram

The phase diagram drawn with the temperature- concentration (*T*,*C*) coordinates of a solid solution under irradiation, is non-equilibrium, but can be stationary. The crystal under irradiation is an open system, and the relationships of equilibrium thermodynamics are inapplicable to it.

The effect of irradiation on the phase diagram is well exemplified by ordering alloys [255]. In the crystal areas covered by the radiation cascades, the long-range order is breaking, but as the temperature is high enough, an increased diffusion mobility of the irradiated atoms provides for reconstruction of the long-range order, while with no irradiation the alloy phase diagram has the form presented in Fig. 2.1a, but being irradiated with the flux density Φ, it takes the form shown in Fig. 2.1b. If $\Phi>\Phi_C$ (Fig.2.1b), the order has no time to be reconstructed at no temperatures.

There were attempts to regard a binary alloy under the irradiation as a quadrupole system *ABvi* [256], however, the results do not agree with the available observations.

The radiation induced decomposition was observed in a number of binary systems *(Al-Zn, Ni-Si, Ni-Be, Ni-Ge, Mg-Cd, W-Re)* [257], as well as in many industrial alloys whose phase diagrams are not known well enough to shurely distinguish between the radiation-induced and radiation-accelerated decomposition surely. The phase diagram obtained during irradiation of the *Al-Zn* alloys by the 1 MeV-energy electrons in a high-voltage microscope is given in Fig. 2.2.

A simple thermodynamic approach according to that the alloy free energy F is changed by δF if the stationary concentration of radiation defects is equal to C_j^+, does not permit to explain the observed shifts of the solubility curve.

Another, kinetic, approach is suggested in Ref. 258 (see also its generalization in Ref. 259). It is based on determining the growth conditions of a spherical incoherent precipitate with a relative volume effect $\delta=(\Omega_P-\Omega_M)/\Omega_M$ (Ω_P, Ω_M are the precipitate and matrix atomic volumes) in a medium of impurity and point-defect concentrations C_a and C_j^+, respectively. Capture of the diffusing impurity provides growth of the precipitate and that of vacancy or interstitial compensates a volume change. The precipitate state is characterized by the number of containing foreign atoms (n_a) and point defects (n_j). Evolution of the precipitate - growth, dissolution or vibration conditions - are described by a plane path ($n_a n_j$). There is a critical point (n_a^*, n_j^*) which may be a steady site, a steady focus, saddle or center, depending on the values of theoretical constants. The number of atoms is fitted by the radius

$$R_P^* = -2\gamma\Omega_P/\Delta\varphi, \tag{2.1}$$

where γ is the surface energy, $\Delta\varphi$ is the radiation-modified potential depending on δ, the generation rate of Frenkel pairs g and other parameters of the problem. When $\Delta\varphi>0$, all the particles dissolve.

The model has accounted for some experimental facts, for example β-phase precipitation (*Zn*) in *Al-Zn* alloys, but failed to do it for incoherent-precipitate formation when $\delta<0$.

It was noticed in Refs. 257, 258 that the precipitate growth under irradiation is substantively facilitated if the foreign atoms form with the vacancies and interstitials some mobile complexes which diffuse toward the precipitates where the vacancies and interstitials recombine. To describe this mechanism it is suggested that on the surface of the growing n_a-atom precipitate there are $z(n_a)$ centers which are capable to trap both the individual vacancies and interstitials and the *va* and *ib* complexes, the bond energy of which with the precipitate being much smaller than that of foreign atoms, so that the concentrations C_v and C_i in the matrix and at the precipitate are balanced.

Analogous reasons permit to obtain an expression for the concentration of the precipitates containing n_a foreign atoms and n_j point defects. In its turn, this gives a chance to write down the flux $J(n_a \longrightarrow n_{a+1})$ in the precipitate-size configuration space. As accepted in the classical nucleation theory [260], this may be treated as particle flux along the axis n_a with the n_a-depending diffusion coefficient in the strength field determined by the effective potential $\tilde{\varphi}(n_a)$, the critical precipitate size n_a^* being fitted by the turning point in the potential relief where $\partial\tilde{\varphi}/\partial n=0$.

Ref. 265 lists the calculation results for the solubility limit under the irradiation when $\Phi=0.02$ and 0.002 dpa/s for various values of δ, ε_i^m, ε_v^m, and volume v_P-precipitate fraction.

When the constant values are reasonable, the calculated curves agree, within the limits of measurement accuracy, with the experimental data presented in Fig. 2.2 although they give a stronger shift of the solubility limit under the irradiation.

The model presented is sufficiently rough, it does not account for various stages of precipitate origination, the deviation of the vacancy and interstitial distributions from the homogeneous equilibrium one due to their capture at the precipitates, the formation of defect clusters, a finite time of their recombination at one precipitate.

As distinguished from the Gibbs phase diagrams whose form is determined by the requirement to minimize the free energy and does not depend on the initial conditions and the peculiarities of kinetic processes leading to the equilibrium, a characteristic property of the diagram of phase equilibrium on irradiation is its essential dependence on the structural state of the alloy (of the type of interphase boundaries, the distribution of extended structural defects, etc.). It is for that reason that it is proposed the radiation-modified phase diagrams in Refs. 261, 262 to be referred to as structural- phase ones. In radiation physics of alloys, such diagrams are called to play a part analogous to that which the diagrams of thermodynamically equilibrium phases do in conventional material science.

The mechanisms of alloy-component separation at the dynamic stage were studied in Refs. 82, 85, 263-266. In so doing, the component separation in Refs. 82, 85, 263, 264 is caused by a difference in the interaction energies of various kinds of interstitial atoms with the elastic stress field created by all point defects. In Refs. 265,266 the reason for the composition inhomogeneities to be formed is a difference in the diffusion mobilities of various-kind interstitial atoms.

In Refs. 80, 81, 261, 262, 265-267 the foundations to kinetic theory of irradiated alloys and solid solutions have been regarded , as well as the conditions of phase equilibrium on irradiation, the nucleation rates of the new-phase precipitates in the cascades of moving atoms.

However, it should be noted that the solid solutions can be in the working state at various decomposition stages, each of these being characterized by its own peculiarities in the radiation processes to proceed.

2.3. Radiation-Induced Nucleation and Dissolution of the Precipitates

As far as the radiation-defect concentrations are much higher than those of vacancies and interstitials, they change kinetics of precipitate nucleation strongly.

The precipitate nucleation can take place both inhomogeneously in the material volume (homogeneous decomposition) [268-270] and close to the linear and triaxial defects, for example dislocations, voids, and grain interfaces [271-273].

Ref. 258 suggested a theoretical model being responsible for the formation of incoherent spherical precipitates. According to Ref. 258, the capture of foreign atoms diffusing into the solid solution by a new-phase nucleus ensures its growth, and the join of the vacancies and interstitials compensates for a change in its volume. In this case, the mechanism of foreign-atom transport toward the precipitate is not concretized.

Ref. 259 theoretically considers the growth of coherent- precipitate nuclei on the assumption that such precipitates grow at the expense of vacancy and interstitial complex diffusion toward them. On the nucleus surface, the complexes decompose, the foreign atoms join the precipitation, and the intrinsic point defects mutually recombine. Thus, in the model of coherent-precipitate formation considered in Ref. 259, these precipitates are zones of an increased recombination of vacancies and interstitials.

The theoretical study of precipitate formation is based on the classical nucleation theory [260]. One seeks in the precipitate-size configuration space for the density of distribution in sizes and the precipitate-nucleus fluxes [200,259,264] knowledge of whose explicit form makes it possible to find (see, e.g., Ref. 259) the critical sizes of steady precipitates (R_{P0}).

If the supersaturation arising as a result of increasing the impurity by the radiation defects is not large enough, the new-phase nuclei do not form and do not grow (see Ref. 88). However, near the radiation-defect sinks (dislocations, voids) there occur impurity seggregates which change the alloy properties, in particular its swelling stability [200]. The radiation-induced impurity seggregation on the radiation-defect sinks was observed in a number of systems, for example in nickel alloys [262].

The seggregation may be caused not only by the diffusion of *av* and *ib* complexes, but a so-called reversed Kirkendall effect [274,275]. A forward Kirkendall effect consists, as known, in that when the solid-solution concentration gradient is given and the coefficients of vacancy diffusion of the components are not equal ($D^a_v \neq D^b_v$), there appears in the system a resulting vacancy flux which leads to shifting the inertial marks, pair formation, and so on.

The reversed Kirkendall effect is the formation of the impurity-concentration gradient ∇C_a in the presence of the vacancy-concentration gradient ∇C_a and their oriented flux. For a concentrated solid solution [275], we have

$$\nabla C_a = \frac{D^b_i D^a_i}{\alpha' (D^b_i D^{\mathrm{irr}}_a + D^a_i D^{\mathrm{irr}}_b)} \left[\frac{D^a_v}{D^b_v} - \frac{D^a_i}{D^b_i} \right] \nabla C_v, \tag{2.2}$$

where $D^{a,b}_j$ is the diffusion coefficient of *a* or *b* atoms in the interstitial (dumbbell) or vacancy mechanisms, $D^{\mathrm{irr}}_{a,b}$ is the bulk diffusion coefficient of *a* or *b* atoms on irradiation, α' is the thermodynamic factor equal to unity for the ideal solutions. A maximum enrichment by the component *a* will take place if *a* is diffused by the interstitial mechanism, and *b* component by the vacancy one only (Fig. 2.3).

The seggregation on a thin-foil surface due to entraining the impurity by the radiation defects was investigated with the method of numerical integration of the balance equations for the foreign-atom, vacancy and interstitial concentrations, in Refs.275-278. The seggregation is maximal when $T=0.45T_{melt}$ in te system *Al-Zn*. The impurity is essentially transported by the mixed dumbbells. The temperature of maximal seggregation becomes higher with the irradiation intensity, which is shown diagrammatically in Fig. 2.4. At low temperatures, the moving complexes have no time to be formed, and the radiation defects recombine. At high ones, the seggregates disaggregate due to the reverse diffusion.

As the binding energy of *av* and *ib* complexes used to be no more than 0.2-0.3 eV [181], the complexes are decomposed at the elevated temperatures, and the mechanism of reverse Kirkendall effect becomes more effective. At moderate temperatures, especially in case of small-range impurity which readily forms the *av* and *ib* complexes, the entrainment mechanism becomes prevalent.

The question of a possibility of spatially homogeneous precipitate nucleation and of its relation to the selective nucleation on sinks was considered as early as in Ref. 279 for the binary systems. However, the temperature dependence of the average impurity concentration necessary for the homogeneous nucleation, was not considered. In Refs. 280, 281, the nucleation on sinks is compared with that in volume. Therefore, for consideration of the nucleation possibilities of precipitates in the material volume (homogeneous nucleation), it is necessary to take into account various types of the heterogeneous nucleation, for example that on dislocations, impurities and so on. Ref. 282 gives some estimates of the impurity concentration C_a^*, which are representative of the transition from the grain-boundary nucleation to the volume one. When the temperature becomes lower, the value C_a^* falls in spite of rising the ratio of C_a^* to the solubility value. However, in this case the solubility is reduced more badly than C_a^* . It follows from Ref. 282 that it is principally possible to reach in experiment a spatially homogeneous classical nucleation. Besides, the concentration C_a^* depends on the grain sizes, and the spatially homogeneous nucleation must be better performed in coarse-grained materials.

The irradiation can stimulate not only growth but dissolution of the precipitates. One of the dissolution mechanisms consists in that the primary and secondary fast particles transfer to the colliding foreign atoms of the precipitate such an energy (more than the threshold one) at which their path exceeds the distance to the precipitate surface, so that they return into the matrix. If the damage rate is equal to K dpa/s, then the flux of the atoms knocked- on of the precipitate through its surface is equal to $\Phi = 10^{18}K$ m^{-2}s^{-1} for typical values of the constants. Then the dissolution rate of a spherical particle P with the volume $V_P = (4/3)\pi R_P^3$ is equal to $dV_P/dt = -4\pi R_P^2 \Phi \rho_{nuc}^{-1}$, where ρ_{nuc} is the number of atoms per unit volume. The atomic mechanism of such a dissolution is simulated in Ref. 283 by computer. Some other mechanisms are discussed in Ref. 284.

2.4. Kinetic Equations for the Point-Defect Concentrations in a Solid Solution

In case of a solid solution with two kinds of substuting impurities, *a* and *b*, we shall believe that the impurities of the *a* kind (a size larger than that of matrix atom) form the complexes *va* with the vacancies (v), and the smaller-sized impurities (b) form the complexes *ib* with the interstitial atom (i). Then the system of equations for the point-defect concentrations has the form [89, 91, 93, 97-100, 285, 286]

$$\frac{\partial C_v(\mathbf{r},t)}{\partial t} = \mathrm{div}\mathbf{J}_v(\mathbf{r},t) + g_v(\mathbf{r},t) - \mu D_i C_i C_v - \alpha_{va} C_v C_a + \chi_{va} C_{va} - \mu_{ib}^v C_{ib} C_v - D_v C_v k_v^2, \tag{2.3}$$

$$\frac{\partial C_a(\mathbf{r},t)}{\partial t} = \mathrm{div}\mathbf{J}_a(\mathbf{r},t) - g_a(\mathbf{r},t) - \alpha_{va} C_v C_a + \chi_{va} C_{va} + \mu_{va}^i C_i C_{va} - D_a C_a k_a^2, \tag{2.4}$$

$$\frac{\partial C_{va}(\mathbf{r},t)}{\partial t} = \mathrm{div}\mathbf{J}_{va}(\mathbf{r},t) + \alpha_{va} C_v C_a - \chi_{va} C_{va} - \mu_{va}^i C_i C_{va} - D_{va} C_{va} k_{va}^2, \tag{2.5}$$

$$\frac{\partial C_m(\mathbf{r},t)}{\partial t} = -g_m-g_i-g_{bm}+2\mu D_iC_iC_v+\alpha_{ib}C_iC_b-\chi_{ib}C_{ib}+\mu^v_{ib}C_{ib}C_v, \qquad (2.6)$$

$$\frac{\partial C_i(\mathbf{r},t)}{\partial t} = \mathrm{div}\mathbf{J}_i(\mathbf{r},t)+g_i(\mathbf{r},t)-\mu D_iC_iC_v-\alpha_{ib}C_iC_b+\chi_{ib}C_{ib}-$$

$$\mu^i_{va}C_iC_{va}-D_iC_ik_i^2, \qquad (2.7)$$

$$\frac{\partial C_b(\mathbf{r},t)}{\partial t} = \mathrm{div}\mathbf{J}_b(\mathbf{r},t)-g_b(\mathbf{r},t)-\alpha_{ib}C_iC_b+\chi_{ib}C_{ib}+\mu^v_{ib}C_{ib}C_v-D_bC_bk_b^2, \qquad (2.8)$$

$$\frac{\partial C_{ib}(\mathbf{r},t)}{\partial t} = \mathrm{div}\mathbf{J}_{ib}(\mathbf{r},t)+g_{ib}(\mathbf{r},t) +\alpha_{ib}C_iC_b-\chi_{ib}C_{ib}-$$

$$\mu^v_{ib}C_{ib}C_v-D_{ib}C_{ib}k_{ib}^2. \qquad (2.9)$$

Here we have $j=v,i,va,ib,a,b,m$ (m is the matrix atom in a lattice site), $g_j(\mathbf{r},t)$ are the generation rates of the defects of j type, $\alpha_{jk}=N_jaD_j$ is the kinetic coefficient describing jk complex formation [246], N_j is the number of positions of intrinsic point defects j in the zone of foreign-atom capture,

$$\chi_{jk}= N_j^{jk}\, \nu_o \exp\,(-\varepsilon^b_{jk}\,/kT) \qquad (2.10)$$

is the kinetic coefficient of dissociation of the jk-complexes of the binding energy ε^b_{jk} ($k=a,b$), N_j^{jk} is the number of j-defect positions which it can reach after the *jk*-complex dissociation, $\mu^n_{jk} = 4\pi r_{jk}(D_{jk}+D_n)$ is the coefficient of $n=i,v$ defect recombination with *jk* complexes, and r_{jk} is the radius of such a recombination.

The balance equations being a constituent of equations (2.3)-(2.9) are listed in many papers (see, e.g., Refs. 15, 201, 275-278, 287, 288) and, as a rule, solved by numerical computations.

Analysis of the literature indicates that studying kinetics of movable point defects in the solid-solution decomposition process is yet paid little attention. should note the monographs [184,185], original works [245,289-292], computations (see, e.g., Refs. 275-278), as well as the results listed in Refs. 15, 200, 293, which were obtained for the stationary problem.

To solve analytically the systems of equations like (2.3)-(2.9) is, of course, failed in virtue of their extraordinary complicity. It is necessary to formulate a series of simplifying assumptions so that this system be reduced to a smaller number of equations, maintaining a fundamental physical sense of the problem considered.

Introduce the following simplifications:

1. The foreign atoms only form the movable complexes with one kind of intrinsic point defects (j): vacancies ($j=v$) or interstitials ($j=i$).
2. Diffusion-mobile are vacancies, interstitials, foreign atoms (a), and ones connected into the *va* and *ia* kind complexes.
3. One neglects in a solid solution the reactions of the type $ja+n \rightarrow a$ (here $n=v,i$ but $n \neq j$).

4. Curvature of the precipitate surface does not affect the near-surface layer defect concentrations.

Then the system of balance equations (2.3)-(2.9) for the point-defect concentrations may be written-down in the following form:

$$\frac{\partial C_j(\mathbf{r},t)}{\partial t} = g\text{-div}\mathbf{J}_q^j(\mathbf{r},t)-\alpha_{ja}C_jC_a+\chi_{ja}C_{ja}-\mu D_iC_jC_n-D_jC_jk_j^2, \tag{2.11}$$

$$\frac{\partial C_a(\mathbf{r},t)}{\partial t} = -\text{div}\mathbf{J}_q^a(\mathbf{r},t)-\alpha_{ja}C_jC_a+\chi_{ja}C_{ja}-D_aC_aK_a^2, \tag{2.12}$$

$$\frac{\partial C_{ja}(\mathbf{r},t)}{\partial t} = -\text{div}\mathbf{J}_q^{ja}(\mathbf{r},t)+\alpha_{ja}C_jC_a-\chi_{ja}C_{ja}-D_{ja}C_{ja}k_{ja}^2, \tag{2.13}$$

$$\frac{\partial C_n(\mathbf{r},t)}{\partial t} = g\text{-div}\mathbf{J}_q^n(\mathbf{r},t)-\mu D_iC_jC_n-D_nC_nk_n^2, \tag{2.14}$$

where j is a complex-forming point defect.

The diagram in Fig. 2.5 illustrates possible solutions of equations (2,11)-(2.14) together with expression (1.31) for the flux density $\mathbf{J}_a^j(\mathbf{r},t)$.

When calculating the concentration profiles of the mobile point defects j near the structural inhomogeneities of the material q one used to play into action the effective-medium method (see, e.g., Refs. 204, 206, 207) whose idea consists in the following. The q sink occupies the region F=I. One cuts out the region F=II of r_q size in a homogeneous absorbing medium (Fig. 2.6). This region is replaced by the same volume of a real medium with the q type sink at the center. The size of r_q is chosen in such a way in order for a correct volume fraction of structural inhomogeneities of the material to be conservedin the absorbing medium. The constituent medium obtained is obeyed the condition that its profiles far from the studed q sink ($r>r_q$) in the region F=III) are the same as those of homogeneous medium. The balance equations are solved in both regions and the obtained concentration profiles are joined at the interface between the real medium and the homogeneous absorbing one ($r=r_q$).

It is to formulate the corresponding boundary and initial conditions for the system of balance equations (2.11)-(2.14) depending on the considered problem. The initial atomic concentrations, in accordance with (1.31) and $\mathbf{J}^j_q(\mathbf{r},t)$ being equal to zero at the zero initial moment (t=0), will be presented in the form (see, e.g., Refs. 184, 185, 204):

$$C_a(\mathbf{r},0)=C^0_a\exp\left[-E^a_q(\mathbf{r},0)/kT\right],$$

$$C_{ja}(\mathbf{r},0)=C^e_{ja}\exp\left[-E^{ja}_q(\mathbf{r},0)/kT\right], \qquad (2.15)$$

$$C_j(\mathbf{r},0)=C^e_j\exp\left[-E^j_q(\mathbf{r},0)/kT\right].$$

Here C^0_j is the initial concentration of the foreign atoms unlinked into the complexes, and C^e_j and C^e_{ja} are the thermodynamically equilibrium concentrations of intrinsic point defects and complexes, respectively, far from the q-type sinks.

As to the boundary conditions, here it is to distinguish between two cases [285]:

a) when one considers kinetics of q-type sink formation (precipitates, voids, interstitial loops) in the material volume whose sizes essentially exceed the separation between them, the boundary conditions for the concentrations must be formulated only for the surface bordering this volume (external surface of the sample, grain boundaries),

b) if one studies the defect kinetics and their distribution affecting a change in some intrinsic sink (e.g. an individual precipitate, $q=P$), then it is necessary to write down the conditions for the corresponding concentrations on the surface of such a sink. In this case, the sink under consideration must be excluded from the sink strength sum (1.26), as far as now the problem treats its surface as an internal one.

Below we shall be interested in the distribution of j-type mobile point defects close to q sinks (i.e. case (b)). The boundary conditions are now representative in a standard form (see, e.g., Ref. 294):

$$\left[\beta_1(R_q)C_j(\mathbf{r},t)+\beta_2(R_q)\frac{\partial C_j(\mathbf{r},t)}{\partial n}\right]\Bigg|_{\Lambda_q(R_q)} = \vartheta(R_q,t),$$

$$\left[\beta_1(L_q)C_j(\mathbf{r},t)+\beta_2(L_q)\frac{\partial C_j(\mathbf{r},t)}{\partial n}\right]\Bigg|_{\Lambda_q(L_q)} = \vartheta(L_q,t). \tag{2.16}$$

Here the values β_1 and β_2 define the type of boundary conditions (2.16), $C_j(\mathbf{r},t)/\partial n$ is the derivative of the mobile j point-defect concentration with respect to the outside normal n to the surfaces $\Lambda_q(R_q)$ and $\Lambda_q(L_q)$; $R_q(\mathbf{r},t)$ and $2L_q(\mathbf{r},t)$ are the sizes of q-type sinks and the average separation between them, respectively, the form of the functions $\vartheta(R_q,t)$ and $\vartheta(L_q,t)$ being determined by a concrete problem.

To solve the balance equations (2.11)-(2.14) analytically with initial and boundary conditions (2.15), (2.16) is a complicated problem too. Such problems used to be solved by numerical computer methods (see, e.g., Refs. 275-278). However, in various limiting cases (see, e.g., the diagram in Fig. 2.5), the system of balance equations (2.11)-(2.14) is essentially simplified. In this case, it becomes possible to calculate analytically the concentration profiles of mobile point defects and hence their fluxes (1.31) toward the structural inhomogeneities of the material.

The aggregates of intrinsic point defects and the precipitates may be regarded as sinks for the vacancies and interstitials. Their evolution in the material exposed to irradiation influences the diffusion fluxes of intrinsic point defects toward some sink. Such influence is effectively accounted in the balance equations (2.11)-(2.14) via the sum of sink strengths. A systematic consideration of such equations will be given in the following Sections of this Chapter.

A solid solution dissolves under the action of temperature and irradiation in a diffusion way. This process is participated by many physical objects described either by equations (2.3)-(2.9) or a simplified system (2.11)-(2.14). Therefore, embedded into the precise solution of the problem is the diffusion coefficient where all diffusion channels with the participation of individual vacancies and interstitials and their various complexes are taken into account (see, e.g., Ref. 88). For the sake of simplicity, we shall regard the pipe diffusion through the dislocations and the surface diffusion through the grain and precipitate boundaries. A relative contribution of various channels depends on the temperature, being determined by proper binding and migration energies. The irradiation effect is reduced to the creation of the inhomogeneous vacancies and interstitials forming new transporters of foreign atoms.

In case when the diffusion of the a-kind impurity may be realized with the participation of vacancies or mixed dumbbells, the effective diffusion coefficient of the impurity has the form .

$$D_a^{\text{eff}} = D_a^v C_v^+ + D_a^i C_i^+,$$

where the migration coefficients are

$$D_a^j = \frac{1}{6} a^2 z^j \nu_a^j .$$

Here z^j is the coordinate number in the vacancy and interstitial mechanisms, ν_a^j is the effective frequency of the atom jump (including the correlation factor).

It is necessary in many problems of radiation physics, e.g. in studying the void-growth mechanism, to know the effective coefficients of vacancy and interstitial diffusion, which if the foreign atoms are present, can be captured by them with the formation of mobile or immobile complexes. In this case, the effective coefficient of vacancy diffusion is determined as follows [210]:

$$D_v^{\mathrm{eff}}=(D_v C_v^+ + D_{va} C_{va}^+)(C_v^+ + C_{va}^+)^{-1}.$$

The dependencies of C_v^+ and C_{va}^+ on the defect-generation rates are derived from the solution of balance equations (2.11)-(2.14). As a rule, only stationary solutions are considered, but in a general case, D^{eff} is time-dependent. The diffusion coefficients of free vacancies, D_v, and complexes, D_{va}, depend on the corresponding migration and binding energies. The relations obtained describe the complicity of kinetics in many-component materials. Used below will be the diffusion coefficients introduced in (1.24).

2.5. Radiation Swelling of Alloys

Of a great spectrum of the manifested radiation damages, we shall only decide for the present upon one of them to be the radiation swelling.

The study of such a phenomenon has been carrying out for a long time both experimentally (e.g. in Refs. 7, 8, 80, 185- 191, 295-302) and theoretically [26-29, 92-100, 193-195, 206, 207, 217, 246]. A great deal of materials has been observed, some models worked out to describe this phenomenon, the foundations of the theory built. However, as correctly noted in Ref. 40, solving key questions of the radiation-swelling theory requires the continued investigations to be made. The theoretical investigations of radiation swelling dealt with the homogeneous materials in the main. It is experimentally known [7, 295-300] that the impurities and secondary-phase precipitates being at various stages of their formation and evolution can lead to the swelling of the materials to be lowered.

The vacancies and interstitials create various elastic fields (see, e.g., Refs. 222, 223), therefore, their elastic interaction with a sink is also various. As a result, there occurs a preference to adsorption of mobile defects of one kind as compared to that of another kind on the same sink. Thus, for example, in the pure crystals, the interstitial adsorption by an edge dislocation is performed more intensively than the vacancy absorption [29, 204]. Besides, as far as the energy of vacancy migration is more than that of interstitial migration [181], the latter diffuses more rapidly in volume of the material, departing toward the sinks or forming new structural inhomogeneities (e.g. interstitial loops). This leads to a hindered mutual recombination of opposite intrinsic point defects, as a result of which the less mobile vacancies also have time either to go away toward their sinks or form the aggregates expanding into voids.

The microstructural changes which take place in the continuous point-defect generation (e.g. interstitial-loop and vacancy-void formation) and in their migration toward the existing sinks, were studied in detail in a number of works (see, e.g., Refs. 29, 203, 204, 206, 207, 246) in connection with the radiation stability problems of metals.

The situation becomes more complicated in solid solutions as far as the evolution of their defect structure is determined by the diffusion of both intrinsic point defects and foreign atoms. When we have a supersaturated solid solution, the precipitates of the secondary phase and the individual foreign atoms which are formed in the irradiation process, change its radiation swelling and radiation creep (see, e.g., Refs. 29, 203, 204, 206, 207, 246).

The recent investigations [8, 93, 96, 254, 299] have shown that it is necessary to account for the elastic fields of coherent precipitates in estimating the radiation-swelling rate. In Refs. 295-300 it was established that the radiation swelling can be as good as suppressed in the many-component chromium-nickel alloys where a developed homogeneous decomposition takes place.

Basic mechanisms and ways of weakening the radiation swelling of structural materials were preferentially concerned to the solid solution implantation by the elements (impurities) causing appreciable dilatation of the crystal lattice. Depending on the sign and value of mismatching the atomic radii of the alloying element and the matrix, there form compressed or extended regions around the impurities. In this case, the small-range alloying elements (e.g. *Si*, *Be*, *B*) are preferentially fixed with the interstitials, and the larger-range atoms (e.g. *Al*, *Ti*, *Mo*, *W*) with the vacancies. Weakening the swelling is explained in this case by delaying the opposite radiation point defects, by reason of which their escaping toward the sinks is prevented from (or decelerated) and *ipso facto* increases the probability of their recombination on meeting [306].

The decelerating defect departure toward the sinks is the basis of the mechanism of radiation point-defect capture by the coherent interfaces. The dislocation and interface "poisoning" by the impurities is indicated to be important as well. Forming impurity atmospheres around the dislocations and lowering the stacking fault energy during implantation are followed by relaxing the elastic stresses around them and decreasing the preference effect for the interaction with interstitials, which can favour an increased recombination of opposite point defects [307].

The mechanisms indicated, especially such as implantation by the elements which create crystal-lattice dilatation or have different diffusibility in a solid solution of a given composition, as well as cause the short-range order in the solid solution or influence the stacking fault energy, must favour the recombination of opposite point defects, and affect the radiation swelling in such a way that it be weak in development [248, 306-311]. However, those mechanisms preferentially sought for the austenitic chromium-nickel steels of 18-8 and 15-5 types (basic structural materials which are used in the fast-neutron atomic power plants) have regard, as a rule, only to the state of initial solid solution ignoring temporary structural changes which develop in them at various temperatures. Such approaches have not provided a substantial increase in the material resistance to the radiation swelling.

Refs. 295-300 experimentally showed that the ability of solids for the opposite radiation point defects to recombine increases mainly not due to the concentration and dimension mismatch which is created in volumes of initial solid solutions on their proper implantation (solid solution hardening), but the mismatch which is created during decomposition of supersaturated (metatable) solid solutions, and the decomposition intensity as well. Therefore, in order to suppress the radiation swelling in steels and alloys it is necessary, using a proper implantation, to provide a developed, continuously homogeneous decomposition of the solid solution which might have a pronounced incubation period and a definite value of volume dilatation at the interface "the forming secondary phase - matrix". The developed strong fields of structural stresses which appear in such a temporal decomposition, prove to be able to redistribute the fluxes of opposite point defects, weaken or suppress the interstitial migration toward the dangerous structural sinks and to provide a possibility of their recombination with the vacancies in the elastically distorted regions of the decomposing solid solutions.

A direct evidence in favour of the elastic field presence near the precipitates in a slightly swelling alloys is given by the electron microscopic patterns in which one can see, around the forming precipitates, a typical extinction contrast [26, 312-314], which vanishes at the separated (losing the coherency) phases. The presence of the high intrinsic stresses even reaching the value of brittle strength is evidenced by the occurrence of microcracks (and brittle fracture) in the decomposition process [315, 316].

The anomalous recombination of opposite point defects in the decomposing solid solutions is experimentally observed in the form of their radiation swelling which is slight in comparison with the solid solutions where the decomposition has been already finished or with those of a close composition and negligible selective decay. This is evidenced by the following data on the austenitic chromium-nickel alloys being the most common ones in reactor material science:

1. Swelling of the alloys where the decomposition proceeds weakly or does not go at all, is very appreciable [295, 296, 299, 300, 317]. Thus, the swelling of pure nickel exposed to the damaging doses up to 100 dpa, is 9-16 %. When irradiated by the 50-1000 dpa fluency, the weakly aging austenitic chromium- nickel alloys of the type 18-8, 18-8T and so on with about the same composition, give the swelling value γ up to 20-50 % [295, 298-300].

2. In the decomposing alloys with the small volume effect ($\Delta a/a$=1 %, Δa is the change in the lattice parameter a) when one deals with the stages of secondary-phase precipitate formation (e.g. the σ phases in steels), the swelling is appreciable: when $\Delta a/a$ = 14-20 %, the swelling is weak [295, 297-300].

3. Aging in whose course there takes place exhaustion of the secondary-phase nucleation sites and loss of coherency by the early separated phases, transits the alloy into the state weakly resistant to the radiation swelling. It has been indicated that the precipitation hardening alloy of the quality ХI5Н35М2БТЮР with a very high radiation swelling resistance in the state of optimal heat treating (γ=1-2 %) proved to be strongly inclined to the swelling ($\gamma \simeq$100 %) after the preliminary aging of 10,000 hr duration at the temperature 750°C and subsequent neutron irradiation of 25,000 hr duration at the temperature 600°C (the dose of about $1 \cdot 10^{13}$ neutron/cm^2). It follows from the diagram

of structural transformations that upon made aged at 600°C and during 25,000 hr, this alloy is in the state of the developed solid-solution decomposition being not so much denuded by the alloying elements which form the γ' phase.

4. The alloys with selective decomposition swell only at the grain boundaries, those with homogeneous decomposition weakly [295-300].

The precipitation hardening alloys are described by the most developed homogeneous precipitates of γ' and β phases of the Ni_3T type, the volume dilatation being appreciable in its formation at the incubation decomposition period. The precipitate density of such phase may reach 10^{22} m^{-3}. Such a situation is observed in the high-nickel solid-solution-hardening alloys. The mentioned positive factors cause a strong resistance to the radiation swelling of the alloys with the homogeneously developed decomposition (γ=1÷3 %).

The viewpoint presented here makes it possible to explain the nature of weak resistance of 18-8 and 15-5 type steels to the radiation swelling. These steels are characterized by the selective precipitation of the secondary phase with a low factor of dimensional mismatch. The steels of the indicated compositions show, as was mentioned, a low resistance to swelling (γ=20-50 %) [295, 297, 299, 300].

5. The swelling may be lowered by increasing the precipitate density by the rare earth element additions becoming nucleation sites for the secondary precipitates or forming an independent redundant phase. Refs. 296, 317 illustrate the influence of scandium and praseodymium

microadditions which effectively decreases the radiation swelling of nickel. Positive influence of such microalloyage is attrubuted to increasing the nucleation-site density of the secondary phase and strengthening the decomposition of the initial matastable solid solution.

In connection with their large atomic sizes, the rare-earth elements in the solid solutions of steels and alloys have a rather limited solubility which sharply falls with lowering the temperature [323]. Scandium and praseodymium incorporated into nickel from about 0.1 and more percents result in an appreciable supersaturation by these elements in the temperature irradiation interval (500-700°C). This provides the processes of secondary redundant phase formation to develop, the volumetric dilatation being essential at the precipitation stage. In consequence, the conditions are created for the processes of anomalous recombination of opposite defects to proceed.

In the selective decomposition (the steels of 18-8 type) there are no secondary phases in the grain body, and the processes of anomalous recombination are developed very slightly, since the barriers for the interstitials to sink toward the dislocations are absent. When the homogeneous decomposition is developed (e.g. the steel Х15Н35М2БТЮР), the interstitial escape toward the dislocations is hindered by the elastic stress fields occuring at the stage of precipitation of the γ' phase of $Ni_3(Ti,Al,Nb)$. In such a structural state (the distance between the dislocations is equal to ~500 nm, the size of coherent γ' phase is ~5-7 nm, the distance between the precipitates is equal to ~50 nm), the conditions are provided for the processes of anomalous recombination of opposite defects to proceed. The concept suggested, as well as the models considered, is validated by the direct experiments [296, 299]. Inhomogeneity in the scandium-alloying nickel leads to radiation swelling in the alloy inhomogeneous from grain to grain. The scandium-denuded grains do not practically contain the secondary phase being given to swelling. In the scandium-enriched grains (0.15-0.25 %), there forms a great number of radiation-induced intermetallic precipitates in the irradiation process and such grains posess a high stability to the radiation swelling.

2.6. Self-Consistent System of Equations to Describe the Radiation Processes in Alloys

The survey of experimental and theoretical data on the study of physical processes in the irradiated materials given in this Chapter and Chapter 1 permits to devide the whole process of radiation damage formation into three stages (see Fig. 1.1 and 2.7).

The dynamic stage is described by equations (1.4) for the distribution functions of moving atoms $\Phi_k(\mathbf{r},\varepsilon,\Omega,t)$ of k type, (1.16), (1.17) for the generation rates $g_j(\mathbf{r},t)$ of j defects and by the conditions of local conservation of particle number (1.15). The laws described by these equations are diagrammatically listed in Fig. 2.7 (Level 1).

The diffusion stage in the substitutional solid solutions is described by the system of nonlinear differential equations (2.3)-(2.9) for the concentrations $C_j(\mathbf{r},t)$ of the point defects j. The corresponding processes described by these equations are shown in Fig. 2.7 (Level II).

The stage of sink evolution. In all available theoretical investigations of the radiation damage problem numbered above, one basic assumption is made on that the growth of defect aggregates (q sinks) is performed slower than the j point-defect diffusion toward these sinks. Such adiabatic assumption allows not to trace out the system of equations in the form (2.3)-(2.9) up to the high cluster multiplicity, but to introduce their size (in case of spherical clusters it is the radius) $R_q(t)$ and the function of distribution in sizes $f_q(R_q,t)$ (dimension is m^{-4}) which are described by the following equations (see Fig. 2.7, Level III)

$$\frac{d}{dt}\left[\frac{4}{3}\,\frac{\pi}{\Omega_q}R_q^3(t)\right] = \sum_{j^+=1}^{n_j^+} I_q^{j^+}(R_q,t,k_{j^+}^2) - \sum_{j^-=1}^{n_j^-} I_q^{j^-}(R_q,t,k_{j^-}^2), \tag{2.17}$$

$$\frac{\partial f_q(R_q,t,k_j^2)}{\partial t} = W_q(R_q) - \frac{\partial}{\partial R_q}\left[f_q(R_q,t,k_j^2)\,\frac{dR_q}{dt}\right]. \tag{2.18}$$

Here W_q is the nucleation rate of q clusters of the given size R_q; Ω_q is the atomic volume in the q sink; j^+ and j^- are the types of j defects causing the q sink increse and decrease, respectively. In case of plane R_q-sized sinks, the left side of equation (2.17) takes the form

$$\frac{d}{dt}\left[\frac{\pi R_q^2(t)}{\Omega_q^{3/2}}\right].$$

In addition to equations (2.17) and (2.18), there are expressions

(1.38), (1.39) for the rates $I^j_q(R_q,t)$ and (1.40) for the efficiencies $K^j_D(r_*)$ of j defect adsorbing by the sinks q and the dislocations D.

It is also necessary to determine the sink strengths S^j_q which appear in expression (1.26) for the sum of sink strengths

$$k^2_j = \sum_{q=1}^{n_j} S^j_q(t,k^2_j). \tag{1.26}$$

The determination of S^j_q value is possible to perform most effectively using the effective medium method (see Section 2.4). In this case the homogeneous j-defect concentration in the matrix will be denoted as $C^m_j(t)$, and the values $I^j_q(R_q,t)$ and $K^j_D(r_*)$ will be sought for with (1.38)-(1.40) using the inhomogeneous solutions for the concentrations $C_j(r,t)$ inside the r_q-sized region near the sinks q to determine the flux densities $\mathbf{J}^j_q(R_q,t)$ according to (1.31).

Then, proceeding from the physical sense of the summands in the equations (2.3)-(2.9), one can write down as follows:

$$D_j C^m_j(t,k^2_j) S^j_q(t,k^2_j) = \int I^j_q(R_q,t) f_q(R_q,t,k^2_j) dR_q, \tag{2.19}$$

$$D_j C^m_j(t,k^2_j) S^j_D(t,k^2_j) = \rho_D K^j_D(t,k^2_j). \tag{2.20}$$

Substituting $S^j_q(t,k^2_j)$ and $S^j_D(t,k^2_j)$ from (2.19), (2.20) intoexpression (1.26), we obtain

$$k^2_j = \rho_D \frac{K^j_D(t,k^2_j)}{D_j C^m_j(t,k^2_j)} + \tag{2.21}$$

$$\left[D_j C_j^m(t,k_j^2)\right]^{-1} \sum_{q=1}^{n_j} \int I_q^j(R_q,t) f_q(R_q,t,k_j^2) dR_q .$$

In fact, the equation (1.26) or (2.21) closes the whole system of equations (1.4), (1.15)-(1.17), (1.38)-(1.40), (2.3), (2.17)-(2.20) (see Fig. 2.7), and makes it self-consistent, since the concentrations $C_j^m(t,k_j^2)$ themselves depend on k_j^2.

In studying the evolution of single-crystal structure under irradiation it is necessary to account for the influence of various structural inhomogeneities on one another, that is, to solve in common the whole complex of interdependent problems. Consider in short the methods allowing to solve this complicated problem in principle.

The evolution of single-crystal defect structure gives rise to changing the sink-strength sums k_j^2 appearing in (2.3)- (2.9). The k_j^2 determination is a combersome problem as requiring knowledge of not only diffusion fluxes of vacancies and interstitials toward their sinks (dislocations, loops, voids, etc.), but the distribution functions of intrinsic point-defect aggregates $f_q(R_q)$ in the sizes R_q. To the recent time this problem has not yet been resolved.

There exist approximate methods which make it possible to estimate the k_j^2 values. Thus, in Refs. 206, 207, 247-250, the approximate values of sink strengths S_q^j are calculated basing on the theory of reaction rates in the approximation of stationary inhomogeneous distribution of intrinsic point defects over the material volume. In Ref. 206, the sink strengths of vacancy voids and edge dislocations are determined in the approximation that the pure crystal does not contain other sinks. A complete k_j^2 determination requires to calculate the sink strengths for all structural inhomogeneities practically found in a pure crystal. Thus, estimated in Ref. 207 are the sink strengths of grain boundaries, interstitial and vacancy loops, dislocation networks. In Refs. 212, 213, the powers are calculated for such sinks as a dislocation dipole, dislocation wall, small-angular tilt boundary.

Among the approximate methods of k_j^2 determination, the cycle of Refs. 247-250 is to be noted. In connection with the problem of radiation swelling, extraordinary consideration was given in these publications to the influence of saturated sinks (secondary-phase precipitates, immobile foreign atoms) onto the strengths of intrisic point defect sinks toward the structural inhomogeneities of the material. Growth of the voids surrounded by a "cloth" of the saturable sinks was considered. Capturing the vacancies and interstitials which migrate toward the void, such a "cloth" effectively flattens their diffusion coefficients, thus increasing the volume recombination of the intrinsic point defects. The results of theoretical investigations [247-250] are expressed by a rather unwieldy formulas. While estimated numerically, they, however, agree well enough with the experimental data for the M316 type steels (see, e.g., Ref. 249).

Refs. 206, 207, 247-250 are deficient in that they do not take into account the influence of elastic fields of structural inhomogeneities on the diffusion mobility of intrinsic point defects.

The processes occurring on irradiation of the solid solutions at various stages of their decomposition can be presented diagrammatically in Fig. 2.8.

The secondary phases are not likely to precipitate in diluted solid solutions, therefore it is advisable in studying the irradiation effect to fix basic attention onto the impurity influence onto:

1) the dynamic stage of damage, i.e. onto descibing the development of moving atom cascades in an impurity material;

2) kinetics of annealing under irradiation with allowance for the impurities.

Of the ordered alloy radiation impurities, the processes of propagating the focusons and dynamic crowdions manifest themselves most effectively, which can result in the order parameter to be changed.

As noted above, two types of Gibbs transformations are possible in the concentrated solid solutions (see Refs. 324, 325), i.e. the solid-solution decomposition can be performed in two ways. First, the nucleation and growth of the secondary-phase precipitates are possible. In this case there forms a small region of a new phase inside the matrix where the nucleation composition takes place, which is different from that of the matrix. Such type of transformations may be otherwise referred to as heterogeneous one. The second way is the composition fluctuations. Initially homogeneous in composition, the region transforms into that containing the concentration wave whose amplitude increases with time. As distinguished from the first (heterogeneous) way, the changes are here limited in their magnitude, but not localized spatially. Such a decomposition is called spinodal one. As a rule, such decomposition does not realized in the materials which are used as structural ones for the needs of atomic energetics (see Section 2.5). Therefore, it is not considered below. When the solid solution decomposes according to the first type (nucleation and growth), it is necessary to separate two basic stages: the stage of coherent precipitate formation and the stage of separated presipitates of the secondary phase.

When the mismatch between the atomic parameters in the matrix and the forming precipitate is small, the difference in the atomic distances can be compensated by the elastic deformation of the two phases, but when the precipitates are large in size, the system can lower its energy with concentrating the deformation, e.g. in the transverse network of edge dislocations [325]. For the small precipitates when the nucleus diameter is smaller than the separation between the dislocations, one should anticipate that the nucleus remains coherent, but will be elastically deformed, since the elastic energy raises that of the precipitated phase. There exist two different types of deformation which may be due to the new-phase formation in the material [325]. The first type is concerned to the mismatch of two coherent phases at the stage of coherent precipitations. The deformation in the precipitate neighbourhood can reach the level which is sufficient to the structural defect (dislications, microcracks) formation removing the lattice distortion. When the deformation is of the second type, the new phase may occupy a volume different from that of the matrix fraction which be found to occupy. This type of deformation can appear in different ways.

For example, the precipitate increases with no changing in the site number, while the atomic volume in two phases differs. Another example is given by the case of nucleation of a less dense bcc ferrite in pure iron from austenite. The volume deformation is also realizes at an appreciable difference in the diffusion rates of two alloy components. For example, when zinc-reach γ-brass precipitates are formed from the β brass, there exists a more intense, inward-directed flux of zinc atoms which are diffusing more rapidly than the copper atoms which compensate this outward-directed flux.

If one takes that such coherent precipitates are spherical (with the radius R_P), then, according to the isotropic theory of elasticity, they create the homogeneous elastic stress fields, the diagonal components of whose stress tensor have the form [237]

$$\sigma_P^{rr\mathrm{I}}=\sigma_P^{\varphi\varphi\mathrm{I}}=\sigma_P^{\vartheta\vartheta\mathrm{I}}=-\frac{2E}{3(1-\nu)}(1-v_P)(\varepsilon_{\mathrm{I}}-\varepsilon_{\mathrm{II}}),$$

$$\sigma_P^{rr\mathrm{II}}=-\frac{2E}{3(1-\nu)}\left[\left(\frac{R_P}{r}\right)^3-v_P\right](\varepsilon_{\mathrm{I}}-\varepsilon_{\mathrm{II}}), \qquad (2.22)$$

$$\sigma_P^{\varphi\varphi\mathrm{II}}=\sigma_P^{\vartheta\vartheta\mathrm{II}}=\frac{2E}{3(1-\nu)}\left[\frac{1}{2}\left(\frac{R_P}{r}\right)^3+v_P\right](\varepsilon_{\mathrm{I}}-\varepsilon_{\mathrm{II}})$$

where r,ϑ,φ are the spherical defect coordinates with respect to the precipitation center, F=I, II are the precipitation (I) and matrix (II) regions, E is the Young modulus, ν is the Poisson coefficient, ε_{I} and $\varepsilon_{\mathrm{II}}$ are the relative deformations inside and outside the precipitate, and the volume fraction v_P of regions I, i.e. the precipitates themselves, is expressed in the form

$$\nu_P = \int V_P(R_P) f_P(R_P) dR_P. \tag{2.23}$$

Here $V_P(R_P)$ is the volume, $f_P(R_P)$ is the function of precipitate distribution in R_P sizes. The components of stress tensor (2.22) satisfy the equilibrium equation over the whole crystal volume

$$\mathrm{Sp}\sigma_P = \nu_P \mathrm{Sp}\sigma_P^{\mathrm{I}} + (1-\nu_P)\mathrm{Sp}\sigma_P^{\mathrm{II}} = 0, \tag{2.24}$$

where

$$\mathrm{Sp}\sigma_P^{\mathrm{I}} = \frac{2E}{1-\nu}(1-\nu_P)(\varepsilon_{\mathrm{I}}-\varepsilon_{\mathrm{II}}); \quad \mathrm{Sp}\sigma_P^{\mathrm{II}} = -\frac{2E}{1-\nu}\nu_P(\varepsilon_{\mathrm{I}}-\varepsilon_{\mathrm{II}}).$$

The material regions and the dependence $\mathrm{Sp}\sigma_P^F$ on r are shown diagrammatically in Fig. 2.9.

The energy of j point-defect interaction with the precipitate P has the form (see, e.g., Refs. 184, 185)

$$E_P^{jF} = \Delta\Omega_j^F \mathrm{Sp}\sigma_P^F, \tag{2.25}$$

where $\Delta\Omega_j^F$ is the volume relaxation of j-type defect in the F region of the material. It follows from (2.22) that $\mathrm{Sp}\sigma_P(r)$ is constant in the material volume and undergoes discontinuity on the precipitate surface (see Fig. 2.9). In this case the elastic field of coherent precipitates does not appear in balance equations (2.11)-(2.14), since the energies of vacancy and interstitial interaction with precipitate (2.25) are constant over the solid solution volume, as $\mathrm{Sp}\sigma_P^F$ does, and undergo discontinuities at the phase interface. Substituting (2.22) into (2.25), we have

$$E_P^{j\mathrm{I}} = -\frac{2E}{1-\nu}\Delta\Omega_j^{\mathrm{I}}(1-\nu_P)(\varepsilon_{\mathrm{I}}-\varepsilon_{\mathrm{II}}); \quad E_P^{j\mathrm{II}} = \frac{2E}{1-\nu}\Delta\Omega_j^{\mathrm{II}}\nu_P(\varepsilon_{\mathrm{I}}-\varepsilon_{\mathrm{II}}), \tag{2.26}$$

i.e. $E_P^{jF}(r)$ is a stepwise function.

The effect of the elastic field of spherical coherent precipitates on the defect concentration profiles manifests itself through the boundary conditions only. In accordance with (1.31) and (2.25), the account for elastic stresses (2.22) in the balance equations (2.11)-(2.14) leads merely to the renormalization of thermodynamically equilibrium concentrations of mobile point defects on the q-sink surface (when $r=R_q$) having the form (see, e.g., Ref. 204):

$$\tilde{C}_{j0}^{eF} = C_j^{eF}\exp\left[-\frac{E_q^{jF}(R_q)}{kT}\right] = C_j^{eF}\chi_j^F, \tag{2.27}$$

where C_j^{eF} is the thermodynamically equilibrium concentration of j defects in the F region.

At the stage of separating the secondary-phase precipitates, the elastic stresses are removed due to the precipitate boundary formation.

The stages of heterogeneous decomposition differ one from another. The precipitates are evolved in different ways, therefore, the change in the defect distribution under the irradiation is of independent interest.

To estimate with correctness various radiation changes in the material properties (see Fig. 2.7) it is necessary to solve the whole system of equations (1.4), (1.15)-(1.17), (1.38)- (1.40), (2.3)-(2.9), (2.17)-(2.20). In this case, equation (2.21) (more correctly, here we have so many equations how many kinds of mobile j defects exist) is that to determine the concentrations $C_j^m(t)$ even independent of k_j^2 [93, 94, 96-99]. A subsequent substitution of thus calculated concentrations C_j^m into the expressions for f_q, R_q, and S_q^j will permit to obtain the expressions for the distribution functions $f_q(R_q,t)$, the sizes $R_q(t)$ and the sink strengths $S_q^j(R_q,t)$, which do not depend on k_j^2. And this makes it possible, in its turn, to calculate the changes in such material properties, as swelling, hardening or creep, occurring under irradiation.

CHAPTER 3
BASIC RADIATION PROCESSES IN DILUTE SOLID SOLUTIONS

3.1. Cascades of Moving Atoms in Impure Crystals

Extremely important question of theory and practice is concerned with the structure and parameters of the atomic displacement cascades and the regions in alloys formed by them, especially in polycomponent alloys. It is at once to say that the progress in experimental investigations of these phenomena leaves behind the theory. It is easy to understand due to great difficulties in calculating all the variety of radiation processes in polycomponent materials.

To study how impurities affect the development of cascades is of an extraordinary interest, as much as such situations can be correlated to the diluted solid solutions.

In the experiments described in Ref. 326, the monocrystals of *Cu* and the *Cu-Al* alloys (up to 10 % of *Al*), oriented along <011>, were irradiated by Cu^+ ions of the energy 30 keV and the exposure dose $2 \cdot 10^{15}$ m^{-2}, and the crystals of *Mo* by Mo^+ ions of the energy 60 keV and the doses $7.4 \cdot 10^{15}$ and $5 \cdot 10^{16} m^{-2}$. When $1.7 \cdot 10^{-4}$ at.% and less doses of nitrogen were introduced into *Mo*, one observed a substantial decrease in defect yield (5 times) and in total number of Frenkel pairs per 1 RKA. The average size of dislocation loops, which was about 2.77, was decreased down to 1.92 nm. The addition of carbon affected more slightly. The same impurities in *Cu* affected the defect parameters little.

In Ref. 327 analogous data for *Mo* are given, and the concentration of vacancy loops at 535°C is indicated to be 10 times as low as that at 20°C.

The study of *Pt* +4 at.% *Au* alloy irradiated by 30 keV- energy ions at the temperature 40 K was carried out in Ref. 107. In this case a detailed autoionomicroscopic analysis of the atomic (vacancy) structure of a lamellar formation was fulfilled. An interesting, if unique, case of denuded zone was described which regenerated into a four-layer lamellar vacancy formation. The microscopic analysis has been run at $T_{irr} \simeq 40$ K (before and after the isochronal annealing which was performed at 100 K after the irradiation), i.e. higher than the temperature of substage II of the return. The atomic structure of the denuded zone was reproduced clearly. The zone-forming vacancies were observed in the region of $(75\bar{1})$ face. The total number of vacancies in the denuded zone was equal to 31, and they were located in the region formed as a 2nm-diameter disk in four planes (220). The atomic vacancy concentration was 35- 44 %. The experimentally observed vacancies were localized at the distance about 9 nm from the irradiated surface of the sample, i.e. at the very tail of the distribution profile. The authors of Ref. 107 compare their result with the data of transmission electron microscopy. They indicate that the number of vacancies united into the dislocation loops in tungsten, iridium and platinum come up to a small part of the total number of vacancies in such irradiated metals. In particular, it is about 10 % in the case of *W*. This fact is even important because of the success in the electron microscopic analysis of the irradiated metals largely depends on the ability of the denuded-zone core rich in vacancies to transform into a dislocation vacancy loop (since such defects are within easy reach of the experimental examination).

Cascade processes in binary systems (nickel-aluminum, platinum-silicon) have been investigated in Ref. 328 by the method of backward Rutherford scattering; the method is described in detail, for example in Ref.329. The same method was used in Ref. 330, where the authors studied the above process in the system of insoluble elements tungsten-copper. In both works the irradiation of the samples was made by the ions Ar^+ of the energies 250 and 30 keV, respectively. The temperature of the samples during the irradiation did not exceed 373 K. Since the diffusion processes in a nonequilibrium solid solution *W* in *Cu* manifested themselves at the temperatures higher than 720 K [331], the interpenetration observed in the system *W-Cu* could be only caused by a dynamic process of cascade mixing. To check this assumption, the authors of Ref. 330 have compared the dose dependence of depth change in the mixed layer which was obtained by them, with the results of analytical evaluations in the scope of the model worked out in Ref. 332. As a result, it was obtained that the depth of a mixed layer depends approximately on the irradiation fluency as a square root, being in a good accord with the dependence calculated.

It follows from the results of Ref. 330 that the dynamic mass transport in the cascades of atomic collisions including those formed by the chains of focused displacements, can essentially exceed the mass transport in the thermal spike attendant the cascade. It is this spike that leads to mixing the insoluble elements, while the atom migration in a thermal spike must favor lamination of the elements. This is borne out by the fact that the creation of the conditions similar to those of the thermal spike by means of irradiating *W-Cu* by a laser with the modulated performance does not result in the material mixing [333].

Note also a very interesting fact of riching the defect regions of single high-energy cascades of atomic displacements by the frontier-zone atoms, which was detected in Ref. 334 with microsond analysis of the neutron-irradiated *W*+10 at.% *Re* and *W*+25 at.% *Re* alloy samples. Previously this effect was found in computerized modeling the cascades in alloys [263].

Interesting physical results on passing the cascades of moving atoms in impure crystals have been obtained in the computerized experiments of Refs. 114, 138, 172.

In the calculations of Ref. 145 the impurity concentration is maintained constant and high enough to increase the probability of the impurity atoms reaching the cascade development zone (C_a=5 at.%). In all alternate calculations the body-centered cubic (bcc) α-*Fe* lattice was stimulated, and a pair potential was used to form it [145]. The impurity mass was varying in a wide limits (m_a/m=1/7-7, where m is the matrix atom mass, and m_a is the impurity atom mass). In a quite number of cases one has succeeded in the choice of impurity - matrix atom interaction potential. This was the case of the carbon-iron [335] and aluminum-iron systems [336]

(m_a/m=0.214 and 0.482, respectively). In the absence of the appropriate potential, the isotopic impurity approximation was used, the mass of impurity atom was changed, but the interaction remained the same as for the matrix atoms. The PKA energy was given equal to ε_0=100 eV, the initial direction of PKA motion was at the angle of $\theta_0=31^\circ$ with respect to <100> in the (001) plane. Such initial conditions have permitted to form a rather simple collision cascade featuring all the typical properties.

As was shown by the modeling calculation, the cascade was most appreciably affected by the impurities located in the path of propagating the cascade branches (dynamic crowdions, focusons). As a result of reactions with the impurities, the cascade essentially changes its shape and the size of propagation zone. This effect is the more appreciable, the more different the atom masses m and m_a. There is a tendency to shrinkage of the region occupied by the cascade, to localization of the cascade. In Fig. 3.1 one can see the trajectories of cascade atoms in various cases: a pure crystal (Fig. 3.1a), a crystal with the impurities whose masses are considerably larger than those of the matrix atoms (Fig. 3.1b). It is seen from Fig. 3.1 that the occurrence of an impurity shortens the cascade, decreases the number of colliding atoms, that of displacements and substitutions. So, for example, when the impurity atoms appear, the number of substitutions falls by 38 % as compared with the case of a pure crystal.

The cascade localization leads to intensification of intra-cascade radiation annealing, as the conditions of separating the opposite defects, vacancies and interstitials, are getting worse. The consequence was that the vacancy concentration in the denuded zone and the cluster dimensions decrease in the aftercascade defect distribution. In the big cascades this leads to decreasing the sizes of the vacancy loops forming as a result of the collapse of the denuded zone. This effect is observed experimentally in the form of decreased average dimension of vacancy loops in the irradiation of the impurity-containing samples [326]. The cascade localization may lead to a complete radiation annealing of the denuded zone, which will at once result in the decrease in a general concentration of loops. This loop concentration decrease is also observed when the impurities are added into the sample [326]. Apart from this, the aftercascade defects are redistributed in this case. Thus in the case of a light impurity the place of vacancy formation migrates. In a number of cases, the formation of mixed dumbbell interstitials is observed (impurity atom - matrix atom, see for example Fig. 3.1b). The occurrence of the latter is found by the method of channeling in the irradiated impurity-containing samples [337].

In a series of Refs. 338-340, the processes of radiation damage in polycomponent materials caused by various ions, protons, neutrons have been considered. Using as an example gadolinium monoaluminate, the conditions of mutating the neutron damages with a proton beam are estimated. The computerized calculations of cascade functions are performed for TaO and the damage profiles of V in V_3Si irradiated by helium atoms of the energy 2 MeV. The program of simulating the atomic collision cascade in polycomponent materials by the method of pair collisions up to 20 eV is worked out. The profiles of vacancy and interstitial distribution are obtained for Fe_3Al irradiated by aluminum and iron ions of the energy 5 keV. A shift by 1 nm of the maximum for the interstitial profile toward more depths with respect to that of the vacancies in the irradiation with both aluminum and iron. However, in the irradiation down to the depth of 4 nm, a strong departure from stoichiometry is observed toward the increase in the iron atom content and there is an *Al*-rich region at the depth of 8 nm. The projective path of iron atoms is less than that of aluminum ones. It is also clear that it is easier for the aluminum atom to knock on an aluminum atom than an iron one, that is why the maxima of profile occurrence are widely different.

In Refs. 341-343 , by computer simulations, the Fourier components of the velocity field for a cascade developing in nickel are obtained, which describe the phonon distribution function. It is shown that generated in the cascade are some nonlinear collective oscillations of definite frequencies. It is shown moreover that when developing the cascade, a phonon shift toward the low- frequency region is observed. It can affect the diffusion processes, and the dynamic rearrangements in the cascade regions. It is shown by the computerized method of pair collision in one- and polycomponent targets, taking into account the thermal atom displacements, that the projective path of an incoming atom decreases with increasing the target temperature, but the number of knocked-on atoms and that of collisions in the cascade increase. A change in the spatial defect distribution with the temperature is revealed.

The theoretical estimations of impurity effect on the cascade characteristics are developed in a series of works. Thus in Refs. 130,134 the possibilities of calculating the cascade functions in the scope of the modified Kinchin-Pease model are considered for two-atom isotropic bodies. The atoms of different kinds were assigned different displacements energies related by the proportion $\varepsilon_{d2}=h\varepsilon_{d1}$. The collisions were assumed elastic and isotropic. The interval of changing the energy was divided into the sections delimited by the energies ε_{d1}, $2\varepsilon_{d1}$, ε_{d2}, $2\varepsilon_{d2}$. The balance equations were set up for the cascade functions $\nu_{kn}(\varepsilon_0)$, i.e. an average number of the *n*-type atoms displaced at the cost of the *k*-type atom with the energy ε_0. The solution to these equations gives the cascade function

$$\nu(\varepsilon_0)=\nu_{12}(\varepsilon_0,h)+\nu_{22}(\varepsilon_0,h)=\nu_{11}(\varepsilon_0,h)+\nu_{21}(\varepsilon_0,h) \tag{3.1}$$

which depends on both ε_0 and *h*.

Using an idea of the "damage energy", one succeeded to develop a method of calculating the cascade functions as applied to the polycomponent materials. The expression for the "damage energy" $E_D(\varepsilon_0)$ of *n* type PKA in the material where it undergoes collisions with various *k*-type particles, was first derived in a general form in Ref. 165. The energy losses for the electron excitation were not taken into account in this case. It is of interest to generalize the involved idea of the "damage energy" onto the case of polycomponent targets. In Ref. 131 the general expressions are listed for the cascade function and the number of substituting collisions. Similar results were obtained in Ref. 137 for two- and three-atomic targets. In this paper studied were the cascades in the energy range 1-1000 keV, where the elastic energy losses are prominent. Here account was taken of the losses ε_d in each collision, and the inelastic losses for electron excitation were allowed for in the continuous deceleration approximation. In Ref. 138, Lindhard's equation for $E_D(\varepsilon_0)$ is numerically integrated, which is written as applied to the case of two- and three-component materials of the type Al_2O_3, MgO, $MgAl_2O_4$, UO_2, and so on. The results are approximated in the form

$$E_D(\varepsilon_0)=\varepsilon_0(1+C_1\varepsilon_0^{0.15}+C_2\varepsilon_0^{0,75}+C_3\varepsilon_0)^{-1}. \qquad (3.2)$$

In the papers indicated, the tables of the C_k constants (k=1, 2,3) are listed for all materials investigated. Compared were the efficiencies of radiation damage defined as the ratio $E_D(\varepsilon_0)/\varepsilon_0$ for each PKA type in a number of two-component materials. The basic result is that in all materials involved the defect formation efficiency is less for small ε_0 and more for large ε_0 as compared with a homogeneous material composed of light atoms. At the same time for the heavy atoms, the damage efficiency at all energies proves to be less than that in a homogeneous material composed of the same atoms. In a number of cases the differences in the corresponding absolute values reach some dozens of percents.

In the calculations of Ref. 344 it was found for copper that the impurity atoms lower the channeling efficiency. However, this does not markedly tell on the cascade dimensions. But the impurity manifests itself more badly on annealing the cascades when there occur vacancy loops bordered with either complete or partial dislocations, or tetrahedra composed of the stacking faults in depending on the stacking fault energy (which is largely decreased by some impurities).

In Ref. 157, the flux of moving atoms $\Phi_k(E)$ is defined with allowance for the inelastic losses on some simplifying assumptions and applied to the calculation in a polycomponent system. Although, as it was noted above, the energy ε_i at which the inelastic losses are comparable to the elastic ones, lies in the range of hundreds keV, for the PKA energy, $\varepsilon_0>\varepsilon_i$ is valid in some cases. Obtained in Ref.15 are simple expressions for the cascade function in the limiting cases of $\varepsilon_0\gg\varepsilon_i$ (a) and $\varepsilon_0\ll\varepsilon_i$ (b) and different ratios of the matrix atom mass m and foreign PKA one m_a. These expressions are applicable to the solid solutions and compounds of the type A=B. In particular, with the notation $\nu_{kn}(\varepsilon_0)$ (see above) we have for the cases (a) and (b)

$$\nu_{nn}(\varepsilon_0)=\frac{2\varepsilon_0}{3\varepsilon_{dn}} \quad (a); \qquad \nu_{nn}(\varepsilon_0)=\frac{8\varepsilon_{in}}{9\varepsilon_{dn}} \quad (b).$$

In the case (b) the cascade function is independent of ε_0. If PKA is of the type k, then when the concentration ratio is $C_k\ll C_n$, we obtain

$$\nu_{kn}(\varepsilon_0)\nu^{-1}_{nn}(\varepsilon_0)=A_{kn}\,\frac{C_k}{C_n}\cdot\frac{3\varepsilon_{dn}}{2\varepsilon_{dn}+h_0\varepsilon_{dk}},$$

where $h_0=(m+m_a)^2(4mm_a)^{-1}$, the constant A_{kn} is approximately equal to 1 and slightly depends on m, m_a and the atomic numbers z, z_a.

Thus the computerized experiments and partially the indirect natural experiments provide information on the influence of the alloying impurity mass on the cascade characteristics.

3.2. Cascade Processes in Crystals with Impurities of Different Masses

The presence of impurities in crystals, i.e. the difference of pure materials from solid solutions, tells on the development of cascade processes, and hence, on an embryo distribution of point defects in the materials irradiated. On the basis of the present state of the subject having presented in Chapter 1, it needs to develop a theory for the cascades of moving atoms for the solid solutions containing foreign atoms of different masses. Since the analytical means to solve the system of equations (1.4), (1.7), (1.15)-(1.17) are limited, we shall perform their generalization for diluted solid solutions, i.e. in condition that the impurity concentration C_a is not large, and consider some limiting cases of heavy impurities of the mass $m_a \gg m$ and light ones of the mass $m_a \ll m$ [6, 103, 148, 345, 346].

3.2.1. Solid Solutions with Heavy Impurities ($m_a \gg m$)

The distribution function of moving atoms, $\Phi_{HI}(\mathbf{r},\varepsilon,\Omega,t)$ in the case of heavy impurities (HI) and on assumptions listed in Section 1.2.1. of Chapter 1, is described by equation (1.4), which here takes the form

$$\Omega\cdot\nabla\Phi_{HI}(\mathbf{r},\varepsilon,\Omega)+\rho_{nuc}\sigma(\varepsilon)\Phi_{HI}(\mathbf{r},\varepsilon,\Omega)+C_a\sigma_a(\varepsilon)\Phi_{HI}(\mathbf{r},\varepsilon,\Omega)=$$

$$\rho_{\text{nuc}} \int_{\varepsilon}^{\varepsilon_0} d\varepsilon_1 \int_{4\pi} d\Omega_1 \Phi_{\text{HI}}(\mathbf{r},\varepsilon_1,\Omega_1)\sigma(\varepsilon_1,\varepsilon_1\text{-}\varepsilon)\mathcal{K}_s(\varepsilon_1,\Omega_1 \rightarrow \varepsilon,\Omega)+ \tag{3.3}$$

$$C_a \int_{\varepsilon}^{\varepsilon_0} d\varepsilon_1 \int_{4\pi} d\Omega_1 \Phi_{\text{HI}}(\mathbf{r},\varepsilon_1,\Omega_1)\sigma_a(\varepsilon_1,\varepsilon_1\text{-}\varepsilon)\mathcal{K}_s(\varepsilon_1,\Omega_1 \rightarrow \varepsilon,\Omega)+$$

$$\rho_{\text{nuc}} \int_{\varepsilon}^{\varepsilon_0} d\varepsilon_1 \int_{4\pi} d\Omega_1 \Phi_{\text{HI}}(\mathbf{r},\varepsilon_1,\Omega_1)\sigma(\varepsilon_1,\varepsilon)\mathcal{K}_d(\varepsilon_1,\Omega_1;\varepsilon,\Omega).$$

Here $\sigma(\varepsilon)=\sigma_{0m}(\varepsilon)$ and $\sigma_a(\varepsilon)=\sigma_{am}(\varepsilon)$ are the total cross-sections of interaction of the matrix atom (m) moving with the energy ε either with the same matrix atom at a lattice point or with a foreign atom (a), respectively (see (1.5)). The same holds for the differential cross-sections of energy transfer $\sigma_{nk}(\varepsilon_1,\varepsilon)$. As noted in Section 1.2.1, to solve the equation (3.3) we shall avail ourselves of the method of distribution function $\Phi_{\text{HI}}(\mathbf{r},\varepsilon,\Omega)$ expanded in infinite series over the systems of orthogonal polynomials [152-156, 164, 166]. The equation (3.3) is rewritten in cylindrical coordinates R,ϑ,Z, in this case the axis Z coincides with the direction of PKA motion Ω_0 starting from the point $\mathbf{R}=0$ and a simmetry is supposed with respect to the polar angle ϑ. Then the distribution function $\Phi_{\text{HI}}(\mathbf{r},\varepsilon,\Omega)$ can be represented in the form of infinite triple series

$$\Phi_{\text{HI}}(Z,R,\varepsilon,\Omega) = \Phi_{\text{HI}}(z,r,\varepsilon,\Omega) =$$

$$\frac{e^{-z}z^{\alpha}e^{-r^2}}{2\pi a^3}\sum_{l=0}^{\infty}\sum_{m=-l}^{l}Y_l^m(\Omega)\sum_{s=0}^{\infty}\sum_{t=0}^{s}\frac{b_{st}^{\alpha}}{a^t\eta_s^{\alpha}}L_s^{\alpha}(z)\times \tag{3.4}$$

$$\sum_{k=0}^{\infty}\sum_{n=0}^{k}\frac{\Delta_{kn}}{a^n h_k}u_k(r)\Phi_{lm}^{tn}(\varepsilon),$$

where $\Phi_{lm}^{tn}(\varepsilon) = 2\pi\int_0^{\infty}z^t dz\int_0^{\infty}x^n dx\Phi_{lm}(z,x,\varepsilon)$ are the momenta of the distribution function; $\Phi_{lm}(z,x,\varepsilon)$ are the factors for the distribution function expansion over the spherical functions $Y_l^m(\Omega)$; $L_s^a(z)$ are the Laguerre polynomials; $Z=az$; $R=ar$; z, r are dimensionless coordinates; a is an interatomic spacing; x, y are the Cartesian coordinates, and $r=x^2+y^2$;

$$\eta_s^{\alpha} = \frac{\Gamma(\alpha+s+1)}{s!}; \quad b_{st}^{\alpha} = \frac{(-1)^t\,\Gamma(\alpha+s+1)}{t!(s-t)!\Gamma(\alpha+t+1)}.$$

The polynomials $u_k(r)=\sum_{n=0}^{k}\Delta_{kn}r^n$ are defined on the interval $0\le r<\infty$ with the weight function $\omega(r)=r\exp(-r^2)$ and are given by the orthogonality relation

$$\int_{\gamma_1}^{\gamma_2}\omega(r)u_k(r)u_{k'}(r)r^m dr = h_k\delta_{kk'}.$$

When $\gamma_1=0$ and $\gamma_2 \longrightarrow \infty$, we have $m=0$, as well as $\Delta_{00}=1$ and $h_0=0.5$ [158].

Substitute the expansion (3.4) together with the Lindhard cross-section (1.10) into equation (3.3) and use the relation $\Phi_{HI}(\varepsilon)=2\sqrt{\pi}\Phi_{00}^{00}(\varepsilon)$ (see Section 1.2.1), and obtain the following equation for the flux of impurity atoms $\Phi_{HI}(\varepsilon)$:

$$\varepsilon^2 \frac{d^2\Phi_{\text{HI}}(\varepsilon)}{d\varepsilon^2} + (2-\zeta)\varepsilon \frac{d\Phi_{\text{HI}}(\varepsilon)}{d\varepsilon} - \beta_0 f \Phi_{\text{HI}}(\varepsilon) = 0, \tag{3.5}$$

where β_0 is explained in (1.12), and the quantity f depends on the impurity composition and the orientation types of atomic motion, taken into account in the consideration. For a diluted isotropic solid solution,

$$f = f_{\text{HI}} = \left[1 + \frac{C_a}{\rho_{nuc}} \cdot \frac{F_{aj}}{F}\right]^{-1} = \left[1 + \frac{C_{rel}}{1 - C_{rel}} \cdot \frac{F_{aj}}{F}\right]^{-1} \simeq 1 - \Delta, \tag{3.6}$$

where F_{aj}, F are defined following (1.8) and (1.11), $C_{rel} = C_a(\rho_{nuc} + C_a)^{-1}$, $\Delta \equiv C_{rel} F_{aj} F^{-1}(1 - C_{rel})^{-1}$. With allowance for the dynamic crowdions (see Section 1.2.1), the quantity f takes the form

$$f = f_{\text{ID}} = f_{\text{HI}}(1 - P_D^d). \tag{3.7}$$

Substituting (3.6) and (3.7) into (3.5), we obtain the following solutions for the fluxes of the atoms moving in the solid solution with heavy impurities, neglecting the dynamic crowdions:

$$\Phi_{\text{HI}}(\varepsilon) = \Psi_0 \nu_0 \chi_{\text{I}} \left[\frac{\varepsilon_0}{\varepsilon}\right]^{\kappa_{\text{I}}} \simeq \Psi_0 \nu_0 \chi_0 \left[\frac{\varepsilon_0}{\varepsilon}\right]^{\kappa_0} \left[1 - \Delta \frac{\chi'}{\chi}\right] \left[\frac{\varepsilon_0}{\varepsilon}\right]^{\delta_0 \Delta}, \tag{3.8}$$

and with allowance for the dynamic crowdions

$$\Phi_{\text{ID}}(\varepsilon) = \Phi_{\text{HI}}(\varepsilon) \left[\frac{\varepsilon}{\varepsilon_F}\right]^{P_D^d \delta_{\text{I}}}, \quad \check{\varepsilon} < \varepsilon < \varepsilon_F; \tag{3.9}$$

where

$$\chi_{\text{I}} = 2^{-1}\left\{1 - (1-\zeta)[4\beta_0 f + (1-\zeta)^2]^{-1}\right\} \simeq \chi_0 - \chi' \Delta;$$

$$\kappa_I = 2^{-1}\left[1-\zeta+\sqrt{4\beta_0 f+(1-\zeta)^2}\,\right] \simeq \kappa_0 - \delta_0\Delta;$$

$$\delta_I = \beta_0 f\left[4\beta_0 f+(1-\zeta)^2\right]^{-1} \simeq \delta_0 - \delta'\Delta;$$

$$\chi' = \frac{\delta_0(1-\zeta)}{4\beta_0+(1-\zeta)^2}; \quad \delta' = \delta_0\frac{2\beta_0+(1-\zeta)^2}{4\beta_0+(1-\zeta)^2}.$$

Using the definition of the cascade function (1.7) with a substitution of (3.8) and (3.9) into (1.7) and taking into account of (1.12), we obtain

$$\nu_{HI}(\varepsilon_0) = \Psi_0\chi_I\left(\frac{\varepsilon_0}{2\varepsilon_d}\right)^{\kappa_I+0.5} \simeq \tag{3.10}$$

$$\nu_0(\varepsilon_0)\left(1-\Delta\frac{\chi'}{\chi}\right)\left(\frac{2\varepsilon_d}{\varepsilon_0}\right)^{\delta_0\Delta}\left(1+\Delta\frac{\delta_0}{\kappa_0-0.5}\right) = \nu_0(\varepsilon_0)H_I(\varepsilon_0),$$

$$\nu_{HID}(\varepsilon_0) = \nu_{HI}(\varepsilon_0)H_D\left(\frac{\varepsilon_F}{2\varepsilon_d}\right)^{p_D^d\delta'\Delta}, \tag{3.11}$$

where H_D is defined in (1.13).

Illustrate the dependencies obtained for the cascade functions using as an example the heavy lead impurity in copper ($m/m_a \simeq 0.3$). The parameters are here as follows: χ_0=0.38, χ'=0.6, κ_0=0.75, δ_0=0.28 at ζ=0.7. The dependence $\nu_0(\varepsilon_0)/\nu_{HI}(\varepsilon_0)$, as well as the function of relative impurity concentration C_{rel}, is shown in Fig. 3.2, from where it is seen that the cascade function decreases in the presence of heavy impurity.

The dependencies of the cascade functions $\nu_{HI}(\varepsilon_0)$ and $\nu_{HID}(\varepsilon_0)$ on the PKA energy ε_0 are also presented in Fig. 1.3 (curves 5 and 6) (see Chapter

1) for copper by (3.10) and (3.11) at the relative lead concentration $C_{rel}=10^{-1}$. It is seen from Fig. 1.3 and formulas (3.10), (3.11) that the general number of defects per cascade in the presence of heavy impurity becomes lower.

In the scope of the above enumerated assumptions, the expressions for the vacancy generation rates (1.16) and the interstitial ones (1.17) can be written as follows:

$$g_{vI}(\mathbf{r})=\rho_{nuc}\int_{\varepsilon}^{\varepsilon_0} d\varepsilon_1\int_{4\pi} d\Omega_1\Phi_{HI}(\mathbf{r},\varepsilon_1,\Omega_1)\int_0^{\varepsilon_1} d\varepsilon''\sigma(\varepsilon_1,\varepsilon'')[p(\varepsilon_1\text{-}\varepsilon'')p(\varepsilon'')+$$

$$p(\varepsilon_1\text{-}\ \varepsilon'')q(\varepsilon'')P_D^d], \tag{3.12}$$

$$g_{iI}(\mathbf{r})=\rho_{nuc}\int_{\varepsilon}^{\varepsilon_0} d\varepsilon_1\int_{4\pi} d\Omega_1\Phi_{HI}(\mathbf{r},\varepsilon_1,\Omega_1)\times$$

$$\int_0^{\varepsilon_1} d\varepsilon''\sigma(\varepsilon_1,\varepsilon'')q(\varepsilon_1\text{-}\varepsilon'')q(\varepsilon'')(1\text{-}P_D^d)+ \tag{3.13}$$

$$C_a\int_{\varepsilon}^{\varepsilon_0} d\varepsilon_1\int_{4\pi} d\Omega_1\Phi_{HI}(\mathbf{r},\varepsilon_1,\Omega_1)\int_0^{\varepsilon_1} d\varepsilon''\sigma_a(\varepsilon_1,\varepsilon'')q(\varepsilon_1\text{-}\varepsilon'')q(\varepsilon'')+$$

$$\rho_{nuc}\mu_{iD}(\mathbf{r},t).$$

Re-establish the distribution function $\Phi_{HI}(x,y,z,\varepsilon,\Omega)$ from its momenta (3.8) or (3.9) (see Refs. 152-156, 164, 166, as well as (3.4)) in the $L_0Y_0^0u_0$-approximation and obtain the expression for $\Phi_{HI}(x,y,z,\varepsilon,\Omega)$ both in the isotropic case and in case when the dynamic crowdions are taken into account in the presence of heavy impurity:

$$\Phi_{HI}(x,y,z,\varepsilon,\Omega) = \frac{\Psi_0 \nu_0 \chi_0}{4\pi^2 a^2 \Gamma(\alpha+1)} e^{-z} z^{\alpha} e^{-(x^2+y^2)} \left[1-\Delta \frac{\chi'}{\chi_0}\right] \left(\frac{\varepsilon_0}{\varepsilon}\right)^{\kappa_I}$$

$$\text{if } \varepsilon_F < \varepsilon < \varepsilon_0, \tag{3.14a}$$

$$\Phi_{HI}(x,y,z,\varepsilon,\Omega) = \frac{\Psi_0 \nu_0 \chi_0}{4\pi^2 a^2 \Gamma(\alpha+1)} e^{-z} z^{\alpha} e^{-(x^2+y^2)} \left[1-\Delta \frac{\chi'}{\chi_0}\right] \times$$

$$\left(\frac{\varepsilon_0}{\varepsilon}\right)^{\kappa_I} \left(\frac{\varepsilon}{\varepsilon_F}\right)^{\delta_I P_D^d} \quad \text{if } \check{\varepsilon} < \varepsilon < \varepsilon_F \,. \tag{3.14b}$$

Substituting (3.14) into (3.12) and (3.13), we obtain the expressions for the corresponding generation rates in the form

$$g_{\nu I}(\mathbf{r}) = g_{\nu}(\mathbf{r}) \left(\frac{\varepsilon_F}{\varepsilon_0}\right)^{\kappa' \Delta} \left[1-\Delta \frac{\chi'}{\chi_0} + \Delta \frac{\delta_0}{\kappa_0+\zeta-1}\right], \tag{3.15}$$

$$g_{iI}(\mathbf{r}) = \left[\tilde{g}_{iI}(\mathbf{r}) + B_{2D} e^{-\eta z} (\eta z)^{\alpha} \cdot e^{-\theta^2 (x^2+y^2) \varphi'_I}\right] \varphi_I(\varepsilon_0), \tag{3.16}$$

where

$$g_{\nu}(\mathbf{r}) = B_{\nu} e^{-z} z^{\alpha} e^{-(x^2+y^2)},$$

$$\tilde{g}_{iI}(\mathbf{r}) = B_i e^{-z} z^{\alpha} e^{-(x^2+y^2)} \left[1+\frac{\Delta}{1-P_D^d}\right],$$

$$\varphi_I(\varepsilon_0) = \left(\frac{\check{\varepsilon}}{\varepsilon_F}\right)^{(\delta_0 - P_D^d \delta')\Delta} \left\{1+\Delta\kappa' \left[\frac{2\kappa_0+3(\zeta-1)}{\kappa_0+2\zeta-(\kappa_0+\zeta-1)(\kappa_0+2\zeta-1)} + \right.\right. \tag{3.17}$$

$$\frac{2\kappa_0+4\zeta-1}{(\kappa_0+2\zeta)(\kappa_0+2\zeta-1)}\Big]\Big(\frac{\varepsilon_F}{\varepsilon_0}\Big)^{\kappa'\Delta}\Big[1-\Delta\frac{\chi'}{\chi}+\Delta\frac{\delta_0}{\kappa_0+\zeta-1}\Big]\Big\},$$

$$\varphi_I'=\varphi_I^{-1}\Big[1-\Delta\frac{\delta_0-P_D^d\delta'}{1-\kappa_0-2\zeta+P_D^d\delta+10a^{-1}}+\Delta\frac{\delta_0}{\kappa+\zeta-1}\Big].$$

Here

$$B_j=\frac{\Psi_0\hat{B}_j}{\pi a^3\Gamma(a+1)},\quad (j=v,i,2D);\quad \hat{B}_v=\hat{B}\Big[1+P_D^d\Big(\frac{\varepsilon_d}{\check{\varepsilon}}\Big)^{\zeta}\Big];$$

$$\hat{B}_i=\hat{B}\hat{\gamma}(1-P_D^d);\quad \hat{B}_{2D}=\hat{B}\gamma';\quad \hat{B}=\frac{\sqrt{2}\,\rho_{nuc}F}{\zeta\sqrt{m}(\kappa_0+\zeta-1)}\varepsilon_0^{\kappa+0.5}\varepsilon_F^{1-\zeta-s}\varepsilon_d^{-\zeta};$$

$$\hat{\gamma}=\frac{\zeta[\kappa_0+2\zeta-(\kappa_0+\zeta-1)(\kappa_0+2\zeta-1)]}{(\kappa_0+2\zeta-1)(\kappa_0+2\zeta)}\varepsilon_F^{\kappa_0+\zeta-1-P_D^d\delta_0}\varepsilon_d^{\zeta}\check{\varepsilon}^{1+P_D^d\delta_0-\kappa_0-2\zeta};$$

$$\gamma'=\zeta P_D^d\chi_0\Big[1-\kappa_0-2\zeta+P_D^d\delta_0+10a^{-1}\Big]^{-1}.$$

List some figures with $\zeta=0.7$, which correspond to the selection of PKA ε_0-energy interval of the order of several keV for copper: $\kappa_0=0.75$, $\chi_0=0.38$ (see (1.12)), $P_D^d=0.32$. These values correspond to $\varepsilon_d=30$ eV, $\varepsilon_F=70$ eV, $\check{\varepsilon}=5$ eV. As a result of integrating in (3.12) and (3.13), we have $\eta=0.28$, $\theta=0.2$, $\hat{\gamma}=0.034$, and $\gamma'=0.012$. For the lead impurity in copper, the estimates $\varphi_I=0.43$ and $\varphi_I'=0.72$ at $C_{rel}=3\cdot10^{-1}$ can be obtained.

The expression (3.15) for the vacancies and the first summand in (3.16) for the interstitials describe the rates of their generation due to the nonorientally moving atoms. The second summand in (3.16) describes sweeping the interstitials from the central cascade region toward the periphery due to the dynamic crowdions, thus favoring the spatial redistribution of vacancies and interstitials. A spatial separation of defects in the cascade zone follows from the energy distribution over the cascade [6]. However, the $L_0 Y^0_0 u_0$- approximation for the function of moving atom distribution is insufficient, in order such separation to be noticeable in the appropriate expressions for $g_j(\mathbf{r})$. For this purpose, it is necessary to take into account higher approximations for the function $\Phi_{HI}(Z,R,\varepsilon,\Omega)$ by (3.4).

It is seen from the expressions (3.15)-(3.17) that the presence in a solid solution the impurity, heavy and in the limit undisplacable, leads to decreasing the vacancy and interstitial generation rates in the cascade of moving atoms as compared with those of a pure metal.

The concentrations of defects (vacancies and interstitials) satisfy the equation (1.14) if the instantaneous recombination described by the rate $g^R_j(\mathbf{r},t)$ (see (1.18)) is taken into account. In the considered case of heavy impurity, we obtain from (1.18) for the instantaneous recombination rate, for example that of vacancies as follows:

$$g^R_{v\mathrm{I}}(\mathbf{r},t) = C_{v\mathrm{I}}(\mathbf{r},t) \int_{\mathbf{r}-\mathbf{r}_{0R}}^{\mathbf{r}+\mathbf{r}_{0R}} d\mathbf{r}'\, g_{i\mathrm{I}}(\mathbf{r},t) \equiv C_{v\mathrm{I}}\,\omega^R_{i\,\max}\,.$$

In this case a complete number of vacancies can be estimated as

$$N^R_{v\mathrm{I}}(t_k) = N_v(t_k)\varphi_{\mathrm{I}}(\varepsilon_0)\left[1\text{-}\exp\left(-\omega^R_{i\,\max}t_k\right)\right]\left(\omega^R_{i\,\max}t_k\right)^{-1},$$

where t_k is the time of cascade development, $N_v(t_k)$ is a complete number of vacancies if the instantaneous recombination in the pure material is left out of account [104]. For C_{rel}, the estimation of N_v/N^R_{vI} gives the vacancy number decreased by about the order of magnitude.

Consequently, in a diluted solid solution with heavy, in the limit undisplaceable, impurities, such impurities only adsorb the energy, increasing the amplitude of their oscillations at a lattice point. As seen from the expressions obtained for the cascade functions (3.10), (3.11) and from the distributions of vacancy and interstitial generation rates in the cascade region (3.15), (3.16), the general defect number in the cascade region becomes lower (see Fig. 1.3). As distinguished from the pure material, the interstitials are formed not only in the periphery of the cascade region, but in its central part, so far as this decreases the dynamic crowdion path (see Sections 3.3 and 3.4). So the mass removal from the central cascade region is hindered. Due to them the instantaneous recombination between the vacancies and interstitials is intensified, which also leads to decreasing the cascade dimensions.

Thus it has been shown that the injection of a heavy impurity into the crystal results in decreasing the cascade efficiency and dimensions, and so favors the localization of the cascade regions, which is in agreement with the data of computerized experiments [172]. Hence, the dimensions of the cascade regions formed under the irradiation may be decreased by means of selecting the composition of a diluted solid solution with allowance for the masses of alloying elements.

3.2.2. Solid Solutions with Light Impurities ($m_a \ll m$)

In a deluted solid solution where the impurity atoms are lighter than the matrix ones ($m_a \ll m$), the moving atom cascade arising under the irradiation will be more complex than in the pure case [152-158] or in a material with heavy impurity (see Section 3.2.1). The problem of such a cascade description becomes more complicated, so long as it needs in this case not only to solve two equations of (1.4)-type for the distribution functions for two types of moving atoms, but to take into account their interaction.

The light impurities also participate in the formation of cascade moving atoms, as the matrix atoms. In the framework of the assumptions listed in Section 1.2.1, write down, using a general expansion of the (3.4)-type distribution function, the equations for the momenta of these functions, which are energy distributions. In this case it is to keep in mind that the cascade is created by a primarily knocked-on matrix atom of the energy ε_0. This is valid for the diluted solid solution when the probability for the incoming particle to come across the matrix atom is more than that to come across the impurity atom with the concentration $C_a < \rho_{\text{nuc}}$. Thus interacting, the moving atoms of the energy ε can be formed both in the processes of scattering and displacement on the fixed matrix atoms and as a result of scattering on the impurity atoms. The whole complex of these mechanisms is reflected in the equation for the distribution function $\Phi_{\text{LI}}(\varepsilon,\varepsilon_0)$ in ε-energy of the moving matrix atoms initiated by the ε_0-energy PKA in a solid solution with light impurities:

$$\Phi_{LI}(\varepsilon,\varepsilon_0)\left[\rho_{nuc}\int_{\varepsilon}^{\varepsilon_0} d\varepsilon''\sigma_{11}(\varepsilon,\varepsilon'')+C_a\int_{\varepsilon}^{\Lambda\varepsilon} d\varepsilon''\sigma_{21}(\varepsilon,\varepsilon'')\right]=$$

$$\rho_{nuc}\int_{\varepsilon}^{\Lambda(\varepsilon_0-\varepsilon)} d\varepsilon''\sigma_{21}(\varepsilon+\varepsilon'',\varepsilon'')\Phi_{LI}(\varepsilon+\varepsilon'',\varepsilon_0)+$$

$$\rho_{nuc}\int_{\varepsilon}^{\Lambda(\varepsilon_0-\varepsilon)} d\varepsilon''\sigma_{11}(\varepsilon+\varepsilon'',\varepsilon)\Phi_{LI}(\varepsilon+\varepsilon'',\varepsilon_0)+ \tag{3.18}$$

$$C_a\int_{\varepsilon}^{\Lambda(\varepsilon_0-\varepsilon)} d\varepsilon''\sigma_{21}(\varepsilon+\varepsilon'',\varepsilon'')\Phi_{LI}(\varepsilon+\varepsilon'',\varepsilon_0),$$

where $\Phi_{LI}(\varepsilon,\varepsilon_0)=\int d\mathbf{r}\int d\Omega\,\Phi_{LI}(\mathbf{r},\varepsilon,\Omega,\varepsilon_0)$; $\sigma_{jk}(\varepsilon,\varepsilon'')$ was defined in (1.10); $j,k=1,2$ (1 is a matrix atom, 2 is an impurity one); $\Lambda=4mm_a(m+m_a)^{-1}$.

The moving impurity atoms are initiated by the impurity PKA atom of the energy ε_{0a} (Fig. 3.3). There takes place an energy transfer both to the impurity atoms and the matrix atoms, and in the latter case there can form a cascade of matrix atoms of the third generation. Such atoms will not be considered in our calculation, as an energy exchange between the atoms of different kinds on condition that $m_a \ll m$ leads to decreasing the efficiency of tertiary cascades. Besides, the analytical calculation was carried out with the use of the Lindhard cross-section (1.10) that sets the role of small energy transfers too high, which leads to an anomalous increase in the displacement number in the periphery cascades. For the distribution function $\Phi_a(\varepsilon,\varepsilon_{0a},\varepsilon_0)$ in the energies ε of the light moving impurities initiated by the ε_{0a}-energy impurity PKA, one can write an equation analogous to (3.18) in the form

$$\Phi_a(\varepsilon,\varepsilon_{0a},\varepsilon_0)\left[\rho_{\rm nuc}\int_{\varepsilon}^{\Lambda\varepsilon}d\varepsilon''\sigma_{12}(\varepsilon,\varepsilon'')+C_a\int_{\varepsilon}^{\varepsilon}d\varepsilon''\sigma_{22}(\varepsilon,\varepsilon'')\right]=$$

$$C_a\int_{\varepsilon}^{\varepsilon_{0a}-\varepsilon}d\varepsilon''\sigma_{22}(\varepsilon+\varepsilon'',\varepsilon'')\Phi_a(\varepsilon+\varepsilon'',\varepsilon_{0a},\varepsilon_0)+$$

$$C_a\int_{\varepsilon}^{\varepsilon_{0a}-\varepsilon}d\varepsilon''\sigma_{22}(\varepsilon+\varepsilon'',\varepsilon)\Phi_a(\varepsilon+\varepsilon'',\varepsilon_{0a},\varepsilon_0)+ \tag{3.19}$$

$$\rho_{\rm nuc}\int_{\varepsilon}^{\Lambda(\varepsilon_{0a}-\varepsilon)}d\varepsilon''\sigma_{12}(\varepsilon+\varepsilon'',\varepsilon'')\Phi_a(\varepsilon+\varepsilon'',\varepsilon_{0a},\varepsilon_0).$$

The cascade function of a composite cascade in a diluted solid solution with light impurities $\nu^0_{\rm LI}(\varepsilon_0)$ can be here represented in the form of the sum of cascade functions: $\nu_{\rm LI}(\varepsilon_0)$ for the main cascade of matrix atoms and $\nu_a(\varepsilon)$ for all subcascades of the impurity atoms created in the main cascade, i.e.

$$\nu^0_{\rm LI}(\varepsilon_0)=\nu_{\rm LI}(\varepsilon_0)+\nu_a(\varepsilon_0). \tag{3.20}$$

The cascade functions $\nu_{\rm LI}(\varepsilon_0)$ and $\nu_a(\varepsilon_0)$ are determined by (1.7) from the functions $\Phi_{\rm LI}(\varepsilon,\varepsilon_0)$ and $\Phi_a(\varepsilon,\varepsilon_{0a},\varepsilon_0)$ which by means of the relations (see (1.3))

$$\Phi(\varepsilon,\varepsilon_0)=\nu(\varepsilon)\Psi(\varepsilon,\varepsilon_0) \tag{3.21}$$

are expressed through the number of moving ε-energy matrix atoms in the cascade, $\Psi_{\rm LI}(\varepsilon,\varepsilon_0)$, and a total number of moving ε-energy impurity atoms in all the cascade created by the PKA with the energy ε_0, i.e. $\Psi_a(\varepsilon,\varepsilon_0)$ is a sum of the moving atoms with the energy ε over all the subcascades created by the impurity atoms with the energy ε_{0a}.

In each individual subcascade with ε_{0a} there form $\Psi_a(\varepsilon,\varepsilon_{0a},\varepsilon_0)$ of moving impurity atoms. This figure may be found from the equation (3.19) using (3.21). In order to find a total number of moving impurity atoms $\Psi_a(\varepsilon,\varepsilon_0)$ of the energy ε, it is also necessary to know the number of cascades with the energy ε_{0a}, i.e. that of knocked-on impurity atoms $N_a(\varepsilon_{0a},\varepsilon_0)$ with the energy ε_{0a}. One should take into account that the value ε_{0a} can range on the interval $2\varepsilon_d<\varepsilon_{0a}<\zeta_0\varepsilon_0$. Then we have

$$\Psi_a(\varepsilon,\varepsilon_0) = \int_{2\varepsilon_d}^{\zeta_0\varepsilon_0} d\varepsilon_{0a} N_a(\varepsilon_{0a},\varepsilon_0)\Psi_a(\varepsilon,\varepsilon_{0a},\varepsilon_0). \tag{3.22}$$

The number of impurity atoms with the energy ε_{0a} can be found from the relation

$$N_a(\varepsilon_{0a},\varepsilon_0)=t_{ka}C_a \int_0^{\zeta_0(\varepsilon_0-\varepsilon_{0a})} d\varepsilon''\Phi_{LI}(\varepsilon_{0a}+\varepsilon'',\varepsilon_0)\sigma_{a0}(\varepsilon_{0a}+\varepsilon'',\varepsilon_{0a}). \tag{3.23}$$

The procedure described here can be extended to the next cascade generation. Such an approach may be used to calculate the dissociation of high-energy cascade into subcascades with allowance for the spatial atom distribution in the cascade.

Solving the equation (3.18) as in Section 3.2.1, using (3.21) and substituting into (1.7), we obtain an expression for the first term in (3.20) in the form

$$\nu_{LI}(\varepsilon_0)=\nu_0(\varepsilon_0)(1-\Delta\eta_{LI})\left(\frac{2\varepsilon_d}{\varepsilon_0}\right)^{\kappa_{LI}\Delta}, \tag{3.24}$$

where $\nu_0(\varepsilon_0)$ is defined in (1.12) and Δ in (3.6) and η_{LI}, κ_{LI} depend on the parameter of Lindhard cross-section intricately.

Let us solve the equation (3.19) and substitute the solutions of the equations (3.18) and (3.19) into (3.22) and (3.23), and next into (1.7). Thus we obtain an expression for the cascade function $\nu_a(\varepsilon_0)$ in the form

$$\nu_a(\varepsilon_0)=\nu_{LI}(\varepsilon_0)\Delta^2\beta(\zeta), \tag{3.25}$$

where $\beta(\zeta)$ depends on ζ only.

Substitute (3.24) and (3.25) into (3.20) and obtain the following expression for the cascade function of a diluted solid solution with light impurities:

$$\nu^0_{LI}(\varepsilon_0)=\nu_0(\varepsilon_0)(1-\Delta\eta_{LI})\left(\frac{2\varepsilon_d}{\varepsilon_0}\right)^{\kappa_{LI}\Delta}(1-\Delta^2\beta). \tag{3.26}$$

It is seen from the formula (3.24) that the matrix atom cascade function slightly depends on the impurity concentration, and $\nu_a(\varepsilon_0)\sim C_a^2$. If one takes, for example, tungsten (m) with copper impurity (m_a), selects the matrix PKA energy equal to ε_0=10 keV, then ν_a/ν_{LI}=0.3 is obtained at C_{rel}= $2\cdot 10^{-2}$, and ν_a/ν_{LI}=0.1 at C_{rel}=10^{-2}.

As a result of the calculation procedure worked out for the cascade functions in diluted solid solutions with light impurities, a simple expression (3.26) is obtained, which shows, as in the case of heavy impurity, a decreased average number of displaced atoms in the cascade developing in a solid solution when compared with a pure metal (cf. $\nu_0(\varepsilon_0)$ according to (1.12)). This approach makes it possible to obtain a spatial distribution of moving atoms in the cascade zone, however, due to difficulties in performing it analytically, it is reasonable to solve the appropriate system of equations with the use of computer.

So it has been shown that, as in the case of heavy impurity (as compared to the matrix atoms), the presence of a light impurity in a diluted solid solution localizes the cascade of moving atoms, which is also in agreement with the results of computerized experiments [172]. It has been thus shown that the efficiency of radiation cascades can be lowered by selecting the alloy components as early as at the dynamic stage of solid solution radiation damage.

3.3 Channeling Effect on the Cascade Function

Influence of the channeling effect if the energies of moving atoms are $\varepsilon>\varepsilon_C$, has been discussed in Chapter 1, Section 1.2.1, where an estimation formula for the cascade function is given which has been obtained in Refs. 112, 149. The atom emissivity from the cascade region center is made easy due to the channeling effects. This leads, on the one hand, to increasing the cascade region size, and, on the other hand, to decreasing the cascade function, since the fast atoms captured into the channel lose almost all their energy for a quasicontinuous deceleration and thus does not create the secondary knocked-on atoms if there will not occur a subcascade in a high-energy region.

Evaluate, in the framework of the developed procedure, the cascade function of the material taking into account the channeling [347]. Write down, as in Section 3.2.2, the equation not for the spatial distribution function of moving atoms, but for the density $\Psi_C(\varepsilon)$ of the moving atoms of the energy ε with the use (1.4), (3.4), and (3.21) and the allowance for the channeling:

$$\Psi_C(\varepsilon)\int_0^{\varepsilon} d\varepsilon''\sigma(\varepsilon,\varepsilon'')=(1-P_C^s)\int_0^{\varepsilon_Q-\varepsilon} d\varepsilon''\sigma(\varepsilon+\varepsilon'',\varepsilon)\sqrt{1+\frac{\varepsilon''}{\varepsilon}}\,\Psi_C(\varepsilon+\varepsilon')+ \tag{3.27}$$

$$(1-P_C^d)\int_0^{\varepsilon_Q-\varepsilon} d\varepsilon'\sigma(\varepsilon+\varepsilon',\varepsilon)\sqrt{1+\frac{\varepsilon'}{\varepsilon}}\,\Psi_C(\varepsilon+\varepsilon').$$

Substituting the energy transfer cross-section in the form of the Lindhard one (1.10), we obtain the following expression:

$$\frac{d\ln\Psi_C(\varepsilon)}{d\varepsilon}=\frac{1}{\varepsilon}\,\frac{\alpha_C-(1-P_C^s)\alpha_s-(1-P_C^d)\alpha_d(\varepsilon)}{(1-P_C^s)J_s(\varepsilon)+(1-P_C^d)J_d(\varepsilon)}, \tag{3.28}$$

where

$$\alpha_C=\frac{1+\zeta}{\zeta};\quad \alpha_s=\frac{2\zeta^2+3.5\zeta}{4\zeta^2-0.25};\quad \alpha_d=\left[\left(\frac{\varepsilon_0}{\varepsilon}\right)^{1.5-\zeta}-1\right](1.5-\zeta)^{-1};$$

$$J_s=\left[\left(\frac{\varepsilon_0}{\varepsilon}\right)^{1.5-2\zeta}-1\right](3-4\zeta)^{-0.5}+2\zeta(4\zeta-1)^{-1}; \tag{3.29}$$

$$J_d=2\left\{\left[\left(\frac{\varepsilon_0}{\varepsilon}\right)^{2.5-\zeta}-1\right](5-2\zeta)^{-1}-\left[\left(\frac{\varepsilon_0}{\varepsilon}\right)^{1.5-\zeta}-1\right](3-2\zeta)\right\}.$$

The channelons are realized at the energy $\varepsilon>\varepsilon_C$(see Chapter 1, Section 1.2.1). However, to obtain the cascade function by the definition (1.7), it is necessary to know the functions $\Psi_C(\varepsilon)$ at the low energies $\varepsilon_d<\varepsilon<2\varepsilon_d$. Therefore, we shall consider two ε-energy interval separated by the critical channeling energy. The equation (3.27) is valid on the first interval when $\varepsilon_C<\varepsilon<\varepsilon_0$. From (3.29) we can evaluate as follows: $J_d(\varepsilon)\ll J_s(\varepsilon)$, $(1-P^s_C)J_s(\varepsilon)\approx1$. Then the expression (3.28) will take the form

$$\frac{d\ln\Psi_C(\varepsilon)}{d\varepsilon}=\frac{1}{\varepsilon}\left[\alpha_C-(1-P^s_C)\alpha_s-(1-P^d_C)\alpha_d(\varepsilon)\right]. \qquad (3.28a)$$

The dependency $\alpha_d(\varepsilon)$ (see (3.29)) on $\varepsilon_0/\varepsilon$ is practically linear, i.e. in the ε-energy interval considered, this coefficient can be estimated as $\alpha_d\simeq0.45$. The solution (3.28a) has the form

$$\Psi_C(\varepsilon)=\Psi_0\left(\frac{\varepsilon_0}{\varepsilon}\right)^{(1-P^s_C)\alpha_s+(1-P^d_C)\alpha_d-\alpha_C}, \qquad \varepsilon_C<\varepsilon<\varepsilon_0.$$

On the second energy interval, where $0<\varepsilon<\varepsilon_C$, the equation (3.5) with $f=1$ is valid for the moving atom flux $\Phi_{C2}(\varepsilon)$ and gives the solution

$$\Phi_{C2}(\varepsilon)=C_1\varepsilon^{\kappa_0+\zeta-1}+C_2\varepsilon^{-\kappa_0}.$$

To determine the constants C_1 and C_2 it needs to join at $\varepsilon=\varepsilon_C$ with the solutions on the first energy interval, i.e.

$$\Phi_{2C}(\varepsilon_C)=\Phi_C(\varepsilon_C),\quad \left.\frac{d\Phi_{2C}(\varepsilon)}{d\varepsilon}\right|_{\varepsilon_C}=\left.\frac{d\Phi_C(\varepsilon)}{d\varepsilon}\right|_{\varepsilon_C}.$$

Therefore, we have for the flux of moving atoms if the channeling on the whole interval of changing the energy $\check{\varepsilon}<\varepsilon<\varepsilon_0$ is taken into account

$$\Phi_C(\varepsilon)=\Psi_0\nu_0\left(\frac{\varepsilon_0}{\varepsilon}\right)^{(1-P_C^s)\alpha_s+(1-P_C^d)\alpha_d-\alpha_C}\quad \text{if } \varepsilon_C<\varepsilon<\varepsilon_0, \tag{3.30a}$$

$$\Phi_C(\varepsilon)=\Psi_0\nu_0\frac{(1-P_C^s)\alpha_s+(1-P_C^d)\alpha_d+\kappa_0+\zeta-\zeta^{-1}-1.5}{2\kappa_0+\zeta-1}\times$$

$$\left(\frac{\varepsilon_0}{\varepsilon}\right)^{\kappa_0}\left(\frac{\varepsilon_0}{\varepsilon_C}\right)^{(1-P_C^s)\alpha_s+(1-P_C^d)\alpha_d+\zeta^{-1}-0.5}\quad \text{if } \varepsilon<\varepsilon_C. \tag{3.30b}$$

Deriving the formulas of this section, it was assumed that the probabilities P_C^s and P_C^d do not depend on ε and Ω.

Substitute (3.30b) into the definition (1.7) and obtain the cascade function with the allowance for the channeling:

$$\nu_C(\varepsilon_0)=\nu_0(\varepsilon_0)H_C(\varepsilon_0), \tag{3.31}$$

where

$$H_C(\varepsilon_0)=\left[1-\frac{P_C^s\alpha_s+P_C^d\alpha_d-\gamma_C-0.5}{\kappa_0+\zeta-1}\right]\left(\frac{\varepsilon_C}{\varepsilon_0}\right)^{P_C^s\alpha_s+P_C^d\alpha_d+\kappa_0-\gamma_C-0.5},$$

$$\gamma_C=\alpha_s+\alpha_d-\alpha_C.$$

It is seen from the expression (3.31) obtained for the cascade function

function $\nu_C(\varepsilon_0)$ that the probabilities of channeling the scattered (P_C^s) and displaced (P_C^d) atoms appear in the expression (3.31) with different weights α_s and α_d. It is understood since the crystals are described by not only different probabilities of channeling the scattered and displaced atoms, but different fractions of the channelons formed by the scattered and displaced atoms.

Fig. 3.4 demonstrates the dependence $\nu_0(\varepsilon_0)\nu_C^{-1}(\varepsilon_0)= H_C^{-1}(\varepsilon_0)$ on the PKA energy ε_0 for the channelons in the direction <110> in silicon and germanium at various channeling probabilities $P_C^s \simeq P_C^d$. Fig. 3.4. shows that when the energy ε_0 increases, the number of the displaced atoms in the cascade decreases as compared with those in the isotropic case if the channeling is taken into account.

3.4. Impurity Effect on the Distribution of Focused Atomic Collisions

The presence of impurities in the alloys of various composition affects the processes of propagating the correlated collisions along the close packed crystallographic directions (see, for example, Refs. 348, 349). Since (see Chapter 1) the structure of defect regions in the materials irradiated is to a considerable extent defined by the nature of the focusing substituting collision chains which transport the atoms out of the denuded zone, it becomes important to establish, and look into, the dependence of the chain length λ_D on the purity, temperature and other factors.

In Ref. 105, the value λ_D estimated for the extrapure tungsten (the impurity level is ~10^{-3} at.%) irradiated by the 20 keV-energy ions at T_{irr}=18 K, was within the limits 4.5-15 nm for the directions <100> and 4.5-8.5 nm for the directions <111>. But the irradiation of tungsten by its own ions at T_{irr}=300 K (the ion energy is 50 keV) has shown considerable shortening of the chains. In the estimation of the authors of Ref. 115, the value λ_D ranges within the limits 1-5 nm. Decreasing in λ_D must be related to both decreasing T_{irr} and terminating the focusing atomic collision chains on the impurities (analyzed in Ref. 115 was technical purity tungsten with the impurity level of about 0.2 at.%). A joint action of these factors is borne out by the data of Ref. 350: an increase in the impurity level up to 1.5 at.% without a considerable change in the irradiation temperature (T_{irr}=400 K), i.e. in both cases the interstitials possess high mobility, but it is the impurities where, as was already said, they are stabilized, and hence fixed in the method of autoionic microscopy) has led to decreasing λ_D down to 1-2 nm.

In Refs. 37, 119, 351, the problem of constructing a static λ_D distribution was realized, and the procedure of autoionomicroscopic analysis of propagating the focused atom collision chains was developed in detail. The experimental data were processed with computer using an especially created program. Essential specificity of the experimental part of the works was that fixed were only the interstitials captured by the impurity atoms. In all cases, the irradiation temperature was well above a threshold temperature of beginning a distant migration of interstitials in tungsten. Use of metals with impurities for the analysis (tungsten of technical purity and its alloy with ThO_2) was a positive fact in such a situation. However, obtaining the information on the value λ_D and its statistic distribution requires a definite recalculation to the idealized situation (the absence of impurities).

Examples of the chain distribution in the length obtained in Refs. 37, 119 are shown in Fig. 3.5. In the case of technically pure tungsten irradiated by the deuterons of the energies within the 5-12 MeV limits, the values λ_D were 4.2 and 3.8 nm for the directions <110> and <111>, respectively. When the alloy *W*-1.5 at. % ThO_2 was irradiated by fast neutrons and fission fragments (T_{irr}=400 K), these values were equal to 2.4 and 2.3 nm.

The data interesting in the quantitative respect can be obtained from the computerized experiments, performed in particular by the dynamic method [106, 108, 114, 142, 145, 168, 172].

The procedure of simulating the reactions of dynamic crowdions in α-iron with the simplest defects suggested in Refs. 352, 353 has permitted to have a good understanding of the nature of the reactions with substitutional impurities. The injection of the impurity was approximately simulated by changing the mass of one of the atoms. The direction of motion for the dynamic crowdion was selected along <100>. The crowdion reactions with the impurity atoms of various mass and various location relative to

PKA were calculated. Shown in Fig. 3.6 is one of the interaction alternatives. The initial PKA energy is selected as ε_0 =35 eV, the initial angle to the chain axis is θ_0=5°. As was shown by the calculations of propagating the dynamic crowdion without the impurity atom, the atom - atom collision chain with such initial parameters terminates by typical formation of a "dumbbell" interstition. The location of impurity atom was changing relative to the first atom in the collision chain, and so did the energy of the chain atom which interacted with the impurity directly. The ratio between the masses of impurity and matrix atoms (m_a/m) varied in wide limits from 0.125 to 4. The analysis has shown that in all cases when the impurity atom is located in the path of advancing the dynamic crowdion, the impurity effect on the path length resolves itself into its decrease.

The question of the thermal atomic displacement effect on the focused collision propagation in crystals was first considered in Ref. 354. The energy losses for the thermal oscillations in propagating the collision chains in copper was first calculated in Ref. 355, where the complimentary scattering angles in the chain averaged with respect to all displacement vectors, were determined. In the Ref. 356 developing the representations of Ref. 355 the calculation was performed for gold, lead and nickel as well. The dependencies of the chain length λ_D and the energy losses on the crystal temperature were calculated. A decrease in λ_D with increasing the temperature and the disappearance of the chains at $T \simeq 10^3$K are shown. In Ref. 357 a model is offered to calculate the paths λ_D of dynamic crowdions in the crystals with nonzero temperature. The calculations of the length λ_D

in copper and α-iron were performed at various temperatures, PKA energies and propagation directions, which have shown that the average lengths are reasonably not so large and slightly dependent on the crystal temperature up to ~500 K.

The impurity effects in the neighboring atomic rows on the process of focusing the momentum were first analyzed for the ionic crystals [132, 358], using the central potentials.

In a number of Refs., the simulation of advancing the substituting collisions was attempted both near the structure defects [359] and in the stressed crystals [360-362]. The chain distributions in path length are obtained in Ref. 363 with computer and are in a qualitative agreement with the results of Ref. 115. In Ref. 364 the assumption on the presence of focused substitutions in the crystal makes it possible to calculate the defect distributions in cascades and the cascade functions reasonably properly with allowance for the instantaneous recombination.

Thus, the focusing of atomic collisions in the crystals with impurities or in the ordered biatomic crystals possesses its own features as compared with that in the pure metals considered in Chapter 1, Section 1.2.2. The focusing of the atomic momentum along the close packed directions including the atoms of different masses is possible in this case. The energy losses are here due to the fact that $\Lambda=4mm_a(m+m_a)^{-2}<1$, and the energy $\varepsilon(1-\Lambda)$ is lost in each collision. The value Λ decreases with increasing the difference in the masses of the colliding atoms, and hence, the energy losses increase in each collision.

3.5. Design Model and the Types of Momentum Propagation in the Alternating-Mass Atomic Chains

Let us construct a theory of focused collisions in crystals for the close packed directions with alternating-mass atoms. Such structures can appear in solid solutions as well as in ordered alloys of Fe_3Al or Cu_3Au type. We shall carry out our investigation, using the hard-sphere model. We shall also base on hot testing of phenomenon [114, 119] and on the results of computerized experiments [357] concerned with not so strong temperature dependence of dynamic crowdion path lengths λ_D that will be neglected in our calculations.

Consider an isolated set of alternating atoms of two kinds with the masses m_1 and m_2 (Fig. 3.7) interacting along the selected direction only with the nearest neighbors of the Born-Meier potential (see Section 1.2.2):

$$V_{\mathrm{BM}}(r)=A\exp(-Br). \tag{3.32}$$

The potential constants are well known for the *Fe-Al* system [336, 371, 372], where the atomic masses m_1 and m_2 are essentially different and A=6550 eV and B=46.4 nm^{-1} are selected. The influence of the neighboring atomic chains in the crystals is neglected in the presence.

To study the chain shown in Fig. 3.7 it is convenient to divide it into separate cells: from the *n*th to (*n*+2)th atoms (*n*=0,1,2,...). Then it is sufficient to consider the momentum propagation over one cell, i.e. the interrelation between the successive angles ϑ_n, ϑ_{n+1}, ϑ_{n+2}. In the cell selected, there take place only two collisions: the first, odd, collision of the nth atom with the $(n+1)$th one and the second collision, even, of the $(n+1)$th atom with the $(n+2)$th one. To be definite, assume the first atom in a chain, and hence in a cell, to have the mass m_1. In this case, the value α_0 in the relation (1.99) must be substituted by

$$\alpha_n = a/R_n. \tag{3.33}$$

Here R_n is the collision radius for the nth atom moving with the energy ε_n with the resting $(n+1)$th atom, which is equal to

$$R_n = B^{-1} \ln \frac{A(1+l_0)}{\varepsilon_n} \sqrt{l_0^{(-1)^n - 1}}, \tag{3.34}$$

where $l_0 = m_1/m_2$.

In order the collision to be realized, two conditions should be fulfilled:

1) the collision radius R_n should not exceed the interatomic separation a, i.e.

$$\alpha_n = a/R_n > 1; \tag{3.35}$$

2) the impacted parameter $a\sin\vartheta_n$ should be less than R_n,

$$\sin\vartheta_n < \alpha_n^{-1}. \tag{3.36}$$

The first condition defines the minimal energy ε_n, and the second one the maximum angle ϑ_n at which the collision is even possible.

Using the laws of energy momentum conservation, a recurrence relation can be obtained in the form

$$\alpha_{n+1}=\alpha_n\left\{1-\frac{\alpha_n}{Ba}\ln\left[4l_0^{1+(-1)^n}(1+l_0)^{-2}(1-\alpha_n^2\sin^2\vartheta_n)\right]\right\}^{-1}. \tag{3.37}$$

In the one-atom case, the successive angles ϑ_n and ϑ_{n+1} were considered in studies of the momentum focusing. In this case, it is appropriate to consider the recoil angles for the atoms of the same kind (ϑ_n and ϑ_{n+2}). We shall say therefore that the focusing along the chain takes place if

$$\vartheta_{n+2}\leq\vartheta_n. \tag{3.38}$$

In order to show clearer a special feature of this effect as differentiated from one-atom case, it is convenient to use the low-angle ϑ_n approximation which appreciably simplifies the analytical expressions when the (3.38)-type conditions are investigated. In this case the relations (1.19) and (3.37) obtain the form

$$\vartheta_{n+1}=(\alpha_n-1)\vartheta_n, \tag{3.39}$$

$$\alpha_{n+1}=\alpha_n\left(1-\frac{2\alpha_n}{aB}\ln\frac{2l_0}{1+l_0}\right)^{-1} \tag{3.40}$$

In (3.40) the fact that n is even for the reasons of dividing into the cells has been taken into consideration. With the relationships (3.39) and (3.40) we obtain the following recurrence relation:

$$\vartheta_{n+2}=(\alpha_{n+1}-1)\vartheta_{n+1}=(\alpha_{n+1}-1)(\alpha_n-1)\vartheta_n. \tag{3.41}$$

Using three sequential angles for the recoil atoms, ϑ_n, ϑ_{n+1}, ϑ_{n+2}, six possible relations can be set up which define the types of pulse propagation through the cell:

1. $\vartheta_{n+2}<\vartheta_{n+1}<\vartheta_{n}$; (3.42a)

2. $\vartheta_{n}<\vartheta_{n+1}<\vartheta_{n+2}$; (3.42b)

3. $\vartheta_{n+1}<\vartheta_{n+2}<\vartheta_{n}$; (3.42c)

4. $\vartheta_{n+1}>\vartheta_{n}>\vartheta_{n+2}$; (3.42d)

5. $\vartheta_{n+1}>\vartheta_{n+2}>\vartheta_{n}$; (3.42e)

6. $\vartheta_{n+1}<\vartheta_{n}<\vartheta_{n+2}$; (3.42g)

Relations (3.42a) and (3.42b) describe the focusons (Φ) and the defocusons (D) already known. The remaining four are those of new types. For example, relation (3.42c) defines an odd focuson (OF), i.e. such a focuson where there takes place a focusing collision ($\vartheta_{n+1}>\vartheta_{n}$) in the first (odd) impact, and a defocusing collision ($\vartheta_{n+2}>\vartheta_{n+1}$) in the second one, although the cell as a whole executes the pulse focusing ($\vartheta_{n+2}<\vartheta_{n}$). Analogously, relation (3.42d) defines an even focuson (EF), (3.42e) an odd focuson (OF), and (3.42f) an even focuson (EF). Consider the conditions in R_n and ε_n which the various relations (3.42) between the successive angles lead to. Use for this purpose the recurrence relations (3.39)-(3.41), as well as (3.34), and write down the obtained limitations on the interaction radius and the energy:

when $\vartheta_{n+1}\leq\vartheta_{n}$,

$$R_n \geq a/2, \quad \varepsilon_n \leq \varepsilon^* = \frac{2l_0}{1+l_0}\varepsilon^0, \tag{3.43}$$

when $\vartheta_{n+2} \leq \vartheta_n$,

$$R_n \geq R_n^0 = \frac{a}{2} + \frac{1}{B} \ln \frac{2A}{1+l_0},$$

$$\varepsilon_n \leq \varepsilon^0 = A(1+l_0)\exp(-BR^0), \tag{3.44}$$

when $\vartheta_{n+2} \leq \vartheta_{n+1}$,

$$R_n \geq R^{**} = R^0 + \frac{1}{B} \ln \frac{2l_0}{1+l_0},$$

$$\varepsilon_n \leq \varepsilon^{**} = \frac{1+l_0}{2l_0}\varepsilon^0. \tag{3.45}$$

Using relations (3.43)-(3.45), we can estimate, depending on the magnitude of l_0, the conditions of propagating various types of focusons and defocusons whose existence limits have

been given schematically in Fig. 3.8. For the low angles ϑ_n, only four types of propagating through the cell are realized which are different depending on l_0. Common are the usual focusons and defocusons. If one is not confined to low angles (see below), then the even focuson and odd defocuson can be also involved in the pulse propagation process at $l_0>1$. For $l_0<1$, the mentioned four types of focusing are maintained for the high angles ϑ_n as well, since in so doing the collision chain has as its origin the lighter nth atom ($m_1<m_2$), and here we have $R_{n+1}>R_n$ (see (3.34) and (3.40)). That is why the pulse will tend to focus after the first (odd) collision, i.e. the relation $\vartheta_{n+2}<\vartheta_{n+1}$ will be valid in the second (even) collision. As a result, the focusing effect occurring in the first collision will be maintained in the second one, and hence there cannot occur both an odd focuson (OF) and an even defocuson (ED), which assume defocusing in each even impact. *Momentum focusing energy.* The focusing energy for one-atom chains is defined by the expression (1.23). In our case the one-atom chain is in agreement with $l_0=1$. Then we obtain from (3.43)-(3.45) $R^0= R^{**}= a/2$,

$$\varepsilon^{*}= \varepsilon^{**}= \varepsilon^{0}= 2A\exp\left(-\frac{aB}{2}\right)=\varepsilon_{F1}.$$

Therefore, as has been obtained in Refs. 109, 111, there take place only usual focusons and defocusons.

It makes sense for a two-atom chain to define the focusing energy ε_{F2} as the energy below that all types of focusons are realized for a given atomic row ($l_0>1$ or $l_0<1$). It is apparent that ε_{F2} agrees with the energy ε^0 (see Fig. 3.8 and expression (3.44)), i.e. we have

$$\varepsilon_{F2}\equiv\varepsilon^{0}=2A\,\frac{(1+l_0)^2}{4l_0}\exp\left(-\frac{aB}{2}\right)=\frac{\varepsilon_{F1}}{\Lambda_{12}},$$

where $\Lambda_{12}=4m_1m_2(m_1+m_2)^{-2}$. As we might expect, ε_{F2} depends only on the relationship between the atomic masses in the chain and the form of the interaction potential. For example, we have $\varepsilon_{F1}=32$ eV for the iron-aluminum system.

We consider how the values of the recoil angles ϑ_n change with increasing the atom number n measured from the original point of the chain, i.e. we construct the dependence $\vartheta_n=\vartheta(n)$. We resorted for this purpose to the recurrence relations (3.39), (3.40) specifying the initial values of the angles ϑ_0 and the energy ε_0 (or α_0). The curves $\vartheta(n)$ thus constructed for the iron-aluminum system are given in Fig. 3.9. Apart from the initial angle $\vartheta_0 = 1^\circ$, the curves for $\vartheta_0 = 2, 5, 10^\circ$ were calculated which coincide qualitatively. There only takes place a difference in the magnitude of a sudden change between ϑ_n and ϑ_{n+1} for the successive collisions.

A typical property of all curves in Fig. 3.9 is that all the collision chains have been eventually focused even at the energies higher than the focusing one ε_{F2} (but not so high) and $\vartheta_0=1^\circ$. Analogous picture is seen for the other values of ϑ_0. Consequently, it is safe to say that the momentum focusing condition is fulfilled beginning from some collision n^*, i.e. there exists a transition region where the defocuson is "reconstructed" into the focuson. This is confirmed by the conclusions of Refs. 109, 111, where the result has been obtained that qualitatively coincides with that shown in Fig. 3.9a.

The presence of the transition region is due in the one-atom case to the fact that ε_n decreases, and R_n increases, with increasing n. In this case such initial values ε_0 and ϑ_0 can be chosen that R_n will grow more "rapidly" than ϑ_n so that the momentum focusing condition is fulfilled beginning from a certain n^*. If the ε_n change with increasing n is neglected, there will be no transition region: realized in this case can be either focuson or defocuson. The dependence $\vartheta(n)$ for $l_0=1$ ($m_1=m_2$) (see Fig. 3.9a) with the same constants in the interaction potential and with the interatomic separation a=0.265 nm has been considered for the purpose of a detailed study of the defocuson-focuson transition. Note that there only takes place either a focuson or defocuson in the limit $\vartheta_n \longrightarrow 0$ at l_0=1. The analysis of the dependence $\vartheta(n)$ obtained for the low, but finite angles shows that the transition region realized in the case occupies only one cell on the whole studied interval of the initial energies ε_0 and the angles ϑ_0, realized in which region is either an odd defocuson or an even focuson depending upon the initial conditions. It must be noted that the value $\vartheta^*=\vartheta(n^*)$ given by the analytical expressions in Ref. 111, leads to the results somewhat too high as compared with those calculated basing on the recurrence relations (3.39), (3.40).

The curves presented in Fig. 3.9b,c involves, as distinguished from the case l_0=1 (Fig. 3.9a), several peaks, i.e. relative maxima. Hence in this case the transition region can consist of several cells. On the investigated interval of the initial energies ε_0, the maximum number of the occupied cells is equal to 3 with the transition region located directly at the very beginning of the collision chain at the energies of the order of the focusing one. It shifts toward the higher n with increasing ε_0 and ϑ_0. If located at the beginning of the collision chain, the transition region occupies 1-2 cells reaching three cells only at the higher ε_0. For l_0=0.48, this region is presented either by an even focuson or its combination with an odd defocuson, depending on ε_0, and for l_0=2.07, either an odd focuson or its combination with an even defocuson. The defocusons in this case correspond to the beginning of the transition region, and the focusons to its end. Consequently, as opposed to the case l_0=1, this region cannot here consist of only defocusons (even or odd). If one does not confine to the low angles, the transition region at l_0=2.07 can consist of either even focuson or its combination with an odd defocuson. In the case l_0=0.48, the type of focusons and defocusons responsible for the existence of this region, does not change with increasing the angle.

Conditions of occurring the focusons and defocusons. Each type of focusing pulse propagation is defined by an appropriate system of two inequalities (3.42). In order to find the conditions of occurrence of each focuson and defocuson type it is necessary to write the system of respective angle inequalities corresponding to the given type, and then, unlike the low-angle approximation considered above (see (3.39) and (3.40)), to reduce the obtained system to two unknowns, ϑ_n and ε_n (or α_n), with the help of the recurrence relations (1.19) and (3.37).

As convenience in the analytical calculation of the fields defining each type of focused motions, it is reasonable to change from the variables ϑ_n and ε_n to ξ_n and α_n ones, where

$$\xi_n=\sqrt{1-\alpha_n^2\sin^2\vartheta_n}\ . \qquad (3.46)$$

It follows from the boundary conditions making it possible to realize collisions (see (3.42)) that the allowed values of the new variables are

$$0 < \xi_n \le 1, \qquad (3.47)$$

$$\alpha_n \ge 1. \qquad (3.48)$$

Apart from this, the existence of the second (even) collision in the cell (the (n+1)th atom collides with the (n+2)th one) gives

$$\sin\vartheta_{n+1} < \alpha_{n+1}^{-1}, \qquad (3.49)$$

$$\alpha_{n+1} \geq 1. \tag{3.50}$$

It is easy to show from the expression (3.37) that (3.50) is fulfilled if

$$\alpha_n \geq \left[1 + \frac{2}{Ba}\ln(\mu_0 \xi_0)\right]^{-1}, \tag{3.51}$$

where

$$\mu = 2\lambda(1+\lambda)^{-1}.$$

If (3.37) is substituted into (3.49) and account is taken of the recurrence relation (1.18), obtained is

$$\alpha_n^2 \beta_1(\xi_n, \mu_0) + 2\alpha_n \beta_2(\xi_n, \mu_0) - \beta_3(\xi_n) < 0, \tag{3.52}$$

where

$$\beta_1(\xi_n, \mu_0) = 1 - \xi_n^2 - 4\ln^2(\mu_0 \xi_n)(aB)^{-2};$$

$$\beta_2(\xi_n, \mu_0) = 2\left[1 + \xi_n\sqrt{1-\xi_n^2}\right]\ln(\mu_0 \xi_n);$$

$$\beta_3(\xi_n) = 2 - \xi_n^2 + 2\xi_n\sqrt{1-\xi_n^2}.$$

As shown by the analysis of inequality (3.52), the condition (3.49) is valid if:

$$1 \leq \alpha \leq \alpha^*(\xi_n, \mu_0), \tag{3.53}$$

where

$$\alpha^*(\xi_n, \mu_0) = \beta_1^{-1}(\xi_n, \mu_0)\left[\sqrt{\beta_2^2(\xi_n, \mu_0) + \beta_3(\xi_n)\beta_1(\xi_n, \mu_0)} - \beta_2(\xi_n, \mu_0)\right].$$

The relations (3.51) and (3.53) apply some additional restrictions on the region of allowed values ξ_n and α_n. The variables ξ_n and α_n which meet conditions (3.47) and (3.48), but at which either (3.51) or (3.53) are not valid, are forbidden, since for such ξ_n and α_n, the collision chain will terminate at the (n+1)th atom. When the conditions (3.51) or (3.50) are violated, the atom becomes embedded in the chain, and when violated are (3.53) or (3.49), leaves it without interaction.

Now, after the estimation of the allowed regions for ξ_n and α_n, we turn to the detection of the conditions of occurring each type of focuson or defocuson. To do this, rewrite the relation between the angles in terms of ξ_n and α_n. 1. The inequality $\vartheta_n \leq \vartheta_{n+1}$ is equivalent to the condition

$$2\arcsin\left[\alpha_n^{-1}\sqrt{1-\xi_n^2}\right] \leq \arcsin\sqrt{1-\xi_n^2} ,$$

or

$$\alpha_n \geq \alpha_1(\xi_n), \tag{3.54}$$

where

$$\alpha_1(\xi_n)=\sqrt{2\xi_n^2+1}.$$

2. The relation $\vartheta_n \geq \vartheta_{n+2}$ yields

$$\arcsin\sqrt{1-\xi_n^2}\left(\sqrt{\alpha_n^2-1+\xi_n^2}-\xi_n\right)\left[1-2\alpha_n(aB)^{-1}\ln(\mu_0\xi_n)\right] \leq$$

$$\arcsin\sqrt{1-\xi_n^2} ,$$

or

$$\alpha_n^2\varepsilon_1(\xi_n,\mu_0)+2\alpha_n\varepsilon_2(\xi_n,\mu_0)-\varepsilon_3(\xi_n) \le 0, \tag{3.55}$$

where

$$\varepsilon_1(\xi_n,\mu_0)= 1-4(aB)^{-1}\ln^2(\mu_0\xi_n);$$

$$\varepsilon_2(\xi_n,\mu_0)=2(1+\xi_n)\ln(\mu_0\xi_n); \quad \varepsilon_3= 2(1+\xi_n).$$

The inequality (3.55) is fulfilled if:

$$1 \le \alpha_n \le \alpha_2(\varepsilon_n,\mu_0), \tag{3.56}$$

where

$$\alpha_2(\xi_n,\mu_0)= \varepsilon_1^{-1}(\xi_n,\mu_0)\left[\sqrt{\varepsilon_2^2(\xi_n,\mu_0)+\varepsilon_1(\varepsilon_n,\mu_0)\varepsilon_3(\xi_n)} - \varepsilon_2(\xi_n,\mu_0)\right].$$

3. The condition $\vartheta_{n+1} \le \vartheta_{n+2}$ yield

$$2\left[\arcsin\sqrt{1-\xi_n^2} + \arcsin\frac{\sqrt{1-\zeta_n^2}}{\alpha_n}\right] \le$$
$$\arcsin\frac{\sqrt{1-\zeta_n^2}\left[\sqrt{\alpha_n^2-1+\zeta_n^2} - \zeta_n\right]}{1-2\alpha_n(aB)^{-1}\ln(\mu_0\zeta_n)}, \tag{3.57}$$

or

$$\alpha_n^4\gamma_1(\xi_n,\mu_0)+\alpha_n^3\gamma_2(\xi_n,\mu_0)+\alpha_n^2\gamma_3(\xi_n,\mu_0)+$$
$$\alpha_n\gamma_4(\xi_n,\mu_0)+\gamma_5(\xi_n,\mu_0) \le 0, \tag{3.58}$$

where

$$\gamma_1(\xi_n,\mu_0)=16\xi_n^2\ln^2(\mu_0\xi_n)(aB)^{-2}-1;$$

$$\gamma_2(\xi_n,\mu_0)=-8(aB)^{-1}(1+\xi_n^2)\ln(\mu_0\xi_n);$$

$$\gamma_3(\xi_n,\mu_0)=4\left[1-4(aB)^{-2}(1-\xi_n^2)\ln^2(\mu_0\xi_n)\right];$$

$$\gamma_4(\xi_n,\mu_0)= 16(aB)^{-1}(1-\xi_n^2)\ln(\mu_0\xi_n);$$

$$\gamma_5(\xi_n,\mu_0)= 4(\xi_n^2-1).$$

A solution to the inequality shows that it is fulfilled when

$$\alpha_n \geq \alpha_3(\xi_n,\mu_0), \tag{3.59}$$

where $\alpha_3(\xi_n,\mu_0)$ is expressed in a sophisticated manner via $\gamma_i(\xi_n,\mu_0)$ (see Ref. 373).

Fulfilling condition (3.59), we obtain that the region of ξ_n and α_n values contains foreign roots. To exclude such non-ambiguity, note that for the left-hand side of (3.57), the following relation should be valid:

$$\arcsin\sqrt{1-\xi_n^2} + \arcsin\frac{\sqrt{1-\zeta_n^2}}{\alpha_n} \leq \frac{\pi}{4},$$

i.e.

$$(1-2\xi_n^2)\alpha_n-2\sqrt{2}\xi_n\alpha_n\sqrt{1-\xi_n^2}-2(1-\xi_n^2) \leq 0. \tag{3.60}$$

The inequality (3.60) is satisfied for $\xi_n \leq \frac{1}{\sqrt{2}}$ if

$$1 \leq \alpha_n \leq \tilde{\alpha}(\xi_n),$$

and when $\xi_n \geq \frac{1}{\sqrt{2}}$ for all $\alpha_n \geq 1$. In this case

$$\tilde{\alpha}_n(\xi_n)= \sqrt{2(1-\xi_n^2)}\left(\sqrt{1-\xi_n^2}-\xi_n\right)^{-1}.$$

The analysis of the $\alpha_3(\xi_n,\mu_0)$ and $\tilde{\alpha}_n(\xi_n)$-dependencies performed in detail in Ref. 373 yield that the relation (3.57) is satisfied under the condition (3.59) only for $\xi_n \geq 1/\sqrt{2}$. For $1/\sqrt{2} \leq \xi_n \leq \xi^0$, it is satisfied if

$$\alpha_3(\xi_n,\mu_0) \leq \alpha_n \leq \tilde{\alpha}(\xi_n),$$

where ξ^0 is a root of the equation

$$\alpha_3(\xi^0,\mu_0)=\tilde{\alpha}(\xi^0).$$

For $\xi_n<\xi^0$, the inequality (3.57) has no solution. Now we consider the conditions (3.47), (3.48), (3.53), (3.54), (3.56), and (3.59) taken as a whole (i.e. presented in one plot), and return to the variables $\xi_n \longrightarrow \vartheta_n$ and $\alpha_n \longrightarrow \varepsilon_n$ again. So we may obtain the diagrams presenting the conditions of arising each type of focuson and defocuson in any case involved. The diagram obtained in this way for the *Fe-Al* system with l_0=2.07 is pictured in Fig. 3.10a, and that with l_0=0.48 in Fig. 3.10b. The idealized case of l_0=1 effectively corresponds to the concentration of the curves B, C, D in Fig. 3.10a so that they intersect the abscissa axis at one point. This fits the fact that in an one-atom crystal there exist at low angles focusons and defocusons only.

It is seen from Fig. 3.10 that possible at l_0<1 is the realization of four focusings: an even defocuson and an odd focuson are here nonexistent. But if l_0>1, there can exist all six types of pulse propagation. Responsible for this is the character of interacting two kinds of atoms in the chains involved.

There exists at all l_0 some transition region in the dependence of ϑ_n

on the collision number n (at least at $\varepsilon_0 \leq 100$ eV). For various l_0 and ϑ_0 (and of course ε_0), this region consists of various types of focusons and defocusons or their combinations. For $l_0<1$, it is realized at all ϑ_0 at the cost of an even focuson or its combination with an odd defocuson, and for $l_0>1$ and small ϑ_0 at the cost of an odd focuson and its combination with an even defocuson. In this case when ϑ_0 increases, the odd focuson is substituted by the even focuson, and the even defocuson by the odd defocuson.

Free paths in isolated chains. Prescribing the initial values of ϑ_0 angle on the interval $0 \div \vartheta_0^{max}(\varepsilon_0)$, where $\vartheta_0^{max}(\varepsilon_0) = \arcsin \frac{1}{\alpha_0(\varepsilon_0)}$, and the energies $\varepsilon_0 \leq 100$ eV for the atom that creates a collision chain, one can obtain, with the recurrence relations (1.19) and (3.37), the dependence of ϑ_n and ε_n (or more specifically α_n) on the collision number n. Studying these dependencies, we note that for each ϑ_0 and ε_0 pair there exists such an $n=N_F$ for which the boundary conditions (3.47),(3.48) are violated, this value of N_F being believed as the free path of a collision chain.

According to the energy conservation law, in the hard sphere model we have for a chain of alternating atoms the following expression instead of (1.20):

$$\varepsilon_{n=1} = 4m_1m_2(m_1+m_2)^{-2}\varepsilon_n(1-\alpha_n^2\sin^2\vartheta_n).$$

The difference in the atomic masses is responsible for large energy losses in the collision as compared with the one-atom case. That is why the free path of a collision chain is steeply impaired and becomes finite as distinguished from the infinite path of focusing chains in this model when the atomic masses are equal.

Fig. 3.11 pictures the differences of the free path length N_F on the angle ϑ_0 obtained at the given initial energy for the cases $l_0<1$ and $l_0>1$. One should note a step-like manner of the N_F change, which is due to the collision chain terminating on a heavier atom in the vast majority of cases. Escape of a lighter atom from the chain is only possible when ϑ_0 increases (at small N_F). It is also seen from Fig. 3.11 that if the collision chain has gone through the origin cells ($N_F>3$), the atom escapes from the chain due to the violation of condition (3.47) is no longer possible. In this case the condition (3.48) is violated. The transition region mentioned above is here observed. The dependence $N_F(\vartheta_0)$ becomes steeper with increasing ε_0, i.e. N_F decreases more rapidly with increasing the angle ϑ_0. In this case at low ϑ_0 , the run of the curves $N_F(\vartheta_0)$ for $l_0<1$ and $l_0>1$ is almost the same, and at high ϑ_0 the path length for $l_0<1$ is less than that for $l_0>1$. Note that in the $l_0>1$ case the steps are located at even N_F, and in the $l_0<1$ case at odd N_F. This means that the collision chain terminates on a heavier atom in both cases. This rule is possibly departed at high angles (small N_F).

The dependence of the path length N_F on the initial energy ε_0 is also interesting at a prescribed initial angle ϑ_0 (see Fig. 3.12). One observes the same step-like change in N_F as in $N(\vartheta_0)$. In the majority of cases, the boundary conditions (3.47) and (3.48) are violated on a heavier atom, and it is the energies $\varepsilon_0>100$ eV when their violation is possible on a lighter atom. As seen from Fig. 3.12, the dependencies $N_F(\vartheta_0)$ at $\vartheta_0=0^\circ$ show essentially the same run for various l_0 and increase with the energy. A decease in the path length N_F is observed with increasing the ε_0 energy when the angle ϑ_0 is increased. At small ε_0, the path lengths for $l_0<1$ and $l_0>1$ coincide. When the energy increases, the chain beginning with a lighter atom will have a shorter length than that beginning with a heavier atom, the difference in path lengths increasing with ϑ_0. It is seen from Fig. 3.12 that the collision chain termination takes place due to a violated boundary condition (3.48), i.e. at the cost of the mass difference between the interacting atoms. Therefore, the path length is largely due to a rapid decrease in the impinging atom energy, rather than an increase in its recoil angle.

It should be noted that the obtained values of focused atomic collision paths are overestimated as a consequence of the isolated chain model consideration. However, when the paths of focused collisions are considered in the isolated chains, possible are the occurrence of extra energy transfer along the chain and an increment of collision paths as a function of the mass retio for the atoms interacting in the chain.

We consider as an example a chain of atomic collisions beginning with the nth atom (see Fig. 3.13) and fitting the following conditions:

$$l_{n,n+1}=\frac{M_n}{M_{n+1}}>1, \quad l_{n+1,n+2}=\frac{M_{n+1}}{M_{n+2}}<1. \tag{3.61}$$

Modifying the definition of the l_0 value and denoting the collision radius as R_n (see Fig. 3.34)), we write

$$R_n = R_{n,n+1} = R^0.$$

The remaining collision radii $R^{(W)}_{n\,,\,n+1}$ are expressed through R^0 as follows:

$$R^{(W)}_{k,\,k+1} = \eta^{(W)}_{k\,,\,k+1} R^0 + \beta^{(W)}_{k\,,\,k+1},\ k=n,n+1,n+2,$$

where

$$\eta^{(W)}_{k\,,\,k+1} = \frac{B_{n,n+1}}{B_{k,k+1}};\quad \beta^{(W)}_{k\,,\,k+1} = B^{-1}_{k,k+1} \ln\left[\frac{A_{k,k+1}}{A_{n,n+1}} f^{(W)}_{k\,,\,k+1}\right];$$

$A_{n\,,\,n+1}$, $B_{n\,,\,n+1}$ are the constants of the Born-Meier potential in (3.32) (A and B) in the interaction of nth and (n+1)th atoms; $f^{(W)}_{k\,,\,k+1}$ depends only on the mass retio between the interacting atoms; W=1,2 indicating the respective main or repeated collision, respectively. In the atomic chain involved (Fig. 3.1) the energy ε_n is imparted to the atom of m_n mass (iron) at the instant of time t_0, and it begins to move toward the atom of m_{n+1} mass (aluminum) and donates it the energy ε_{n+1} at the instant of time t_1. In virtue of relations (3.61), the (n+1)th atom executes oscillations between the neighboring, heavier (n+2)th and nth atoms in the chain. Reflected backward as a result of such a motion at the instant of time t_2 from the (n+2)th atom, this atom interacts with the nth atom at the instant t_3. This collision again changes the direction of its motion, and it catches up the (n+2)th atom with the velocity v''_{n+1} to the instant t_4, donates it the additional energy $\Delta\varepsilon_{n+2}$, thus increasing its energy up to the value $\varepsilon'_{n+2} = \varepsilon_{n+2} + \Delta\varepsilon_{n+2}$. If the repeated interaction is lacking, then the (n+2)th atom is subject to collision with (n+3)th one at the instant t_5.

Consequently, in order the repeated collision to be realized, it needs that

$$t_5 \geq t_4 \tag{3.62}$$

and what is more

$$v''_{n+2} > v_{n+2} \,. \tag{3.63}$$

with the use of the energy and momentum conservation laws, the condition can be rewritten in the form of the following limitation on the atomic mass relations:

$$l_{n+1,n+2} < \frac{l_{n,n+1} - 1}{l_{n,n+1} + 1}, \tag{3.64}$$

The combined account of inequality (3.62),(3.64), as well as that of positivity of all interaction radii and all time intervals $\Delta t_j = t_j - t_{j-1}$, permits to determine the threshold of allowed energy values ε_n for realizing a repeated collision in the many-atom chain.

A relative increase in the $(n+2)$th atom energy at the cost of additional energy transfer in the repeated collision with the $(n+1)$th atom can be represented by the expression

$$\frac{\Delta\varepsilon_{n+2}}{\varepsilon_{n+2}} = \frac{8l_{n,n+1}[l_{n,n+1}(1-l_{n+1,n+2})-1-l_{n+1,n+2}]}{(1+l_{n,n+1})^2(1+l_{n+1,n+2})^2}, \tag{3.65}$$

which depends only on the mass retio between the interacting atoms.

If the collision chain begins with a lighter atom, then its oscillatory motion leads to an additional transfer of the energy $\Delta\varepsilon_{n+1}$ to the atom with the mass m_{n+1}.

The transfer of additional energy along the chain is associated with a considerable increase in the path length of a sequence of atomic collisions, since the repeated interactions can be realized along the whole length of the collision chain.

By way of illustration of the effect of additional energy transfer let us consider a chain of alternating atoms of two kinds, i.e. the condition (3.61) will be written in the form

$$l_{n,n+1}=\frac{m_n}{m_{n+1}}=l_0>1;\quad l_{n+1,n+2}=\frac{m_{n+1}}{m_{n+2}}=\frac{1}{l_0}<1.$$

The fulfillment of the above mentioned conditions of realizing the repeated collisions permits to obtain, depending on the value l_0, the regions of allowed energy ε_n, which are shown in Fig. 3.14. The region of allowed ε_n-energy values is divided into two subregions separated by the range of l_0 values at which the inequality (3.62) is violated. These subregions are exemplified by different behavior of the threshold energy ε_{thr} determined from the condition $\Delta t_5=0$: in the left subregion it falls with increasing l_0, and in the right one has a bell-like shape. Such a behavior of ε_{thr} can be understood ready. In the second subregion when the energy ε_n is constant, the interval Δt_5 first increases with increasing the mass of the nth atom up to some value $l^*(\varepsilon_n)$ and then falls. In this case $l^*(\varepsilon_{thr})=l_{min}$, where l_{min} corresponds to a minimum point of the $\varepsilon_{thr}/\varepsilon_m$ -dependence on l_0 (Fig. 3.14). But with increasing the energy (when l_0 is constant) the interval Δt_5 increases at all times up to $\varepsilon_n/\varepsilon_{thr}=e^2$, and further decreases. Therefore, Δt_5 increases with l_0 and then decrea.

Therefore, Δt_5 increases with increasing l_0, i.e. the repeated collisions can be realized even at a lower energy. Further increase in l_0 ($l_0 > l_{min}$) leads to a decrease in Δt_5, i.e. rises ε_{thr}.

In the left subregion, the value ε_{thr} decreases, in so far as Δt_5 all the time increases with l_0 and ε_n. In addition, this subregion features a narrow interval $\delta\varepsilon_n$ of allowed values ε_n. Such a situation leads to a sharp decrease in the repeated collision efficiency, since for the given l_0, they cannot take place all the way from the beginning to the end of the chain , but only in its minor part. That is why the path length of collision sequences either do not increase at all, or their maximum increment is equal to two interatomic separations.

In the right subregion (Fig. 3.14) for large values of l_0, there is only a rigid limitation on the lower value of the allowed energy ε_n, which is due to the requirement to satisfy the condition (3.62). In this case, the maximum value of the allowed energy ε_n proves rather large: $\varepsilon_n/\varepsilon_m \sim 10^5 \div 10^6$, which leads, naturally, to a substantial increase in the interval $\delta\varepsilon_n$ in this subregion, and hence in the value of path length increment ΔN_F of the collision sequence as a function of ε_n (Fig. 3.15). The presence of an asymptote when l_0=4.23, corresponds to the fact that the energy increment (3.65) at the cost of repeated collisions completely compensates the energy losses due to the mass difference between the atoms interacting in the chain.

3.6. Crowdion Free Path in Two-Atom Crystals

In the scope of the focused collision model for the two-atom isolated rows described in Section 3.5, we account for the effect of surrounding atomic rows on propagating the chains of successive knock-on collisions.

For this purpose, it is necessary to describe not only the interaction along the direction OX of the chain of alternating p and q-kind atoms (with the masses m_p and m_q) (see Fig. 3.16) in the form of the Born-Meier potential, but to construct a potential outer with respect to the chain. This potential set up by the surrounding atoms which are fitted to their lattice points, has the form

$$U_p(x)=\sum_{\Psi=-\infty}^{\infty}\sum_{q=1}^{k} V_{pq}^{(\Psi)}(r), \tag{3.66}$$

where Ψ is an index of a lens of surrounding-row q-atoms, k is the number of atoms in common lens, and the potential (3.32) is written in the form

$$V_{pq}(r)=A_{pq}\exp(-B_{pq}r). \tag{3.32a}$$

Combining (3.32a) and (3.66) and expanding $U_p(x)$ into the Fourier series in the parameter x/a, we have

$$U_p(x)=\sum_{l=-\infty}^{\infty} (-1)^l\chi_p^{(l)}\cos(2\pi l\frac{x}{a}), \tag{3.67}$$

where $\chi_p^{(l)}$ are the expansion coefficients depending on the constants A_{pq}, B_{pq}, and the lens structure.

Considering as an example a collision sequence along the direction

<100> in the crystal Fe_3Al (Fig. 3.16) and using the constants for the interatomic potential from Refs. 371, 372, we obtain from the expansion (3.67) with a precision of 10^{-4} under the condition $U_p(0)=0$

$$U_p(x)=1.5003-1.5742\cos\left[2\pi\frac{x}{a}\right]+0.0755\cos\left[4\pi\frac{x}{a}\right]-$$

$$0.0016\cos\left[6\pi\frac{x}{a}\right]. \qquad (3.68)$$

Since (3.67) and (3.68) are the series which lead to problems, the expression (3.68) can be approximated by the following function

$$\tilde{U}_p(x)=U_p(1-\eta_*)\left[1-\cos\left[2\pi\frac{x}{a}\right]\right]\left[1+\eta_*^2+2\eta_*\cos\left[2\pi\frac{x}{a}\right]\right]^{-1}, \qquad (3.69)$$

where $U_0=2\sum_{l=-\infty}^{\infty}\chi_p^{(2l+1)}$ is the height of the potential barrier, i.e. the energy difference between the atoms at the lens center and at a point in the unit cell, and the parameter η_* is determined from the condition of a minimum deviation $\tilde{U}_p(x)$ from $U_p(x)$, being the potential shape parameter $(|\eta_*|<1)$. The values U_0 and η_* are unambiguously related to the lens parameters, their shape and the constants of atom-atom interactions. In particular, we have for the system *Fe-Al*: $\tilde{A}_p=1.5788$ eV, $\eta_*=5\cdot10^{2}$. The potential (3.69) at $\eta_*=0$ transforms into the Frenkel-Kontorova potential [374].

Writing the energy balance in a knock-on atomic collision along the row selected with consideration for the outer potential (3.69) which is set up by the surrounding atomic rows, we have

$$m_p(m_p + m_q)^{-1}\varepsilon_0\left[1- \tilde{U}_p(R_0)\varepsilon_0^{-1}\right]= V_{pq}(R_0), \tag{3.70}$$

where R_0 is the collision radius in view of the outer field. Using the definition of the collision radius without considering the outer field (3.34), we write it in the form

$$R^0(\varepsilon_0)=B_{pq}^{-1}\ln\left(\frac{m_p}{m_p + m_q}\cdot\frac{A_{pq}}{\varepsilon_0}\right).$$

Inserting a relative deviation of the radius from R^0 with allowance made for the surrounding effect $\zeta_0=(R_0-R^0)(R^0)^{-1}$, we obtain from (3.70) an equation to determine ζ_0 in the form

$$1- \varepsilon_0^{-1}\tilde{U}_p[(1+\zeta_0)R^0]=\exp\left[-aB_{pq}z_0\zeta_0\right], \tag{3.71}$$

where $z_0=R^0(\varepsilon_0)a^{-1}$. Estimate (3.71) numerically and obtain that the value $\zeta_0(\varepsilon_0)\ll 1$ at the energies ε_0 corresponding to the pulse focusing. We now obtain from (3.71) with (3.69) the following analytical expression for $\zeta_0(\varepsilon_0)$ as a linear approximation in ζ_0 and when $\eta_0\simeq 0$:

$$\zeta_0(\varepsilon_0)\simeq\frac{\tilde{A}_p}{z_0(\varepsilon_0)}\cdot\frac{1- \cos[2\pi z_0(\varepsilon_0)]}{B_{pq}a\varepsilon_0-2\pi\tilde{A}_p\sin[2\pi z_0(\varepsilon_0)]}. \tag{3.72}$$

It is seen from the expression (3.72) that the value ζ_0 increases with the outer field ($\tilde{A}_p$ increases), i.e. the atomic collision radius in the selected row increases with increasing the height of the outer-field potential barrier. It can be readily understood, since if any atom of the selected row moves in the field of surrounding atoms, its kinetic energy decreases by the value $\tilde{U}_p$, which leads to increasing R_0. Having a chance of finding the dependence of the collision radius on the energy ε_0 of the incident atom, $R_0(\varepsilon_0)$, one can write the recurrence relations for the energies sequentially transferred along the atomic row. This permits to calculate the length of substituting collisions in the two-atom chains, i.e. to determine a path length of the dynamic crowdion. To perform an event of substituting one atom by the other it needs that $R_0 < 0.5a$. As R_0 depends on the mass relation between the interacting atoms, the energy necessary for overcoming the potential barrier proves to be different for the atoms of different kinds. In our case, this energy for iron atom is equal to about 46 eV, and for aluminum atom to about 23 eV. Interacting with a heavy atom (iron), a light atom (aluminum) can obtain a momentum in the opposite direction. If the energy difference between the incident atom and that knocked-on from a lattice point, proves to be more than $\tilde{U}_p$, then the light atom will return into its potential well and interact with the foregoing iron atom already presenting here. Again overcoming the barrier, this atom will "heal" a vacancy at the iron atom lattice point. However, if the aluminum atom energy is not enough to overcome the barrier repeatedly in the direction of propagating the substitutions, it forms an interstitial configuration with the iron atom.

One can see in Fig. 3.17 the results of calculating the substituting collision lengths for various initial energies of the iron atoms. After the path of dynamic crowdion has been ended, the energy will be scattered over the lattice in the form of focusons.

The dynamic crowdion path lengths obtained in the calculations for alternating close packed directions, were compared with the experimental data on the order parameter of the irradiated ordered alloys [375]. Processing these experiments by the theory [376] yields the path lengths which are in a sufficiently good accordance with the dependence shown in Fig. 3.17.

CHAPTER 4

KINETICS OF POINT DEFECTS UNDER IRRADIATION IN SOLID SOLUTIONS WITH COHERENT PRECIPITATES

4.1. Concentration of Radiation Point Defects in Diluted Solid Solutions

In the process of irradiating the solid solution at the temperature T_{irr}, there will take place some redistribution of the material in line with the balance equations (2.3)-(2.9) or in a simplified form (2.11)-(2.14). However, it is felt for a diluted solid solution that the impurity concentration (C_a or C_b) is not so large and distributed in volume not inhomogeneously. When the irradiation intensities are high enough to fit reactor ones, one can consider a homogeneous stationary problem on the basis of the value t_* by (1.45). Reasoning that there are impurities of two kinds (a and b) in the solid solution, we write equations (2.3),(2.5), (2.7), and (2.9) in the following form with due regard for an additional effect of bivacancies [377]:

$$\frac{dC_v(t)}{dt} = g - \mu D_i C_i^+ C_v^+ - \alpha_{va} C_v^+ (C_a^+ - C_{va}^+) + \chi_{va} C_{va}^+ -$$

$$\mu_{ib}^v C_v^+ C_{ib}^+ - D_v C_v^+ k_v^2 + \delta_2 D_i C_i^+ C_{2v}^+ - \alpha_2 D_v (C_v^+)^2 + 2\chi_{2v} C_{2v}^+ = 0, \tag{4.1}$$

$$\frac{dC_{va}(t)}{dt} = \alpha_{va} C_v^+ (C_a^+ - C_{va}^+) - \chi_{va} C_{va}^+ - \mu_{va}^i C_i^+ C_{va}^+ - D_{va} C_{va}^+ k_{va}^2 = 0, \tag{4.2}$$

$$\frac{dC_i(t)}{dt} = g - \mu D_i C_i^+ C_v^+ - \alpha_{ib} C_i^+ (C_b^+ - C_{ib}^+) + \chi_{ib} C_{ib}^+ -$$

$$\mu_{va}^i C_i^+ C_{va}^+ - D_i C_i^+ k_i^2 - \delta_2 D_i C_i^+ C_{2v}^+ = 0, \tag{4.3}$$

$$\frac{dC_{ib}(t)}{dt} = \alpha_{ib} C_i^+ (C_b^+ - C_{ib}^+) - \chi_{ib} C_{ib}^+ - \mu_{ib}^v C_{ib}^+ C_v^+ - D_{ib} C_{ib}^+ k_{ib}^2 = 0. \tag{4.4}$$

With regard to bivacancies (2_v) we have the following equation for their concentrations C_{2v}:

$$\frac{dC_{2v}(t)}{dt} = 0 = \alpha_2 D_v C_v^{+2} - \delta_2 D_i C_i^+ C_{2v}^+ - \chi_{2v} C_{2v}^+ - D_{2v} C_{2v}^+ k_{2v}^2. \tag{4.5}$$

The terms of the form $\alpha_{jk} C_j^+ (C_k^+ - C_{jk}^+)$ describe the saturated impurity traps $k=a,b$ for the defects $j=i,v$. Even in such an algebraic form the system of equations (4.1)-(4.5) cannot be solved analytically for the homogeneous stationary point- defect concentrations. Therefore, we investigate a solution to those balance equations in various simplified versions when some terms may be neglected. Follow also how the solutions obtained change if the omitted terms are accounted for.

In a pure case (the bivacancies are left out of account), the system of equations (4.1)-(4.5) is degenerated into two equations (1.24a), whose solutions are given in the form (1.29) and (1.29a) for true absolute concentrations.

We now consider how the solutions (1.29) will change in the presence of impurity atomic traps of the same kind (without bivacancies). The account for the trap saturation leads to extra terms in the balance equations. First consider one at a time the cases when there are impurities of only one kind. Then the system of balance equations will contain three equation (4.1)-(4.3) and (4.2) or (4.4).

Vacancy traps. If the mobility of *va* complexes is ignored and their dissociation is believed small, then a solution to the system (4.1)-(4.3) has the form

$$C_j^{(a)+} = C_0 H_j \left[\sqrt{1+\xi_v^2} - \xi_v\right]\left[1+\frac{\mathcal{P}_v}{2g}\right], \qquad (4.6)$$

where

$$\mathcal{P}_v = \kappa_{va}\chi_{va} D_i k_i^2 D_v k_v^2; \quad \xi_v = \eta_v \left[1+\frac{\mathcal{P}_v}{g}\right]^{-1};$$

$$\kappa_{va} = \mu_{ia}^i C_a (\mu_{va}^i D_v k_v^2 + \alpha_{va} D_i k_i^2)^{-1}; \quad \eta_v = \eta_0 (1+\kappa_{va}\alpha_{va}).$$

When $\xi_v \ll 1$, the expression (4.6) takes the form

$$C_j^{(a)+} \simeq C_{j0}^+ - C_0 H_j \left[\eta_0 \kappa_{va} \alpha_{va} - \frac{\mathcal{P}_v}{2g}\right]. \qquad (4.7)$$

From this point on we shall only write the equations for the *j=i,v* defect concentrations with the omitted formulas for the complex concentrations C_{jk} which are too cumbersome.

Interstitial traps. Reasoning the dissociation of the complexes to be small and neglecting their mobility, we obtain

$$C_j^{(b)+}=C_0H_j\left(\sqrt{1+\xi_i}-\xi_i\right)\left(1+\frac{\mathcal{P}_i}{2g}\right),$$

where

$$\mathcal{P}_i=\kappa_{ib}\chi_{ib}D_ik_i^2D_vk_v^2;\quad \xi_i=\eta_i\left(1+\frac{\mathcal{P}_i}{g}\right)^{-1};$$

$$\kappa_{ib}=\mu_{ib}^{v}C_b(\mu_{ib}^{v}D_ik_i^2+\alpha_{ib}D_vk_v^2);\quad \eta_i=\eta_0(1+\kappa_{ib}\alpha_{ib}).$$

When $\xi_i \ll 1$ and $\dfrac{\mathcal{P}_i}{2g} \ll 1$, we have

$$C_j^{(b)+}=C_{j0}^{+}-C_0H_j\left[\eta_0\kappa_{ib}\alpha_{ib}-\frac{\mathcal{P}_i}{2g}\right]. \tag{4.8}$$

There is under way a decrease in the *j* defect concentrations as compared with the pure case (1.29) and (1.29a) (the first term in the parentheses of expressions (4.7) and (4.8)). The account for the dissociation of *jk* complexes, i.e. clearing the traps, leads to the reduction of this effect (the second term in the parentheses). It is to be noted that the traps of only one kind (*a* or *b*) diminish the concentration of both vacancies and interstitials. This is concerned to the fact that, for example, the vacancy traps diminish the concentration of free vacancies, but simultaneously the complexes *va* are sinks for the interstitials. Of course, the interstitial annihilation on the complexes *va* is only beneficial in case when the impurity atom dimension r_a not too much exceeds the matrix atom dimension r_0. Since a solid solution is formed, as known [325], when the inconsistency $|r_a-r_0|/r_0<0.15$ takes place, then the condition for the annihilation involved is fulfilled.

Two kinds of impurities. To obtain visible solutions in this case, we completely ignore the dissociation of *va* and *ib* complexes, the saturation of impurity traps and the migration of complexes. In this case the solution to the system of equations (4.1)-(4.4) retains the structure (1.29) and has the form

$$C_j^{(\mathrm{I})+} = C_0 H_j(\sqrt{1+\xi_\mathrm{I}^2} - \xi_\mathrm{I}), \tag{4.9}$$

where

$$\xi_\mathrm{I} = \eta_0(1+\kappa_\mathrm{I}), \quad \kappa_\mathrm{I} = \alpha_{va} C_a (D_v k_v^2)^{-1} + \alpha_{ib} C_b (D_i k_i^2)^{-1}.$$

If $\xi_\mathrm{I} \ll 1$, then we obtain from (2.102)

$$C_j^{(\mathrm{I})+} \simeq C_0 H_j(1-\xi_\mathrm{I}) = C_{j0}^+ - (\alpha_{va} C_a H_v + \alpha_{ib} C_b H_i) H_j (2\mu D_i)^{-1} \equiv \tag{4.10}$$

$$C_{j0}^+ - \Delta C_j^{(\mathrm{I})},$$

where $\Delta C_j^{(\mathrm{I})}$ is a change in the stationary concentration C_{j0}^+ with allowance for two kinds of impurities simultaneously. Therefore, in the diluted solid solution with two kinds of impurities, the concentrations of both vacancies and interstitial should be lower as compared with the pure case.

Dissociation of complexes. In addition to the above considered processes, we take into account the dissociation of *va* or *ib* complexes treating this effect as a small correction to the earlier solutions. The saturation of impurity traps is here neglected. Now conserved in the system of equations (4.1)-(4.4) are all terms apart from those concerned with the bivacancies. Now we have for the stationary concentrations

$$C_j^{(D)+}= C_0H_j\left[\sqrt{1+\xi_D^2}-\xi_D\right]\left[1+\frac{\mathcal{P}_D}{2g}\right], \qquad (4.11)$$

where

$$\mathcal{P}_D= \frac{\alpha_{va}D_ik_i^2}{\mu^i_{va}D_vk_v^2}\chi_{va}C_a+ \frac{\alpha_{ib}D_vk_v^2}{\mu^v_{ib}D_ik_i^2}\chi_{ib}C_b,$$

$$\xi_D= \xi_I\left[1-\frac{\mathcal{P}_D}{g}\right].$$

When $\xi_D \ll 1$, we derive from (4.11)

$$C_j^{(D)+}= C_j^{(I)+}+ C_0H_j\frac{\mathcal{P}_D}{2g} \equiv C_j^{(I)}+ \Delta C_j^{(D)}, \qquad (4.12)$$

where $\Delta C_j^{(D)}$ is a change in the concentration $C_j^{(I)+}$ due to the *va* and *ib* complex dissociation. The complex dissociation produces, as expected, some increase in the defect concentration as compared to $C_j^{(I)+}$.

Account for the bivacancies. Let us solve a complete system of equations (4.1)-(4.5). The inequality $C_{2v}^+ \ll C_v^+$ will be considered to be valid at not so high temperatures. If so, the solution sought can be considered as a perturbation to (4.12) in the form

$$C_j^{(2v)+}= C_j^{(D)} + \Theta_j\Delta C_j^{(2v)}, \qquad (4.13)$$

where Θ_i=1, Θ_v=-1. First we assume that the condition $\delta_2D_iC_i^+ \gg k_{2v}^2D_{2v}+ \chi_{2v}$ is fulfilled (i.e. the bivacancy decrease is mainly due to the interstitial annihilation on them), that the probabilities of complex dissociation are small, as above, and the complex mobility is ignored. Now we obtain

$$C_j^{(2v)+} = C_j^{(I)+}\left[1+ \Theta_j \frac{\alpha_2}{2k_v^2}(1-\varphi)C_v^{(I)+}\right], \tag{4.14}$$

where $\varphi = 2(D_{2v}k_{2v}^2 + \chi_{2v})(\delta_2 D_i C_i^{(I)+})^{-1}$. It is seen from (4.14) that the bivacancy formation causes the single vacancy concentration to decrease and the interstitial concentration to increase. The bivacancy concentration is equal to

$$C_{2v}^{+} = C_v^{(D)+}H_v^2\left(1-\frac{\varphi}{2}\right)\frac{\alpha_2 D_v}{\delta_2 D_i}\left[1-\frac{3\alpha_2}{2k_v^2}\left(1-\frac{\varphi}{2}\right)C_v^{(D)+}\right].$$

In another limiting case when the bivacancy adsorption on sinks takes prevalence over their dissociation and capture on the interstitials ($D_{2v}k_{2v}^2 \gg \delta_2 D_i C_i + \chi_{2v}$), we obtained instead of (4.14) an analogous expression in which φ is substituted for $\varphi' = 2(\delta_2 D_i C_i^{(D)+} + \chi_{2v})(D_{2v}k_{2v}^2)^{-1}$.

We substitute formula (4.14) into the expression (4.11) for $C_j^{(D)+}$ and have

$$C_j^{(2v)+} = C_0 H_j\left(\sqrt{1+\xi_D^2} - \xi_D\right)\left(1+\frac{P_D}{2g}\right)(1+\Theta_j \Delta C_j^{(2v)}),$$

where

$$\Delta C_j^{(2v)} = \frac{\alpha_2(1-\varphi)}{2k_v^2}C_j^{(D)+}.$$

The system of balance equations close to that involved here was solved with computer for the nonstationary case in Ref. 276. Considered here was the impurity of only one kind, which forms several kinds of complexes with vacancies and interstitials, the vacancies being left out of account. As in the above given calculation, the stationary concentrations depend only on the temperature and energy which are included in the parameters analogous to α_j, χ_j, μ_j, δ_j. The setting time of stationary concentrations is determined by the same parameters.

When the currents are weak and the point defect fluxes are intensive, $\eta_0 \ll 1$ (and hence $\xi_D \ll 1$, $\kappa_I \ll 1$), and the interstitial and vacancy concentration has the form

$$C_j^{(2\nu)+} = C_{j0}^{+} - \Delta C_j^{(I)} + \Delta C_j^{(D)} + \Theta_j \Delta C_j^{(2\nu)},$$

where $\Delta C_j^{(2\nu)}$ can be presented as

$$\Delta C_j^{(2\nu)} = \alpha_2 D_\nu (1-\varphi) \left\{ C_{j0}^{+} \left[C_{\nu 0}^{+} + \Delta C_\nu^{(D)} - \Delta C_\nu^{(I)} \right] + C_{\nu 0}^{+} \left[\Delta C_j^{(D)} - \Delta C_j^{(I)} \right] \right\}.$$

The signs of the corrections to the concentrations C_{j0}^{+}, which occur if various effects are additionally accounted for, are listed in the following table.

Table 4.1

Signs of corrections to point defect concentrations when various reaction channels are taken into account

Effects		ΔC_v	ΔC_i
Vacancy traps	a	-	-
Interstitial traps	b	-	-
Nonsaturated traps		-	-
Complex dissociation	D	+	+
Bivacancies	$2v$	-	+

4.2. Point Defect Concentration in Solid Solutions with Coherent Precipitates

The coherent precipitates forming in consequence of the decay of a solid solution (see Fig. 2.7) create the fields of inner stresses and influence the defect fluxes. This manifests itself either in the drift terms of equations (2.11)-(2.14) or in the sink-describing terms. When considering a microscopic model, as in Section 2.6 (see Fig. 2.8), divide the whole material into the regions of two types, F= I and II. Let the coherent precipitates with $\varepsilon_I>0$ be formed in the solid solution under the irradiation. Then, in accordance with (2.22), the vacancies and interstitial will interact with the regions I and II differently [378-381].

For the complex calculations which were in detail discussed in Chapters

I and II, it is necessary to know the defect distribution $\hat{\tilde{C}}_j(\mathbf{r})$ near the coherent precipitates. In order further to estimate the rates of adsorption on these sinks by (1.38), one needs to know the defect flux densities (1.31) depending on $\hat{\tilde{C}}_j(\mathbf{r})$.

An analytical solution to the equations (2.11) and (2.14) is only possible to find by the effective medium method with some simplifying assumptions :

A. The point defects do not form clusters within the precipitates and have no sinks other than leaving the matrix (F=II).

B. The sinks for the point defects are distributed homogeneously in the region F=II.

C. The reactions with the impurities and the volume recombination are neglected. The equations (2.11) and (2.14) for the supersaturations $\tilde{C}^F_j(\mathbf{r},t)=\hat{\tilde{C}}^F_j(\mathbf{r},t)-\tilde{C}^{eF}_j$ take the form

$$\frac{\partial \tilde{C}^{\mathrm{I}}_j(\mathbf{r},t)}{\partial t} = g^{\mathrm{I}}_j + D_j\Delta\tilde{C}^{\mathrm{I}}_j(\mathbf{r},t)+\frac{D_j}{kT}\nabla\left[\tilde{C}^{\mathrm{I}}_j(\mathbf{r},t)\nabla E^{j\mathrm{I}}_P\right], \tag{4.15}$$

$$\frac{\partial \tilde{C}^{\mathrm{II}}_j(\mathbf{r},t)}{\partial t} = g^{\mathrm{II}}_j + D_j\Delta\tilde{C}^{\mathrm{II}}_j(\mathbf{r},t)+\frac{D_j}{kT}\nabla\left[\tilde{C}^{\mathrm{II}}_j(\mathbf{r},t)\nabla E^{j\mathrm{II}}_P\right]- \tag{4.16}$$

$$D_j\tilde{C}^{\mathrm{II}}_j(\mathbf{r},t)k^2_j.$$

Let the equations (4.15), (4.16) be solved in the stationary case and spherical geometry. We shall use the following transformation for the precipitates (see Ref. 204):

$$\tilde{C}^F_j(r)=\chi^F_j(r)\left[\varphi^F_j(r)-C^{eF}_j\right], \tag{4.17}$$

where $\chi_j^F(r)=\exp\left[-E_P^{jF}(r)(kT)^{-1}\right]$ (see also (2.27)). Now the expressions (4.15), (4.16) can be written for the stationary case in the form

$$\frac{d^2\varphi_j^F(r)}{dr^2}+\frac{2}{r}\frac{d\varphi_j^F(r)}{dr}-\frac{1}{kT}\frac{d\varphi_j^F(r)}{dr}\frac{dE_P^{jF}}{dr}+\frac{g_j^F}{D_j\chi_j^F}=$$

$$\Psi_j^F k_j^2\left[\varphi_j^F(r)-C_j^{eF}\right], \qquad (4.18)$$

where $\Psi_j^{\mathrm{I}}=0$, $\Psi_j^{\mathrm{II}}=1$.

The interaction energy $E_P^{jF}(r)$ has the form (2.25) and is a step function of r. Because of this one can write for dE_P^{jF}/dr

$$\frac{dE_P^{jF}(r)}{dr}=\frac{\Delta\Omega_j^F(r)\sigma_0^F}{R_P}\,\delta\left(1-\frac{r}{R_P}\right). \qquad (4.19)$$

Basing on (4.19), the functions $\varphi_j^F(r)$ will be described in various regions by the equations

$$\frac{d^2\varphi_j^F(r)}{dr^2}+\frac{2}{r}\frac{d\varphi_j^F(r)}{dr}+\frac{g_j^F}{D_j\chi_j^F}=\Psi_j^F k_j^2\left[\varphi_j^F(r)-C_j^{eF}\right]. \qquad (4.20)$$

The boundary conditions for the functions $\varphi_j^F(r)$ will be chosen in accordance with (2.16) in the following form:

$$\left.\frac{d\varphi_j^F(r)}{dr}\right|_{r>0}=0, \qquad (4.21)$$

$$\hat{C}_j^{\mathrm{I}}(R_P)=\chi_j^{\mathrm{I}}\varphi_j^{\mathrm{I}}(R_P)=\chi_j^{\mathrm{I}}C_j^e,\quad \varphi_j^{\mathrm{I}}(R_P)=C_j^e,$$

$$\hat{C}_j^{\text{II}}(R_p)=\chi_j^{\text{II}}\varphi_j^{\text{II}}(R_p)=\chi_j^{\text{II}}C_j^e, \quad \varphi_j^{\text{II}}(R_p)=C_j^e, \tag{4.22}$$

$$\hat{C}_j^{\text{II}}(L_p)=\hat{\tilde{C}}_j^{+}=\chi_j^{\text{II}}\varphi_j^{\text{II}}(L_p), \quad \varphi_j^{\text{II}}(L_p)=\hat{\tilde{C}}_j^{\text{II}}/\chi_j^{\text{II}}.$$

where $2L_p$ is an average separation between the precipitates.

For the determination of the average stationary concentration of point defects $\hat{\tilde{C}}_j^{+}=\hat{\tilde{C}}_j^{+\text{II}}$ in region II in the presence of coherent precipitates, the equations (2.11) and (2.14) will take the form (1.29a) if the assumptions A, B, C are taken into account. In this case one may think that due to the mobile defects (interstitials and vacancies) leaving one of the regions F=I, II, the rates of their generation are not equal, i.e. $g_j^{\text{I}} \neq g_j^{\text{II}}$ and $g_i^F \neq g_v^F$. If the tensor components of the stresses produced by the coherent precipitates have the form (2.22), then the generation rate g in equation (1.24a) should be substituted by g_j^F. In the case of compressed precipitates, we have

$$g_i^{\text{I}}=g^{\text{I}}, \quad g_v^{\text{I}}=g^{\text{I}}+\delta_v; \quad g_i^{\text{II}}=g^{\text{II}}+\delta_i, \quad g_v^{\text{II}}=g^{\text{II}},$$

where g^F is the generation rate for the Frenkel pairs in region F without allowance for the coherent precipitates, and

$$\delta_i=(1-\nu_p)^{-1}\int I_p^{j\text{I}}(R_p)f_p(R_p)dR_p;$$

$$\delta_v=\nu_p\int I_p^{v\text{II}}(R_p)f_p(R_p)dR_p. \tag{4.23}$$

In this case we obtain for the region F= II from the stationary equation

(1.24a)

$$\hat{\tilde{C}}_i^{+II} = C_0^{II} H_i^{II} (\sqrt{1+\tilde{\eta}^2} - \tilde{\eta}) + \tilde{C}_i^e,$$

(4.24)

$$\hat{\tilde{C}}_i^{+II} = \frac{D_i k_i^2}{D_v k_v^2} \left[\hat{\tilde{C}}_i^{+II} - \tilde{C}_i^e \right] - \frac{\delta_i}{D_v k_v^2},$$

where

$$k_j^2 = k_{jII}^2; \quad C_0^{II} = \sqrt{g_i^{II}/\mu D_i}; \quad \tilde{\eta} = \eta_0 \sqrt{g^{II}/g_i^{II}} \left[1 - \frac{\mu \delta_i}{D_v k_v^2 k_i^2} \right],$$

and the values $\tilde{C}_j^e$ are defined in (2.27).

If the defect recombination efficiency is low, we have from (4.24)

$$\hat{\tilde{C}}_j^{+II} \simeq \frac{g_j^{II}}{D_j k_j^2} + \tilde{C}_j^e. \qquad (4.24a)$$

The expression (4.24a) without allowance for the point defect interaction with the coherent precipitates ($E_P^j = 0$) is changed to a known expression (1.29a) for $\tilde{C}_{j0}^+$. Therefore, one can write for the stationary concentration in our model

$$\hat{\tilde{C}}_j^{+II} = \tilde{C}_{i0}^+ (1 + \frac{\delta_i}{g^{II}}) + \tilde{C}_i^e; \quad \hat{\tilde{C}}_j^{+II} \simeq \tilde{C}_{v0}^+ + \tilde{C}_v^e. \qquad (4.25)$$

A solution to equation (4.20) for the region F=1 has here the form

$$\varphi_j^I(r) = C_j^e + \frac{g_j^I}{6 D_j \chi_j^I} (R_P^2 - r^2),$$

which yields the following expression for the j-defect concentration within the precipitates:

$$\hat{C}^{\mathrm{I}}_j(r) = \chi^{\mathrm{I}}_j C^e_j + \frac{g^{\mathrm{I}}_j}{6D_j}(R^2_P - r^2). \qquad (4.26)$$

To obtain the solution to equation (4.20) in region F=II, we first consider a homogeneous equation for the functions $\varphi^{\mathrm{II}}_{j0}(r)$:

$$\frac{d^2\varphi^{\mathrm{II}}_{j0}(r)}{dr^2} + \frac{2}{r}\frac{d\varphi^{\mathrm{II}}_{j0}(r)}{dr} - k^2_j\varphi^{\mathrm{II}}_{j0}(r) = 0,$$

whose solutions are representable in the form

$$\varphi^{\mathrm{II}}_{j0}(r) = C_1 r^{-1}\mathrm{ch}(rk^2_j) + C_2 r^{-1}\mathrm{sh}(rk^2_j), \qquad (4.27)$$

where C_1 and C_2 are the constants. The solution to the inhomogeneous equation (4.20) with the use of the homogeneous solution (4.27) and the notation $\mathcal{A}_j = g^{\mathrm{II}}_j(D_j\chi^{\mathrm{II}}_j)^{-1} + k^2_j C^e_j$ can be written as follows (see Ref. 382):

$$\varphi^{\mathrm{II}}_j(r) = \mathcal{A}_j k^{-2}_j + r^{-1}\mathrm{sh}(rk_j)\left[C_2 + \mathcal{A}_j R_P k^{-2}_j \mathrm{sh}(R_P k_j) - \mathcal{A}_j k^{-3}_j \mathrm{ch}(R_P k_j)\right] +$$

$$r^{-1}\mathrm{ch}(rk_j)\left[C_1 - \mathcal{A}_j R_P k^{-2}_j \mathrm{ch}(R_P k_j) + \mathcal{A}_j k^{-3}_j \mathrm{sh}(R_P k_j)\right].$$

Using the boundary conditions (4.22) we find the values of C_1 and C_2 constants and obtain, taking into account (4.17), the following expression for the defect concentration in the F=II region:

$$\hat{C}^{\mathrm{II}}_j(r) = \left[\chi^{\mathrm{II}}_j C^{e\mathrm{II}}_j + g^{\mathrm{II}}_j(D_j k^2_j)^{-1}\right]\left\{1 - \frac{R_P}{r}\,\frac{\mathrm{sh}[(R_P + L_P - r)k_j]}{\mathrm{sh}(L_P k_j)}\right\} +$$

$$\chi_j^{\mathrm{II}} C_j^e \frac{R_P}{r} \frac{\mathrm{sh}(rk_j) - \mathrm{sh}[(r-R_P)k_j]}{\mathrm{sh}(R_P k_j)} . \tag{4.28}$$

It can be seen from formulas (4.26) and (4.28) not only the dependence $\hat{C}_j^F$ on r, but on the interaction energy E_P^{jF} through χ_j^F. So, for example, if considered is the case of a compressed precipitate, i.e. $\varepsilon_{\mathrm{I}}<0$, there take place $\mathrm{Sp}\sigma_P^{\mathrm{I}}>0$, $\mathrm{Sp}\sigma_P^{\mathrm{II}}<0$. In this case we have $E_P^{v\mathrm{I}}<0$, $E_P^{i\mathrm{I}}>0$, $E_P^{v\mathrm{II}}>0$, $E_P^{i\mathrm{II}}<0$, and hence,

$$\chi_v^{\mathrm{I}}>1, \quad \chi_i^{\mathrm{I}}<1, \quad \chi_v^{\mathrm{II}}<1, \quad \chi_i^{\mathrm{II}}>1.$$

Therefore, when the solid solution in whose decomposition the coherent compressed precipitates to form, are irradiated, incoming to the precipitates will be chiefly the vacancies, and outcoming the interstitials, thus increasing the interstitial supersaturation in regions II and favoring, for the future, the intensification of opposite-defect recombination in these regions. Use the same equations (4.15), (4.16) to calculate the relaxation of the defect profiles after the irradiation due to of the diffusion and with ignoring the sink terms in (4.16). Now, using the transformation (4.17), we obtain for $\varphi_j^F(r,t)$ the following equation instead of (4.18)

$$\frac{1}{D_j}\frac{\partial\varphi_j^F(r,t)}{\partial t} = \frac{\partial^2\varphi_j^F(r,t)}{\partial r^2} + \frac{2}{r}\frac{\partial\varphi_j^F(r,t)}{\partial t}$$

$$\frac{1}{kT}\frac{\partial\varphi_j^F(r,t)}{\partial r}\frac{dE^{jF}}{dr} . \tag{4.29}$$

With allowance for (4.19), we have actually in each region F an ordinary diffusion equation with the following boundary and initial conditions in accordance with (2.15) and (2.16):

$$J_j^{\mathrm{I}}(r,t)\Big|_{r=0}=0,\quad J_j^{\mathrm{I}}(r,t)\Big|_{R_P}=J_j^{\mathrm{II}}(r,t)\Big|_{R_P},$$

$$J_j^{\mathrm{II}}(r,t)\Big|_{r=L_P}=0,\quad C_j^{\mathrm{I}}(r,t)/C_j^{\mathrm{II}}(r,t)=\chi_j^{\mathrm{I}}/\chi_j^{\mathrm{II}},\tag{4.30}$$

$$\varphi_j^F(r,t_0)=\varphi_j^F(r),$$

where the fluxes $J_j^F(r,t)$ have been defined in (1.31), t_0 is the initial time, $\varphi_j^F(r)$ is the defect distribution taking shape under the irradiation.

Solving equation (4.19) in regions F by the Fourier method with (4.30), we obtain the dependence $\varphi_j^F(r,t)$. Analytical expressions for the concentrations can be written leaving only two terms of the Fourier series with respect to the eigenvalue λ in the form

$$\hat{C}_j^{\mathrm{I}}(r,t)=C_j^{eF}+\left[\hat{C}_j^{\mathrm{I}}(R_P,t_0)-C_j^{e\,\mathrm{I}}\right]\frac{\sin\lambda r}{\lambda r}\,e^{-\lambda^2 D_j(t-t_0)},\tag{4.31}$$

$$\hat{C}_j^{\mathrm{II}}(r,t)=C_j^{e\mathrm{II}}+\left[\hat{C}_j^{\mathrm{II}}(R_P,t_0)-C_j^{e\mathrm{II}}\right]\left\{\left[1-(\chi_j^{\mathrm{II}}-1)\frac{\lambda^2R_P^2}{3}\right]\frac{\sin\lambda r}{\lambda r}+(\chi_j^{\mathrm{II}}-1)\frac{\lambda^3R_P^3}{3}\,\frac{\cos\lambda r}{\lambda r}\right\}e^{-\lambda^2 D_j(t-t_0)},\tag{4.32}$$

where λ is determined from the condition $\lambda R_P\simeq\mathrm{tg}(\lambda R_P)$ with $\lambda_1=0$, $\lambda_2=\lambda$ being the two first solutions of this equation.

4.3. Stationary Profiles of Intrinsic Point Defects near an Edge Dislocation in the Formation of Coherent Precipitates

The calculation of concentration profiles for the internal point defects close by an edge dislocation can be handled analogously [204, 225] as the approximation of stationary inhomogeneous vacancy and interstitial distribution in the volume of a solid solution [383]. The balance equations for the concentration $C_{jD}(\mathbf{r},\vartheta)$ of the intrinsic point defects j in the elastic fields of the edge dislocation and precipitates can be now presented in the stationary case as (1.30), (1.31) (see Fig. 4.1, Diagram 2-2.1). As this takes place, the energy $E^j_D(r,\vartheta)$ of j defect interaction with the dislocation is defined in (1.32). In line with the results of Section 4.2, the elastic field of the coherent precipitate does not appear in the balance equation (1.32) implicitly.

Simplifying assumptions. To find the vacancy and interstitial concentrations in the neighborhood of an edge dislocation, we fall back on the method of effective medium (see, for example, Ref. 248 and Chapter 2, Section 2.4). We divide the material volume accounted for one dislocation by three coaxially-cylindrical regions F_D = I, II, III. Such subdivision of the solid solution is shown in Fig. 4.2 as a projection on the plane (r,ϑ). For convenience in further references, the assumptions taken into considerations in such volume subdivision have been numbered as 4.1, 4.2,...

The region F_D=I: $r_* \leq r \leq R$.

(4.1) Here all sinks, but the dislocation itself ($k^2_{jI}=0$), are missing. We are reminded that k^{-1}_j is the free path length of an intrinsic point defect before its adsorption on a sink of any kind.

(4.2) The radius R of the region I is selected as equal to a half separation between the precipitates surrounding the given edge dislocation (i.e. $4\pi R^3/3 \approx C^{-1}_P$, where C_P is the absolute concentration of precipitates in the solid solution).

(4.3) We consider that

$$\frac{|E^j_D(R,\vartheta)|_{\max}}{kT} = \frac{A^j_D}{kTR} \equiv \frac{R^j_0}{R} < 1. \tag{4.33}$$

The region F_D=II: $R \le r \le r_D$.

(4.4) There exist sinks for intrinsic point defects (e.g. precipitates, voids, dislocation loops), $k^2_{jII} \neq 0$.

(4.5) We consider that

$$\frac{|E^j_D(r_D,\vartheta)|_{\max}}{kT} = \frac{R^j_0}{r_D} \ll 1. \tag{4.34}$$

The region F_D=III: $r_D \le r \le L_D$.

(4.6) We may here ignore the elastic field of the edge dislocation as compared with the thermal background.

(4.7) The radius L_D in the region III is selected equal to a half separation between the dislocations (i.e. $2L_D \approx \rho_D^{-0.5}$), and $k^2_{jIII} \neq 0$.

Apart from this, we have:

(4.8) In all F_D regions the vacancy and interstitial recombination is neglected.

(4.9) The sums of sink strengths in II and III obey the relations

$$k_{j\mathrm{II}}R \gg 1, \quad k_{j\mathrm{III}}R \gg 1. \tag{4.35}$$

The boundary conditions for the intrinsic point defect concentration $C^{F_D}_{jD}(r,\vartheta)$ will be defined analogously to Ref. 204. We shall consider that a thermodynamically equilibrium point defect concentration is maintained in the neighborhood of the dislocation (of the radius r_*), and midway between the dislocations ($r{=}L_D$) their diffusion fluxes are equal zero due to the symmetry, and hence $C^{\mathrm{III}}_{jD}(r,\vartheta)$ reaches its maximum value. So we have

$$C^{\mathrm{I}}_{jD}(r_*,\vartheta)= \tilde{C}^e_j \exp\left[-\frac{E^j_D(r_*,\vartheta)}{kT}\right]= C^e_j e^{-\frac{E^j_D+E^j_P}{kT}}, \tag{4.36}$$

in accord with (1.33) and (2.27), but

$$C^{\mathrm{III}}_{jD}(L_D,\vartheta)=\hat{C}^{+}_{j0}= \text{const.}$$

Here C^e_j is the thermodynamically equilibrium concentration of the intrinsic point defects of j type remote of a dislocation in a pure crystal, and the value of average stationary concentration $\hat{C}^{+}_{j0}$ of j kind defects is defined by expressions (1.29) and (4.24), where $g{=}g^{\mathrm{III}}$, $k^2_n{=}k^2_{n\mathrm{III}}$, $k^2_j{=}k^2_{j\mathrm{III}}$.

For the concentration profiles of intrinsic point defects in the neighborhood of an edge dislocation being defined completely, we need to include the conditions of their joining on coaxial cylindrical surfaces of the radii R and r_D:

$$C^{\mathrm{I}}_{jD}(R,\vartheta)= C^{\mathrm{II}}_{jD}(R,\vartheta),$$

$$\left.\frac{\partial C^{\mathrm{I}}_{jD}(r,\vartheta)}{\partial r}\right|_{r=R} = \left.\frac{\partial C^{\mathrm{II}}_{jD}(r,\vartheta)}{\partial r}\right|_{r=R}, \tag{4.37}$$

$$C^{\mathrm{II}}_{jD}(r_D,\vartheta) = C^{\mathrm{III}}_{jD}(r_D,\vartheta),$$

$$\left.\frac{\partial C^{\mathrm{III}}_{jD}(r,\vartheta)}{\partial r}\right|_{r=r_D} = \left.\frac{\partial C^{\mathrm{III}}_{jD}(r,\vartheta)}{\partial r}\right|_{r=r_D}.$$

Consequently, the problem on finding the spatially inhomogeneous vacancy and interstitial distribution in the elastic fields of an edge dislocation and coherent precipitates is reduced to solving the balance equations (1.30) with appropriate boundary conditions (4.36) and joining conditions (4.37) in the F_D regions in the neighborhood of the dislocation (see Fig. 4.1, Diagram 2-2.1).

To solve the problem in hand, we substitute (see, e.g., Ref. 204)

$$C^{F_D}_{jD}(r,\vartheta) = \tilde{C}^{e}_{j} e^{-U^{j}_{D}(r,\vartheta)} + \Phi^{F_D}_{jD}(r,\vartheta) e^{-U^{j}_{D}(r,\vartheta)/2}, \tag{4.38}$$

where

$$U^{j}_{D}(r,\vartheta) = \frac{E^{j}_{D}(r,\vartheta)}{kT} = \frac{R^{j}_{0}}{r}\sin\vartheta. \tag{4.39}$$

Upon straightforward transformations [383,384] we obtain from (1.30) the following equation for the function $\Phi^{F_D}_{jD}(z_{F_D},\vartheta)$:

$$\Delta_{z_{F_D}} \Phi^{F_D}_{jD}(z_{F_D},\vartheta) - \Phi^{F_D}_{jD}(z_{F_D},\vartheta) + \Psi^{F_D}_{jD}(x,\vartheta) = 0, \tag{4.40}$$

where

$$x=\frac{R_0^j}{2r};\quad z_{\mathrm{I}}=x;\quad z_{\mathrm{II}}=k_{j\mathrm{II}}r;\quad z_{\mathrm{III}}=k_{j\mathrm{III}}r,$$

$$\Psi^{\mathrm{I}}_{jD}(x,\vartheta)=0.25D_j^{-1}g^{\mathrm{I}}R_0^{j2}x^{-4}\exp(x\sin\vartheta);$$

$$\Psi^{\mathrm{II}}_{jD}(x,\vartheta)=D_j^{-1}k_{j\mathrm{II}}^{-2}g^{\mathrm{II}}\exp(x\sin\vartheta)-\tilde{C}_j^{\mathit{e}}\exp(-x\sin\vartheta);$$

$$\Psi^{\mathrm{III}}_{jD}(x,\vartheta)=\Psi^{\mathrm{III}}_{jD}=D_j^{-1}k_{j\mathrm{III}}^{-1}g^{\mathrm{III}}-\tilde{C}_j^{\mathit{e}}=\mathrm{const.}$$

The boundary conditions for $\Phi^{F_D}_{jD}(z_{F_D},\vartheta)$ are determined from (4.36), (4.37). A solution to the equation (4.40) for the functions $\Phi^{F_D}_{jD}(z_{F_D},\vartheta)$ has been obtained in Refs. 383,384 with the following expansion:

$$\Phi^{F_D}_{jD}(z_{F_D},\vartheta)=\varphi^{F_D}_{0D}(z_{F_D})+$$
$$2\sum_{n=1}^{\infty}(-1)^n\left[\varphi^{F_D}_{2n,D}(z_{F_D})\cos 2n\vartheta-\varphi^{F_D}_{2n-1,D}(z_{F_D})\sin(2n-1)\vartheta\right]. \tag{4.41}$$

In an exact solution (4.41) of the differential equation (4.40) for F_D=I, one may confine himself to the first terms of the series. Then the concentration profiles of intrinsic point defects are described in the neighborhood of an edge dislocation by expression (4.38) when F_D=I, where

$$\Phi^{\mathrm{I}}_{jD}(x)\equiv\varphi^{\mathrm{I}}_{0D}(x)=\left[\eta_{01,D}+\eta_{02,D}\right]V_0(x)-\alpha^{\mathrm{I}}_{0D}P^{\mathrm{I}}_0(x). \tag{4.42}$$

We have

$$V_0(x)= \frac{I_0(x)K_0(x_D)-I_0(x_D)K_0(x)}{I_0(x_R)K_0(x_D)-I_0(x_D)K_0(x_R)};$$

$$\eta_{01,D}\approx \frac{C_j^+(g^{\text{III}})+\alpha_{0D}^{\text{III}}\left\{\sqrt{r_D/L_D}\ \text{ch}[k_{j\text{III}}(L_D-r_D)]-1\right\}}{\Lambda_0\Gamma_2\sqrt{R/L_D}} + \frac{\alpha_{0D}^{\text{II}}\Gamma_3}{\Lambda_0\Gamma_2};$$

$$\eta_{02,D}\simeq \frac{\alpha_{0D}^{\text{I}}}{\Lambda_0}\left[P_0(x_R)- \frac{x_R\Gamma_1}{k_{j\text{II}}R\ \Gamma_2}\left.\frac{dP_0(x)}{dx}\right|_{x=x_R}\right];$$

$$\Lambda_0\simeq 1+ \Gamma_1[k_{j\text{II}}R\ \Gamma_2\ln(1/x_R)]^{-1};$$

$$x_D= \frac{R_0^j}{2r_*};\qquad x= \frac{R_0^j}{2R};$$

$$\alpha_{0D}^{\text{I}}= \frac{g^{\text{I}}R_0^{j2}}{4D_j};\qquad \alpha_{0D}^{F_D}= \frac{g^{F_D}}{D_jk_{jF_D}^2};\qquad F_D= \text{II,III}.$$

$$\Gamma_1= \frac{k_{j\text{II}}}{k_{j\text{III}}}\text{ch}\left[k_{j\text{II}}(r_D-R)\right]\text{sh}\left[k_{j\text{III}}(L_D-r_D)\right]+$$

$$\text{sh}\left[k_{j\text{II}}(r_D- R)\right]\text{ch}\left[k_{j\text{III}}(L_D-r_D)\right];$$

$$\Gamma_2= \frac{k_{j\text{II}}}{k_{j\text{III}}}\text{sh}\left[k_{j\text{II}}(r_D-R)\right]\text{sh}\left[k_{j\text{III}}(L_D-r_D)\right]+$$

$$\text{ch}\left[k_{j\text{II}}(r_D- R)\right]\text{ch}\left[k_{j\text{III}}(L_D-r_D)\right];$$

$$\Gamma_3 = \Gamma_2 - \sqrt{r_D/R}\ \mathrm{ch}\left[k_{j\mathrm{III}}(L_D - r_D)\right];$$

$$P_0(x) = K_0(x)\int_x^{x_R} Q_{01}(x')dx' - I_0(x)\int_x^{x_D} Q_{02}(x')dx';$$

$$Q_{01}(x) = x^{-3} I_0^2(x); \quad Q_{02}(x) = x^{-3} I_0(x) K_0(x);$$

$I_0(x)$, $K_0(x)$ are the Bessel functions of pure imaginary argument (the same as in (1.35)).

In the limiting case of a pure crystal ($E_P^j=0$) and in the absence of sinks interposed between the dislocations (i.e. $L_D=r_D=R$). If is seen from (2.27) and the above listed formulas that

$$\tilde{C}_j^e = C_j^e; \quad \Gamma_1 = \Gamma_{03} = 0; \quad \Gamma_2 = 1; \quad \Lambda_0 = 1, \tag{4.43}$$

and the expression (4.42) is essentially simplified:

$$\Phi_{jD}^{\mathrm{I}}(x) = (\hat{C}_j^{+} - C_j^e)V_0(x) + \alpha_{0D}^{\mathrm{I}}\left[P_0^{\mathrm{I}}(x_R)V_0(x) - P_0^{\mathrm{I}}(x)\right]. \tag{4.44}$$

It follows from (4.38), (4.42)-(4.44) that in the special case $g^{\mathrm{I}}= 0$ the relation (4.38) with $F_D=1$ turns into a known expression (1.33) derived in Ref. 204.

4.4. Stationary Inhomogeneous Concentrations of Intrinsic Point Defects in the Elastic Field of Small-Sized Dislocation Loops in the Presence of Coherent Precipitates

The calculation of the concentration profiles of intrinsic point defects in the elastic field $C_{jL}(\mathbf{r})$ of small-sized dislocation loops can be performed as the approximation of stationary inhomogeneous vacancy and interstitial volume distribution [385]. In so doing we neglect the angular dependence in the expression $E_L^j(r)$ (see (1.36)) for the energy of interaction of the intrinsic point defects with the dislocation loop with a base on the results of Ref. 24. We shall solve the problem in the spherical geometry. Now the concentrations $C_{jL}(r)$ are described by equations (1.30),(1.31) (see Fig. 4.1, Diagram 2-2.1 in the stationary case, where q=L).

Calculating the vacancy and interstitial concentration profiles near the dislocation loop, we fall back on the effective medium method, as in the case of edge dislocation. As E_L^j decreases essentially more steeply than the energy r (see, e.g. Ref. 324), we divide, in accord with Ref. 248, the material volume accounted for one dislocation loop, into two spherical regions F_L=I, II (see Fig. 4.3).

The region F_L=I: $R_L \le r \le r_L$.

(4.10) There are no other sinks, but the dislocation loop itself ($k^2_{j\,\mathrm{I}}=0$).

The region F_L=II: $r_L \le r \le L_L$.

(4.11) We consider that

$$\frac{E^j_L(r_L)}{kT} \ll 1.$$

(4.12) There exist sinks for the intrinsic point defects (e.g. precipitates, voids), $k^2_{j\,\mathrm{II}} \neq 0$.

(4.13) One may neglect the elastic field of the dislocation loop in (1.30).

(4.14) The radius L_L of the F_L=II region is selected equal to a half loop separation.

Apart from this, we consider that the condition (4.8) is fulfilled in both F_L regions.

(4.15) The diffusion path length ($k^{-1}_{j\,\mathrm{II}}$) of vacancies and interstitial before their adsorption by a q-type sink is less than the average separation between the dislocation loops, $k_{j\mathrm{II}} L_L \gg 1$.

The boundary conditions in this process are written analogously [204,225]. We consider that maintained in the neighborhood of dislocation loop on the spherical R_L-radius surface is a thermodynamically equilibrium concentration of intrinsic point defects, the diffusion fluxes of vacancies and interstitials throughout a spherical R_L-radius surface being zero in accordance with assumption (4.14). Thus,

$$C^{\mathrm{I}}_{jL}(r_L)=\tilde{C}^e_j\exp\left[-\frac{E^j_L(R_L)}{kT}\right];\ \nabla C^{\mathrm{II}}_{jL}(r)\Big|_{r=L_L}=0. \qquad (4.45)$$

For a complete definition of the concentration profiles of intrinsic point defects in the neighborhood of the dislocation loop $C^{F_L}_{jL}(r)$, it is necessary, analogously to (4.37), to include the joining conditions for $C^{\mathrm{I}}_{jL}(r)$ and $C^{\mathrm{II}}_{jL}(r)$ on a spherical r_L-radius surface

$$C^{\mathrm{I}}_{jL}(r_L)=C^{\mathrm{II}}_{jL}(r_L),$$

$$\frac{dC^{\mathrm{I}}_{jL}(r)}{dr}\Big|_{r=r_L}=\frac{dC^{\mathrm{II}}_{jL}(r)}{dr}\Big|_{r=r_L}. \qquad (4.46)$$

So the problem on finding the stationary point defect distribution in the elastic field of a short-radius dislocation loop is reduced, as for the edge dislocation, to solving equation (4.45) and to joining conditions (4.46) in the F_L regions close to the loops (Fig. 4.1, Diagram 2-2.1).

The concentration profiles of intrinsic point defects $C^{F_L}_{jL}(r)$ close to the dislocation loop can be calculated going on in the balance equations (1.30) at q=L to a dimensionless interaction energy,

$$\frac{E^j_L(r)}{kT}=U^j_L(r). \qquad (4.48)$$

Basing on the above included assumptions (4.8), (4.10)-(4.15) and the relations (1.31), (4.48), we find a specific form of equation (1.30) at q=L in the F_L=I, II regions (see Fig. 4.3).

In the F_L=1 region ($R_L\le r<r_L$), the balance equation (1.30) is, in accord with assumptions (4.8) and (4.10), essentially simplified and takes the

form

$$g^{\mathrm{I}} + D_j \Delta C^{\mathrm{I}}_{jL}(r) + D_j C^{\mathrm{I}}_{jL}(r) \Delta U^j_L(r) +$$
$$D_j \nabla C^{\mathrm{I}}_{jL}(r) \nabla U^j_L(r) = 0. \tag{4.49}$$

In the F_L=II region ($r_L \le r \le L_L$), the balance equation (1.30) at q=L is transformed to the form

$$g^{\mathrm{II}} + D_j \Delta C^{\mathrm{II}}_{jL}(r) - D_j k^2_{j\mathrm{II}} C^{\mathrm{II}}_{jL}(r) = 0 \tag{4.50}$$

on a basis of assumptions (4.8) and (4.12), (4.13). Solving the differential equations (4.49), (4.50) with the boundary conditions (4.45) and the joining conditions (4.46), we find the concentration profiles of intrinsic point defects in the neighborhood of the dislocation loop at an arbitrary value of the interaction energy $U^j_L(r)$.

The difference between equations (4.46) and those considered in Section 4.3 consists in that the problem is spherically symmetric (the dependence on the angle ϑ is absent).

The calculation of the concentration profiles of intrinsic point defects in the solid solution F=I region adjacent to the dislocation loop is made in the same way as in the edge dislocation case. For this purpose, we use the transformation

$$C^{\mathrm{I}}_{jL}(r) = \Phi^{\mathrm{I}}_{jL}(r) \exp\left[-U^j_L(r)\right],$$

where $\Phi^{\mathrm{I}}_{jL}(r)$ is the function to be determined. Now the expression for the density of the j-type defect flux toward the dislocation loop (see (1.31)) is transformed to

$$\mathbf{J}^j_L(r)=-D_j \frac{d\Phi^{\mathrm{I}}_{jL}(r)}{dr} \exp\left[-U^j_L(r)\right]\mathbf{i}_r, \tag{4.51}$$

where $\mathbf{i}_r$ is the unit vector of the normal to the spherical r-radius surface, and the equation (4.49) is essentially simplified and takes the form

$$\frac{d^2\Phi^{\mathrm{I}}_{jL}(r)}{dr^2}+\left[\frac{2}{r}-\frac{dU^j_L(r)}{dr}\right]\frac{d\Phi^{\mathrm{I}}_{jL}(r)}{dr}+\frac{g^{\mathrm{I}}}{D_j}e^{U^j_L(r)}=0. \tag{4.52}$$

in the spherical geometry case.

The equation (4.52) can be considered as a linear inhomogeneous differential equation of the first order with respect to $d\Phi^{\mathrm{I}}_{jL}(r)/dr$. Its solution is representative in a standard form:

$$\frac{d\Phi^{\mathrm{I}}_{jL}(r)}{dr}= \tag{4.53}$$

$$\exp\left[-\int_{R_L}^{r} P_L(r')dr'\right]\left\{B^{\mathrm{I}}_L+\int_{R_L}^{r} f(r')\exp\left[\int_{R}^{r'} P_L(r'')dr''\right]dr'\right\},$$

where the notations

$$P_L(r)=\frac{2}{r}-\frac{dU^j_L(r)}{dr}, \tag{4.54}$$

$$f_L(r)=-\frac{g^{\mathrm{I}}}{D_j}\exp\left[U^j_L(r)\right] \tag{4.55}$$

are used, and B^{I}_{L} is the integration constant. Substitute (4.54) and (4.55) into (4.53) and have

$$\frac{d\Phi^{\mathrm{I}}_{jL}(r)}{dr} = \left(\frac{R_L}{r}\right)^2 \exp\left[U^j_L(r)-U^j_L(R_L)\right]\times$$

$$\left\{B^{\mathrm{I}}_L - g^{\mathrm{I}}R_L(3D_j)^{-1}\exp\left[U^j_L(R_L)\right]\left[\left(\frac{r}{R_L}\right)^3-1\right]\right\}. \tag{4.56}$$

Integrating expression (4.56) with account for (4.45) and (4.51), we find

$$\Phi^{\mathrm{I}}_{jL}(r)=\tilde{C}^{e}_j+R^2_L\left[B^{\mathrm{I}}_L e^{-U^j_L(r)}+g^{\mathrm{I}}R_L(3D_j)^{-1}\right]P^j_L(r)-g^{\mathrm{I}}(3D_j)^{-1}Q^j_L(r)$$

$$C^{\mathrm{I}}_{jL}(r)=\left\{\tilde{C}^{e}_j+R^2_L\left[B^{\mathrm{I}}_L e^{-U^j_L(r)}+g^{\mathrm{I}}R_L(3D_j)^{-1}\right]P^j_L(r)-\right. \tag{4.57}$$

$$\left.g^{\mathrm{I}}(3D_j)^{-1}Q^j_L(r)\right\}\exp\left[-U^j_L(r)\right].$$

Here

$$P^j_L(r)=\int_{R_L}^{r}(r')^{-2}e^{-U^j_L(r')}dr'; \quad Q^j_L(r)=\int_{R_L}^{r}r'\,e^{-U^j_L(r')}dr'.$$

To determine an explicit form of $C^{\mathrm{I}}_{jL}(r)$, we find the intrinsic point defect concentrations $C^{\mathrm{II}}_{jL}(r)$ in the $F_L=$ II solid solution region and join the solution obtained to the expression (4.57). Solving the differential equation (4.50) with the boundary condition (4.45), we obtain

$$C^{\mathrm{II}}_{jL}(r)=\hat{C}^{+}_{j0}+B^{\mathrm{II}}_{L}e^{-k_{j\mathrm{II}}L_{L}}r^{-1}\left[\frac{k_{j\mathrm{II}}L_{L}+1}{k_{j\mathrm{II}}l_{L}-1}e^{-k_{j\mathrm{II}}(L_{L}-r)}+e^{k_{j\mathrm{II}}(L_{L}-r)}\right].$$

In the vicinity of a r_L-radius sphere, we have, basing on assumption (4.15),

$$C^{\mathrm{II}}_{jL}(r)\simeq \hat{C}^{+}_{j0}+B^{\mathrm{II}}_{L}r^{-1}\exp(-k_{j\mathrm{II}}r). \tag{4.58}$$

Substitute expressions (4.57) and (4.58) into the joining conditions (4.47), we find the integration constant B^{I}_{L} in (4.57), thus determining the intrinsic point defect profiles in the F_L=I solid solution region adjacent to the dislocation loop. We have ultimately

$$C^{\mathrm{I}}_{jL}(r)=e^{-U^{j}_{L}(r')}\left[\tilde{C}^{e}_{j}+B^{j}_{L}P^{j}_{L}(r)-\frac{g^{\mathrm{I}}}{3D_{j}}Q^{j}_{L}(r)\right]. \tag{4.59}$$

Here

$$B^{j}_{L}=R^{2}_{L}\left[B^{\mathrm{I}}_{L}e^{-U^{j}_{LR}}+\frac{g^{\mathrm{I}}R_{L}}{3D_{j}}\right]=$$

$$R_{L}z^{j}_{L}\left[C^{+}_{j0}-\mathrm{V}_{1L}\tilde{C}^{e}_{j}e^{-U^{j}_{LC}}+\frac{g^{\mathrm{I}}r^{2}_{L}}{6D_{j}}\mathrm{V}_{1L}(\mathrm{V}_{2L}-H^{j}_{L}),\right. \tag{4.60}$$

with

$$z^{j}_{L}=\left[\mathrm{V}_{1L}R_{L}P^{j}_{L}(r_{L})e^{-U^{j}_{LC}}+R_{L}r^{-1}_{L}(k_{j\mathrm{II}}r_{L}+1)^{-1}\right]^{-1}, \tag{4.61}$$

$$H^{j}_{L}=1-2r^{-2}_{L}Q^{j}_{L}(r_{L})\exp(-U^{j}_{LC}),$$

$$V_{1L}=1-\frac{r_L}{k_{jII}r_L+1}\cdot\frac{dU^j_L(r)}{dr}\bigg|_{r=r_L}, \tag{4.62}$$

$$V_{2L}=\frac{k_{jII}r_L+3-r_L\dfrac{dU^j_L(r)}{dr}\Big|_{r=r_L}}{k_{jII}r_L+1-r_L\dfrac{dU^j_L(r)}{dr}\Big|_{r=r_L}},$$

and for simplicity the notations

$$U^j_{LC}\equiv U^j_L(r_L),\quad U^j_{LR}\equiv U^j_L(R_L)$$

have been introduced.

We find a specifical form of $C^{\mathrm{I}}_{jL}(r)$ in a particular case of the power dependence $U^j_L(r)$ on the coordinate r, i.e.

$$U^j_L(r)=\frac{A^j_L}{kTr^n}, \tag{4.63}$$

where A^j_L takes both a positive and negative value depending on the type of j point defect. Insofar as the energy of interacting the intrinsic point defects with the dislocation loop $E^j_L(r)$ is falling with the distance more rapidly than $E^j_D(r)$ [324], we consider that n is any peal-valued number more than, or equal to, two ($n \geq 2$). In this case, we have

$$V_{1L}=1+nU^j_{LC}(k_{jII}r_L+1)^{-1},$$

$$\tag{4.64}$$

$$V_{2L}=(3+k_{jII}r_L+nU^j_{LC})(1+k_{jII}r_L+nU^j_{LC})^{-1},$$

$$P^j_L(r)=R^{-1}_L\varphi^j_L(R_L)-r^{-1}\varphi^j_L(r),$$

$$Q^j_L(r)=\frac{R^2_L}{2}\left\{\left(\frac{r}{R_L}\right)^2 e^{-U^j_L(r)} - e^{U^j_{LR}(r)} + \right. \tag{4.65}$$

$$\left. U^j_{LR}\left[\Psi^j_L(r_L)-\Psi^j_L(r)\right]\right\},$$

where

$$\varphi^j_L(r)={}_1F_1\left(\frac{1}{n},\frac{n+1}{n},U^j_L(r)\right),$$

$$\Psi^j_L(r)=\begin{cases}\mathrm{Li}\left\{\exp[U^j_L(r)]\right\}, & n=2,\\ \dfrac{n}{n-2}\left(\dfrac{R_L}{r}\right)^{n-2}{}_1F_1\left(\dfrac{n-2}{n},2\dfrac{n-1}{n},U^j_L(r)\right), & n>2.\end{cases}$$

Here Li(z) is the integral logarithm.

Substituting (4.65) into expressions (4.61) and (4.62),we obtain

$$z^j_L=\left\{V_{1L}e^{-U^j_{LC}}\left[\varphi^j_L(R_L)-\frac{R_L}{r_L}\varphi^j_L(r_L)\right]+R_Lr^{-1}_L\left(k_{jII}r_L+1\right)^{-1}\right\}^{-1},$$

$$H^j_L=\left(\frac{R_L}{r_L}\right)^2 \mathrm{e}^{-U^j_{LC}}\left\{\mathrm{e}^{-U^j_{LR}}-U^j_{LR}\left[\Psi^j_L(R_L)-\Psi^j_L(r_L)\right]\right\}. \tag{4.66}$$

The relations (4.59), (4.60), (4.64)-(4.66) define the distribution of intrinsic point defects $C^{\mathrm{I}}_{jL}(r)$ near the dislocation loop (F_L=I) in the case of the power dependence of $U^j_L(r)$ on r. However, as it is difficult to operate analytically with expressions (4.66), we consider their approximate representations in case when the conditions

$$U^j_{LC} \ll 1 \ll U^j_{LR} \tag{4.67}$$

are fulfilled which follow from the assumption (4.11) and the expression (4.63). In this case the conditions (4.67) define, together with (4.63), the dependence of the quantity r_L on the form of $U^j_L(r)$. Now we have

$$V_{1L} \simeq 1; \quad V_{2L} = (k_{jII} r_L + 3)(k_{jII} r_L + 1)^{-1}; \quad e^{-U^j_{LC}} \simeq 1. \tag{4.68}$$

when n is not too large.

Using the asymptotic behavior of ${}_1F_1(a,b,z)$ and Li(z) for $z \gg 1$ and $z \ll 1$ (see, e.g., Ref. 386), we find from (4.61) and (4.66) approximate values of z^j_L and H^j_L for the opposite signs of the interaction energy between the dislocation loop and the point defect

$$z^{j(-)}_L \simeq \frac{\dfrac{n}{\Gamma(1/n)} \left(|U^j_{LR}|\right)^{1/n}}{1 - \dfrac{n}{\Gamma(1/n)} \left[|U^j_{LC}|\right]^{1/n} k_{jII} r_L \left[k_{jII} r_L + 1\right]^{-1}}, \quad U^j_L(r) < 0.$$

$$z^{j(+)}_L \simeq nU^j_{LR} \exp(-U^j_{LR}) \quad \text{if} \quad U^j_L(r) > 0, \tag{4.69}$$

$$z^{j0}_L \simeq 1 \quad \text{if} \quad U^j_L(r) = 0,$$

$$H^{j(\pm)}_L \simeq \begin{cases} U^j_{LC} \ln[2|U^j_{LC}|], & n \longrightarrow 2, \\ \dfrac{n}{n-2} U^j_{LC}, & n > 2, \; U^j_L(r) > 0, \\ |U^j_{LC}|^{2/n} \left[\Gamma\left(\dfrac{n-2}{n}\right) - \dfrac{n}{n-2} |U^j_{LC}|^{(n-2)/n}\right], & n > 2, \; U^j_L(r) < 0. \end{cases} \tag{4.70}$$

The substitution of the relations (4.68)-(4.70) into (4.59) permits the expression for the intrinsic point defect concentration in the neighborhood of the dislocation loop to be simplified. Further we shall need to know only the values $z_L^{j(-)}$ and $z_L^{j(0)}$, in consequence we shall omit the indices (-) and (0) and consider only interstitial attraction to the interstitial dislocation loop and the vacancy diffusion close by the loop.

In the $U_L^j(r)=U^j$=const case, the expressions (4.59), (4.60) are still further increased, taking the form

$$C_j^{\rm I}(r) \simeq \tilde{C}_j^e e^{-U^j} + B_{\rm V}^j\left(\frac{1}{R_{\rm V}} - \frac{1}{r}\right) - \frac{g^{\rm I} R_{\rm V}^2}{6D_j}\left[\left(\frac{r}{R_{\rm V}}\right)^2 - 1\right], \tag{4.71}$$

$$B_{\rm V}^j \simeq \frac{C_{j0}^{+} - \tilde{C}_j^e e^{-U^j} + \dfrac{g^{\rm I} r_{\rm V}^2}{6D_j}\left[\left(3+k_{j\rm II} r_{\rm V}\right)\left(1+k_{j\rm II} r_{\rm V}\right)^{-1} - \left(\dfrac{R_{\rm V}}{r_{\rm V}}\right)^2\right]}{\dfrac{1}{R_{\rm V}} - \dfrac{1}{r_{\rm V}} + r_{\rm V}^{-1}(1+k_{j\rm II} r_{\rm V})^{-1}}. \tag{4.72}$$

The relations obtained describe, in particular, the intrinsic point defect distribution in the neighborhood of a spherical void with which the *j*-type defects do not practically interact ($U^j=U_{\rm V}^j\simeq 0$). The radius $r_{\rm V}$ defines the region $F_{\rm V}$=I where there are no other sinks but that with q=V (i.e. $k_{j\rm I}^2=0$).

CHAPTER 5

CONCENTRATION PROFILES OF POINT DEFECTS NEAR THE ISOLATED PRECIPITATES OF A SECONDARY PHASE UNDER IRRADIATION

5.1. Diffusion of Impurities and Their Complexes

If the lifetime of a complex consisting of an impurity atom and an intrinsic point defect is of the order of its jump for one interatomic separation, the concentrations of impurity atoms, intrinsic point defects (j type), and complexes are in a local equilibrium, i.e.

$$\alpha_{ja} C_j(r,t) C_a(r,t) = \chi_{ja} C_{ja}(r,t). \tag{5.1}$$

In this situation the system of balance equations (2.11)- (2.14) is broken into two systems being divorced from each other (see Fig. 2.5). One of these describes the evolution of vacancy and interstitial concentration profiles (see (1.30)), and another the diffusion of single impurity atoms toward their sinks

$$\frac{\partial C_a(\mathbf{r},t)}{\partial t} = -\mathrm{div}\mathbf{J}_q^a(\mathbf{r},t) - D_a C_a k_a^2, \tag{5.2}$$

$$\frac{\partial C_{ja}(\mathbf{r},t)}{\partial t} = -\mathrm{div}\mathbf{J}_q^{ja}(\mathbf{r},t) - D_{ja} C_{ja} k_{ja}^2. \tag{5.3}$$

As the impurity concentration in actual solid solutions is essentially higher than the concentration of complexes, the system of balance equations (5.2), (5.3) degenerates into the diffusion equation (5.2).

A solution to the diffusion equation (5.2) with the initial and boundary conditions (2.15), (2.16) has been obtained in many monographs and original publications (see, e.g., Refs. 184, 185, 387-391).

If the lifetime of *ja* complex is of the order of the time of its movement toward the precipitate, then the precipitate growth is limited by the diffusion of complexes in the solid solution. The concentrations of impurities, intrinsic point *j* defects and complexes are no longer in the local equilibrium, the condition (5.1) is not fulfilled and goes into the inequality

$$\alpha_{ja} C_j(r,t) C_a(r,t) > \chi_{ja} C_{ja}(r,t). \tag{5.4}$$

Assuming that the impurity atoms outside the complexes are fixed (i.e. the diffusion of the impurity atoms unbounded into complexes does not contribute essentially to their total flux to the precipitate), write the system of balance equations (2.11)-(2.14) as follows:

$$\frac{\partial C_j(\mathbf{r},t)}{\partial t} = g - \mathrm{div}\mathbf{J}^j_q(\mathbf{r},t) - \alpha_{ja} C_j C_a + \chi_{ja} C_{ja} - \mu D_i C_j C_n - D_j C_j k_j^2, \tag{5.5}$$

$$\frac{\partial C_a(\mathbf{r},t)}{\partial t} = -\alpha_{ja} C_j C_a + \chi_{ja} C_{ja}, \tag{5.6}$$

$$\frac{\partial C_{ja}(\mathbf{r},t)}{\partial t} = -\mathrm{div}\mathbf{J}^{ja}_q(\mathbf{r},t) - \alpha_{ja} C_j C_a - \chi_{ja} C_{ja} - D_{ja} C_{ja} k_{ja}^2, \tag{5.7}$$

$$\frac{\partial C_n(\mathbf{r},t)}{\partial t} = g - \mathrm{div}\mathbf{J}^n_q(\mathbf{r},t) - \mu D_i C_j C_n - D_n C_n k_n^2. \tag{5.8}$$

The system (5.5)-(5.8) does not split into two independent subsystems of the type (1.30) and (5.2), (5.3). The calculation of the concentration profiles of mobile point defects in the neighborhood of their sinks requires that the nonlinear differential equations (5.5)-(5.8) be solved simultaneously. In Ref. 392 the solution to similar balance equations was found as the approximation of a homogeneous volume distribution of the mobile point defects.

The inhomogeneous distribution of mobile point defects in the balance equations (5.5)-(5.8) was first taken into account in Refs. 289, 292 in the g=0, E_P^j=0, k_j^2=0 and μ=0 cases in studying the the decomposition of oversaturated substitutional solid solutions *Al-Zn*, where the energy of bonding the impurity atoms (a) with the vacancies j=v is so much higher that the complexes migrate toward the precipitates without dissociation in the solid solution volume (χ_{ja}=0). In such a way the authors of Refs. 289, 290 explained a rapid growth of precipitates out of keeping with an ordinary diffusion impurity transfer. The calculation of concentration profiles and diffusion fluxes of complexes close by a void can be calculated following the diagram offered in Ref. 207, since the energy of interacting the mobile point defects with the given type sink (q=V) does not depend on the coordinate r [237].

As distinguished from voids, the edge dislocations produce inhomogeneous fields of elastic stresses in the material volume [222, 324], and, as a consequence, an exact solution to the balance equations (5.5)-(5.8) close by a dislocation is impossible to employ (see, e.g., Ref.

292).

In Ref. 292 the diffusion fluxes of complexes in the elastic fields of edge dislocations and voids have been thus calculated merely to the approximation of the stationary inhomogeneous volume distribution of mobile point defects. However, one considered therewith that the intrinsic point defects of *j* type and the complexes are in a local thermodynamic equilibrium with each other (see (5.1)). The mechanism of vacancy and interstitial transport toward the edge dislocations and voids is similar to that of vacancy pump (see Refs. 289, 290). It was shown in Ref. 292 that if the diffusion rate of impurity atoms (in a bound state) toward a structural defect is more than that of their reverse diffusion into the solid solution, then the secondary-phase precipitates can nucleate in the neighborhood of the structural defects. The segregation formation nearby the dislocations and voids diminishes swelling in the material and the dislocation climbing rate, which is in a good agreement with the results of Refs. 206, 207, 247-250.

5.2. General Equation for Local Nonstationary Concentrations

In many metallic systems the bound energy ε^b_{ja} of *ja* complexes (intrinsic *j*-defect - impurity *a* atom) proves to be sufficiently high [181]. This causes the complexes *ja* to be steady and the impurity transport toward the growing precipitate to take place due to the diffusion of *ja* complexes rather than that of single impurity atoms (see the condition (5.4)). Compared to the case of single impurity atom migration toward the precipitate, the evolution of the latter due to *ja* complex diffusion complicates the physical pattern of solid solution decomposition at this stage. It is well to bear in mind that for elucidating the peculiarities of physical processes, one needs to know how the concentration of the mobile point defects depends on the time *t* and the coordinates. For this purpose, one must solve the system of balance equations (5.5)-(5.8) (see Fig. 3.1, Diagram 3).

To derive the equations (5.5)-(5.8) from a more general system of equations (2.11)-(2.14) one would formulate the following condition instead of the simplifying assumption 2 in Section 2.4.

The bound energy ε^b_{ja} of *ja* complexes and the activation energy ε^m_a for the migration of single impurity atoms would be more than the activation energy ε^m_{ja} for the migration of *ja* complexes in the whole volume of a solid solution, i.e.

$$\varepsilon^b_{ja},\ \varepsilon^m_a > \varepsilon^m_{ja}. \tag{5.9}$$

The condition (5.9) means that only the vacancies, interstitials and *ja* complexes are diffusion-mobile point defects, while the off-complex single impurity atoms remain stationary.

We further believe that the activation energies for the migration of intrinsic point defects ε_j^m and complexes ε_{ja}^m differ little[1], i.e.

$$\varepsilon_{ja}^m \simeq \varepsilon_j^m . \qquad (5.10)$$

As initial conditions for the mobile point defects (*j*, *ja*) nearby the precipitates are taken their thermodynamically equilibrium values, which , in accord with (2.15) and (2.27), are representable in the form

$$C_j(r,0)=C_j^e, \quad C_{ja}(r,0)=C_{ja}^e, \qquad (5.11)$$

and

$$C_a(r,0)=C_a^0,$$

where C_a^0 is the initial impurity concentration in the solid solution.

Let us study, in spherical geometry, without consideration of the precipitate nucleation, a distribution of mobile point defects nearby the isolated precipitates under irradiation [91, 285, 393-395]. Assume that the R_P-radius precipitates are separated by the average distance $2L_P$, so that the growth of each of these takes place inside the sphere of L_P radius. In this case the boundary conditions, in accord with (2.16), have the form

$$\lim_{r \to R_P} C_j(r,t)=C_j(R_P,0); \quad \nabla C_j(r,t)\Big|_{r=L_P}=0, \qquad (5.12a)$$

$$\lim_{r \to R_P} C_{ja}(r,t)=C_{ja}(R_P,0); \quad \nabla C_{ja}(r,t)\Big|_{r=L_P}=0. \qquad (5.12b)$$

An analytical solution to the system of equations (5.5)-(5.8) is

---

[1] In slow depletion of the solid solution under irradiation (see further Section 5.3) the condition (5.10) need not be fulfilled.

impossible. However, taking advantage of the results of Ref. 205, where the nonstationarity period t_* is determined (see (1.27)), one can essentially simplify the balance equations (5.5)-(5.8) on various time intervals. Depending on the values of the energies ε^b_{ja}, ε^m_j, and other parameters of the problem, two limiting cases are possible to be recognized (see Fig. 5.1, Diagram 3):

(a) rapid depletion of the solid solution in the critical time $t_C<t_*$;

(b) slow depletion of the solid solution in $t_C>t_*$.

We shall further demonstrate that the system (5.5)-(5.8) can be reduced to a unified differential equation for some r- and t-dependent function expressed through the sought-for defect concentrations.

Rapid depletion ($t_C<t_*$). In this situation most of the defects j form ja complexes, which virtually do not dissociate in their path toward the precipitate. The decomposition goes rapidly, the solid solution being depleted on the background of the nonstationary concentrations of intrinsic point defects

$$\left.\frac{\partial C_a(r,t)}{\partial t}\right|_{t<t_*}<0, \qquad \left.\frac{\partial C_a(r,t)}{\partial t}\right|_{t>t_*}\approx 0.$$

If the structural inhomogeneities have not managed to be formed in the material volume in the time $t\le t_*$, one would consider that there are no other sinks midway the precipitates. With the recombination of intrinsic point defects the early irradiation stages ($t\le t_*$) and the complex dissociation neglected, one obtains from (5.5)-(5.8) a simplified system of nonlinear differential equations (Fig. 5.1, Diagram 3-3.2):

$$\frac{\partial C_{j0}(r,t)}{\partial t} = g + D_j \Delta C_{j0}(r,t) - \alpha C_{j0} C_a, \tag{5.13}$$

$$\frac{\partial C_a(r,t)}{\partial t} = -\alpha C_{j0} C_a, \tag{5.14}$$

$$\frac{\partial C_{ja0}(r,t)}{\partial t} = D_{ja} \Delta C_{ja0}(r,t) + \alpha C_{j0} C_a. \tag{5.15}$$

Here $C_{j0}(r,t)$ and $C_{ja0}(r,t)$ are the concentrations of intrinsic point defects (j) and complexes (ja) in the absence of the recombination of opposite point defects and complex dissociation of the solid solution volume. In virtue of the condition (5.10), this system is succeeded to be reduced to an ordinary diffusion equation (Fig. 5.1, Diagram 3-3.3)

$$\frac{\partial \varphi(r,t)}{\partial t} = D_j \Delta \varphi(r,t) \tag{5.16}$$

for the function

$$\varphi(r,t) = \alpha \int_0^t C_{j0}(r,\tau) d\tau + \alpha t (C_a^0 - C_j^e) - \frac{\alpha g t^2}{2} =$$

$$\alpha t (C_a^0 + C_{ja}^e) - \alpha \int_0^t C_{ja0}(r,\tau) d\tau, \tag{5.17}$$

satisfying the initial and boundary conditions

$$\varphi(r,0) = 0,$$

$$\lim_{r \to R_P} \varphi(r,t) = \alpha C_a^0 t, \quad \nabla \varphi(r,t) \Big|_{r=L_P} = 0. \tag{5.18}$$

With a knowledge of the solution to the diffusion equation (5.16), one can determine, by (5.17), the point-defect concentration profiles: $C_{j0}(r,t)$, $C_{ja0}(r,t)$, $C_a(r,t)$.

If the complexes dissociate in their path to the precipitate, the solid solution is depleted more slowly, the concentrations $C_i(r,t)$ and $C_v(r,t)$ have a chance to increase substantially. Now one needs to take into account the complex dissociation and the intrinsic point defect recombination in the balance equations (5.5)-(5.8) (see Fig. 5.1, Diagram 3-3.4)

$$\frac{\partial C_j(r,t)}{\partial t}=g+D_j\Delta C_j(r,t)-\alpha_{ja}C_jC_a+\chi_{ja}C_{ja}-\mu D_iC_jC_n, \qquad (5.19)$$

$$\frac{\partial C_a(r,t)}{\partial t}= -\alpha_{ja}C_jC_a+\chi_{ja}C_{ja}, \qquad (5.20)$$

$$\frac{\partial C_{ja}(r,t)}{\partial t}= D_{ja}\Delta C_{ja}(r,t)+\alpha_{ja}C_jC_a-\chi_{ja}C_{ja}, \qquad (5.21)$$

$$\frac{\partial C_n(r,t)}{\partial t}=g+D_n\Delta C_n(r,t) -\mu D_iC_jC_n. \qquad (5.22)$$

For simplicity of mathematical treatment, we shall, however, consider that the concentration profiles of the point defects *l=j, n, a, ja* have a little deviation from $C_{l0}(r,t)$, i.e.

$$C_l(r,t)=C_{l0}(r,t)-C_{l1}(r,t), \quad |C_{l1}(r,t)|\ll C_{l0}(r,t). \qquad (5.23)$$

In this case the system of nonlinear differential equations (5.19)-(5.22) reduces to an ordinary diffusion equation (5.16) for a new function (Fig. 5.1, Diagram 3-3.5)

$$\varphi(r,t)=\mu D_i\int C_{j0}(r,t)C_{n0}(t)dt-C_{j1}(r,t)-C_{ja1}(r,t) \qquad (5.24)$$

satisfying the initial and boundary conditions

$$\varphi(r,0)=\ \mu D_i\left[\int C_{j0}(r,t)C_{n0}(t)dt\right]_{t=0}, \qquad (5.25)$$

$$\nabla\varphi(r,t)\Big|_{r=z_P}=\mu D_i\int C_{n0}(t)\nabla C_{j0}(r,t)\Big|_{r=z_P}dt. \qquad (5.26)$$

$(z_P=R_P,\ L_P)$

Substituting the solution to the differential equation (5.1) for the functions $C_{j0}(r,t)$, $C_{ja0}(r,t)$ from (5.17) and $C_{j1}(r,t)$, $C_{ja1}(r,t)$ from (5.34) into (5.23), one may determine the concentration profiles of mobile point defects $C_j(r,t)$ and $C_{ja}(r,t)$ in the neighborhood of the secondary-phase precipitates.

Slow depletion $(t_C>t_*)$. If the complexes dissociate readily, there takes place a slow depletion of the solid solution, which is performed on the background of actually stationary concentrations of intrinsic point defects $\hat{C}_j^{+II}\equiv\hat{C}_j^{+}$ (see (1.29)), i.e. to the moment of their approaching the stationary values

$$\left.\frac{\partial C_a(r,t)}{\partial t}\right|_{t\leq t_*}\approx 0. \qquad 5.27)$$

In this situation one need to take account of occurring in the solid solution new dislocation loops and voids which are sinks for the mobile point defects, the vacancy and interstitial recombination, as well as the complex dissociation. An analytical solution to the system of balance equations (5.5)-(5.8) remains a complicated problem, as before. However, it may be solved by stages (Fig. 5.1, Diagram 3-3.6-3.8):

(1) Pass on from the equations with a spatial concentration dependence (5.5)-(5.8) (inhomogeneous problem) to the equations for the medium volume concentrations (homogeneous problem, e.g. (1.24)), and then to homogeneous stationary point defect concentrations (see (1.24a) and Fig. 5.1, Diagram 3-3.6).

(2) Using solutions (1.29), we obtain the expressions for the medium impurity volume concentrations $C_a(t)$ and the complexes $C_{ja}(t)$ (Fig. 5.1, Diagram 3-3.7).

(3) Reasoning that the impurity atoms are distributed in the solid solution homogeneously, and the intrinsic point defect concentrations have been taken their stationary values (1.29), the equation (5.7) for the concentration $C_{ja}(r,t)$ can be reduced to a differential equation of the form (Fig. 5.1, Diagram 3-3.8)

$$\frac{\partial C_{ja}(r,t)}{\partial t}=D_{ja}\Delta C_{ja}(r,t)-(\chi_{ja}+D_{ja}k^2_{ja})C_{ja}(r,t)+\alpha_{ja}C^+_jC_a(t) \qquad (5.28)$$

with the initial and boundary conditions analogous to (5.11) and (5.12).

In both limiting cases of rapid and slow solid solution depletion under irradiation, the problem on finding the concentration profiles of mobile point defects is reduced to solving the differential equations (5.16) and (5.28) with the initial and boundary conditions (5.18), (5.25), (5.26) and (5.12), (5.13), respectively. Rather than to solve equation (5.16) and (5.28) in each special case in its own right, we consider a more general differential equation of the form (Fig. 5.1, Diagram 3-3.9)

$$\frac{\partial\varphi^{XY}(r,t)}{\partial t} = D\Delta\varphi^{XY}(r,t) - a\varphi^{XY}(r,t) + bf(r,t) \tag{5.29}$$

with the initial and boundary conditions

$$\varphi^{XY}(r,t)\Big|_{t=0} = \varphi^{XY}(r,0), \tag{5.30}$$

$$\lim_{r\to R_P}\left[\frac{1-X}{R_P}\varphi^{XY}(r,t) + X\nabla\varphi^{XY}(r,t)\right] = \vartheta^{XY}_{R_P}(t), \tag{5.31}$$

$$\lim_{r\to L_P}\left[\frac{1-Y}{L_P}\varphi^{XY}(r,t) + Y\nabla\varphi^{XY}(r,t)\right] = \vartheta^{XY}_{L_P}(t), \tag{5.32}$$

where we have the indices $X,Y=0,1$, but $X\neq Y$; a, b are the constants; and the form of the functions $\vartheta^{XY}_{R_P}(t)$, $\vartheta^{XY}_{L_P}(t)$, as in (2.16), is determined by a specific form of the problem (see, e.g. (5.18) or (5.26)).

If we know solutions to the equation (5.29), we can find the concentration profiles of mobile point defects nearby the secondary-phase precipitates for the considered mechanisms of a solid solution decomposition under irradiation using (5.17), (5.24) or straight-forward from (5.28).

The solution to the equation (5.29) is obtained by the Laplace transformation with respect to time [386, 387, 396] in the form [285, 393]

$$\varphi^{XY}(r,t) = \int_0^t \left\{\vartheta^{XY}_{R_P}(\tau) - e^{-a\tau}\left[\frac{b(1-X)}{R_P}\int_0^\tau f(\tau',r)e^{a\tau'}\,dt' + \right.\right.$$

$$\left.\left.\vartheta^{XY}_{R_P}(0)\right]\right\}\Omega^{XY}_{R_P}(r,t-\tau)\,d\tau -$$

$$\int_0^t \left\{ \vartheta_{L_P}^{XY}(\tau) - e^{-a\tau}\left[\frac{b(1-Y)}{L_P} \int_0^\tau f(\tau',r)e^{a\tau'}\,d\tau' + \vartheta_{L_P}^{XY}(0) \right] \right\} \Omega_{L_P}^{XY}(r,t-\tau)\,d\tau +$$

$$e^{-at}\left[b\int_0^t f(\tau,r)e^{a\tau}\,d\tau + \varphi^{XY}(r,0) \right],$$

where $\Omega_{z_P}^{XY}(r,t)$ are defined by the expression obtained in Refs. 285, 393.

If the solution (5.33) to the equation (5.29) is available, we can write the appropriate expressions for the defect concentrations in the limiting cases of rapid and slow solid solution depletion we are interested in.

5.3. Slow Depletion of Solid Solutions

The analytical calculation of the concentration profiles of mobile point defects near the secondary-phase precipitates, when the solid solution is depleted under irradiation slowly, is realized by stages (see Section 5.2, Fig. 5.1, Diagram 3-3.6-3.8).

Turning to the equations for the average concentrations $C_l(t)$, we consider an approximation of the q-type sinks inhomogeneously distributed in the solid solution volume. Now we have

$$\nabla\left[D_l \nabla C_l(r,t)\right]=0.$$

As this takes place, the system of balance equations (5.5)-(5.8) becomes

$$\frac{dC_j(t)}{dt} = g - \alpha_{ja} C_j C_a + \chi_{ja} C_{ja} - \mu D_i C_j C_n - D_j C_j k_j^2, \tag{5.34}$$

$$\frac{dC_n(t)}{dt} = g - \mu D_i C_j C_n - D_n C_n k_n^2, \tag{5.35}$$

$$\frac{dC_{ja}(t)}{dt} = \alpha_{ja} C_j C_a - \chi_{ja} C_{ja} - D_{ja} C_{ja} k_{ja}^2, \tag{5.36}$$

$$\frac{dC_a(t)}{dt} = -\alpha_{ja} C_j C_a + \chi_{ja} C_{ja}. \tag{5.37}$$

The equations of the system (5.34)-(5.37) are split into two pairs of bound equations (for C_i, C_v and C_a, C_{ja}) in case when the concentrations C_i, C_v change far rapidly than C_a, C_{ja} (i.e. when $t_C > t_*$). We shall indicate later under which conditions this inequality is valid. The decomposition is performed slowly, the concentration C_a falls substantially in the time $t > t_*$; the stationary concentrations of intrinsic point defects ($\hat{C}_i^+$, $\hat{C}_v^+$) always have a chance to build on to the value $C_a(t)$.

Homogeneous stationary concentrations of intrinsic point defects. We consider the balance equations (5.34)-(5.37) reasoning that the vacancy and interstitial concentrations have gone to their stationary values $\hat{C}^+$, i.e. the left-hand sides of equations (5.34) and (5.35) are equal to zero, and C_a=const during the time interval involved. Substituting now (5.37) into (5.34) with allowance for (5.27), we obtain the stationary equation (1.24a) whose solutions have the form (1.29) and when the generation rates are high and the sinks are weak, the form (1.29a). The expressions (1.29) demonstrate that the stationary concentrations of intrinsic point defects depend neither on C_a nor C_{ja} in our model. This takes place because the intrinsic point defect capture followed by the formation of the complexes $(\alpha_{ja}C_jC_a)$ is compensated by the dissociation of the complexes $(\chi_{ja}C_{ja})$ in virtue of the stationarity of the impurity concentration when $t \simeq t_*$ (see (5.27)).

Average nonstationary concentrations of impurity atoms and complexes. In a decomposing solid solution, the concentrations of intrinsic point defects and complexes go on their stationary values, strictly speaking, just when $t \longrightarrow \infty$:

$$\hat{C}^+_j = \lim_{t \to \infty} C_j(t), \quad C^+_{ja} = \lim_{t \to \infty} C_{ja}(t).$$

The equation for C^+_{ja} can be readily obtained if one substitutes the value of $\hat{C}^+_j$ from (1.29) into (5.37) at $dC_a(t)/dt=0$ and takes into account that at $t \longrightarrow \infty$ the solid solution has been depleted completely and the impurity concentration reached its equilibrium value C^e_a:

$$C^+_{ja} = \alpha_{ja}\chi^{-1}_{ja}C^+_jC^e_a.$$

If the rate of changing the impurity concentration is low, i.e.

$$\frac{dC_a(t)}{dt} = \chi_{ja}C_{ja} - \alpha_{ja}C_jC_a \simeq 0, \tag{5.38}$$

then we assume, analogously to a pure crystal [205] that if $t \geq t_* = (D_v k^2_v)^{-1}$ (see (1.27)), the intrinsic point defect concentrations $C_j(t)$ differ moderately from $\hat{C}^+_j (C_j(t)/\hat{C}^+_j \simeq 1)$. So we are considering a slow depletion of the solid solution on the background of stationary $\hat{C}_i^+$ and $\hat{C}_v^+$ concentrations when $t \geq t_*$. Now the equations (5.36) and (5.37) can be written in the form

$$\frac{dC_{ja}(t)}{dt}=\alpha_{ja}C_j^+C_a(t)-\chi_{ja}C_{ja}(t)-D_{ja}C_{ja}(t)k_{ja}^2, \tag{5.39}$$

$$\frac{dC_{ja}(t)}{dt}=-\alpha_{ja}C_j^+C_a(t)+\chi_{ja}C_{ja}(t). \tag{5.40}$$

Taking $C_{ja}(t)$ from (5.40) and substituting into (5.39), we have the following equation for $C_a(t)$:

$$\frac{d^2C_a(t)}{dt^2}+A\frac{dC_a(t)}{dt}+BC_a(t)=0 \tag{5.41}$$

with the initial conditions

$$C_a(t_*)\simeq C_a^0, \quad \left.\frac{dC_a(t)}{dt}\right|_{t=t_*}=0, \tag{5.42}$$

where

$$A=\alpha_{ja}C_j^+ + D_{ja}k_{ja}^2+\chi_{ja},$$

$$B=\alpha_{ja}C_j^+D_{ja}k_{ja}^2<\alpha_{ja}C_j^+A.$$

The solution to the equation (5.40) has the form

$$C_a(t)=\frac{C_a^0}{A}\left(1-\frac{4B}{A^2}\right)^{-1/2}\left\{t_2^{-1}\exp\left[-\frac{t-t_*}{t_1}\right]-t_1^{-1}\exp\left[-\frac{t-t_*}{t_2}\right]\right\}, \tag{5.43}$$

where

$$t_{1,2}^{-1}=\frac{A}{2}\left[1\mp\left(1-\frac{4B}{A^2}\right)^{1/2}\right]. \tag{5.44}$$

Substitute (5.43) into the equation (5.4) and find

$$C_{ja}(t)=C^0_{ja}A^{-1}\sqrt{1-4BA^{-2}}\left\{t_2^{-1}\left[1-(\alpha_{ja}C^+_j t_1)^{-1}\right]\exp\left[-\frac{t-t_*}{t_1}\right]-\right.$$

$$\left. t_1^{-1}\left[1-(\alpha_{ja}C^+_j t_2)^{-1}\right]\exp\left[-\frac{t-t_*}{t_2}\right]\right\}, \tag{5.45}$$

where

$$\tilde{C}^0_{ja}=\alpha_{ja}\chi^{-1}_{ja}C^+_j C^0_a=C^0_{ja}\exp\left(-\frac{E^{ja}_P}{kT}\right) \tag{5.46}$$

is the complex concentration at the instant $t=t_*$.

At low rates and when the solid solution has been depleted (see (5.38)), little complexes are going on to the q sinks (this smallness being reached due to that of $D_{ja}k^2_{ja}$):

$$D_{ja}C_{ja}k^2_{ja}\geq\alpha_{ja}C_jC_a-\chi_{ja}C_{ja}. \tag{5.47}$$

Therefore, in view of the fact that $C_{ja}\ll C_a$ and the summands of the right-hand side of (5.39) are of the same order, we have

$$\frac{4B}{A^2}=\frac{4\alpha_{ja}C^+_jD_{ja}k^2_{ja}}{(\alpha_{ja}C^+_j+D_{ja}k^2_{ja}+\chi_{ja})^2}\approx\frac{4D_{ja}k^2_{ja}}{\alpha_{ja}C^+_j}\left[\frac{\tilde{C}^0_{ja}}{C^0_a}\right]^2\times$$

$$\left(1-\frac{2\tilde{C}^0_{ja}}{C^0_a}\right)\ll 1.$$

It is now seen from (5.44) that

$$t_1\gg t_2,$$

and it is suffice to retain in the expressions (5.43) and (5.45) only the first terms, which gives

$$C_a(t)\simeq C_a^0(At_2)^{-1}\sqrt{1-\frac{4B}{A^2}}\exp\left[-\frac{t-t_*}{t_1}\right]\simeq C_a^0\exp\left[-\frac{t-t_*}{t_C}\right], \quad (5.48)$$

$$C_{ja}(t)\approx \tilde{C}_{ja}^0(At^2)^{-1}\sqrt{1-\frac{4B}{A^2}\left[1-(\alpha_{ja}C_j^+t_1)^{-1}\right]}\exp\left[-\frac{t-t_*}{t_1}\right]\simeq \tilde{C}_{ja}^0\exp\left[-\frac{t-t_*}{t_C}\right], \quad (5.49)$$

because $D_{ja}k_{ja}^2 < \alpha_{ja}C_j^+ + D_{ja}k_{ja}^2 + \chi_{ja}$. In this case

$$t_1^{-1}\equiv t_C^{-1}=\frac{A}{2}\left[1-\sqrt{1-\frac{4B}{A^2}}\right]\simeq\frac{B}{A}=\alpha_{ja}C_j^+D_{ja}k_{ja}^2\left[\alpha_{ja}C_j^+ + D_{ja}k_{ja}^2+\chi_{ja}\right]^{-1}$$

takes place, and hence, the characteristic time t_C of solid solution denudation in the impurity or of the e-fold decrease of the impurity and complexes concentrations is defined as

$$t_C=\frac{\alpha_{ja}C_j^+ + D_{ja}k_{ja}^2+\chi_{ja}}{\alpha_{ja}C_j^+D_{ja}k_{ja}}=\left[D_{ja}k_{ja}^2\right]^{-1}+\frac{k_j}{\alpha_{ja}k_n}\sqrt{\frac{\mu D_iD_j}{gD_n}}\left[1+\frac{\chi_{ja}}{D_{ja}k_{ja}^2}\right].$$

The expression (5.49) shows that the solid solution decomposition (i.e. the growth of the secondary-phase precipitates) becomes slower with increasing the complex dissociation frequency, the recombination of opposite point defects, the frequency of going on the intrinsic defects j

(forming mobile complexes *ja*) to the sinks, and quicker with increasing the defect generation rate, the possibility of mobile complex formation, the frequency of outgoing the intrinsic *n*-type defects which do not form complexes with impurity atoms, to the sinks, and the frequency of *ja*-complex adsorption on the sinks. It can be shown that in the absence of irradiation the point defect concentrations are equal to their thermodynamically equilibrium values

$$\lim_{g \to 0} t_C(g) = \chi_{ja}(\alpha_{ja} C^e_j D_{ja} k^2_{ja})^{-1} » t_C(g).$$

The formula (5.50) can be simplified if one takes into account (5.46) and (5.48), which gives

$$t_C = \frac{\theta_C}{\sqrt{g}}; \quad \theta_C = \frac{\chi_{ja}}{\alpha_{ja} D_{ja} k^2_{ja}} \frac{k_j}{k_n} \sqrt{\mu \frac{D_i D_j}{D_n}}. \tag{5.51}$$

In accordance with the initial conditions (5.42), the obtained solutions (5.43) and (5.44) are valid when $t>t_*$. In particular, the condition of slow depletion $t_C>t_*$ of the solid solution should be fulfilled. In the opposite case, a marked drop of the impurity concentration takes place during the nonstationary period t_* of the intrinsic point defects $t_C<t_*$, and the equations of (5.34)-(5.37) system should be solved simultaneously.

Consider as an example a change with time in the concentration vacancies C_v, interstitials C_i, impurity atoms C_a, and complexes C_{va} in a system with the constants characteristic of a solid *Ni*-based solution

(Table 5.1).

Table 5.1

Characteristic constants for a *Ni*-based alloy

1.	Energy of vacancy formation (1.4 eV)	ε_v^f
2.	Energy of interstitial formation (3.0 eV)	ε_i^f
3.	Energy of vacancy migration (1.4 eV)	ε_v^m
4.	Energy of interstitial migration (0.15 eV)	ε_i^m
5.	Energy of vacancy - impurity atom complex migration (0.6 eV)	ε_{va}^m
6.	Bound energy of the vacancy - impurity atom complex (1.0 eV)	ε_{va}^b
7.	Poisson coefficient (1/3)	ν
8.	Shear modulus ($7.1 \cdot 10^{11}$dn/cm^2)	G
9.	Local change in the volume nearby a vacancy	$\Delta\Omega_v$=-0.23 Ω
10.	Local change in volume nearby an interstitial	$\Delta\Omega$=1.6 Ω
11.	Atomic volume ($1.56 \cdot 10^{-23}$cm^3)	Ω
12.	Initial impurity concentration ($4 \cdot 10^{20}$cm^{-3})	C_a^0
13.	Generation rate of intrinsic point defects (10^{16}cm^{-3}s^{-1})	g
14.	Irradiation temperature (600 K)	T

The dependencies $C_v(t)$, $C_i(t)$, $C_a(t)$, $C_{va}(t)$ obtained for $t>t_*$ by formulas (1.29), (5.48) and (5.49) and those found for $t<t_*$ as a result of numerical integration of the system of balance equations (5.34)-(5.37), are shown in Fig. 5.2.

The dependencies $C_v(t)$, $C_i(t)$, $C_a(t)$ and $C_{va}(t)$ obtained by formulas (1.29), (5.48) and (5.49) and when $t<t_*$ found by the numerical integration of the balance equations (5.34)- (5.37), are shown in Fig. 5.2.

It is seen from Fig. 5.2 that the condition $t_C>t_*$ is fulfilled at the parameter values in choice, and the curves $C_v(t)$ and $C_i(t)$ have, when $t<t_*$, the same form as in the case of a pure crystal (see Ref. 205).

The values of the characteristic times t_C of the solid solution depleted and going on the intrinsic point defects to their stationary values (see the relations (5.51) and (1.27), respectively) are shown in Table 5.2 at the generation rates g and the irradiation temperature T different from those in Table 5.1.

Table 5.2

Characteristic times t_c and t_* for a mock alloy (see Table 5.1) at various temperatures $T(K)$ and generation rates g ($cm^{-3}s^{-1}$) of intrinsic point defects, $k_j^2 \simeq k_n^2 \simeq 10^{10} cm^{-2}$

T	t_c (10^6s)			t_*
	$g=10^{14}$	$g=10^{15}$	$g=10^{16}$	(10^3 s)
600	25.20	7.97	2.52	23.1
650	16.20	5.12	1.62	2.69
700	11.00	3.48	1.10	0.42
750	7.75	2.45	0.78	0.08
800	5.70	1.80	0.57	0.02

It is seen from Table 5.2 that the characteristic time t_c decreases with increasing the generation rate and the irradiation temperature, and the obtained distributions of mobile point defects (see (5.48), (5.49)) are valid in a wide interval of the temperature T and the generation rate g of intrinsic point defects.

Inhomogeneous complex distribution nearby the precipitates. We revert now to the system of balance equations (5.5)-(5.8) and consider the alterations in the solutions (1.29), (5.58), and (5.49) what are made due to the abandonment of the approximation of homogeneous q-sink volume distribution.

As seen from the solution to the problem on precipitate growth in the absence of irradiation [387, 388], the most alterations in the impurity concentrations C_a, and hence in the complex ones C_{ja}, should be forthcoming nearby the precipitate surface, while away from that, the expressions (5.40), (5.58), and (5.49) change little. On the other hand, upon the condition $t_C > t_*$, the stationary intrinsic point defect distribution $C_j(r,t) = C_j^+$ having been found in Ref. 205, manages to be set in the solid solution to the instant t_* (at $C_a \simeq C_a^0$, see (5.42)) with taking place the relation

$$C_{ja}(r,t) \ll C_j^+ \ll C_a(t). \tag{5.57}$$

The first inequality in (5.52) is valid on the condition $\chi_{ja}(\alpha_{ja} C_a^0)^{-1} > 1$ (see (5.46)), the second one is fulfilled, as a rule, for the substitutional impurities at the concentrations of practical importance.

As one did not succeed in solving the system (5.5)-(5.8) analytically, we consider an approximation where the inhomogeneous distribution of the complexes *ja* is only taken into account which performs the impurity transfer toward the precipitates. At the same time the concentrations of the remaining defects are believed as homogeneous and described by the expressions (1.29) and (5.43). We show that such an approach is warranted if a local equilibrium of intrinsic point defects, impurity and complexes nearby the precipitates is warranted. Let us present their concentrations in the form

$$C_j(r,t) = \hat{C}_j^+ \left[1 - \frac{\Delta C_j(r,t)}{C_j^+} \right],$$

$$C_{ja}(r,t)=C_{ja}(t)\left[1-\frac{\Delta C_{ja}(r,t)}{C_{ja}(t)}\right],$$

$$C_{a}(r,t)=C_{a}(t)\left[1-\frac{\Delta C_{a}(r,t)}{C_{a}(t)}\right],$$

where $\hat{C}^{+}_{j}$, $C_{ja}(t)$ and $C_{a}(t)$ are the corresponding solutions to the homogeneous problem (1.29), (5.48), (5.49); and $\Delta C_{ja}(r,t)$, $\Delta C_{j}(r,t)$, $\Delta C_{a}(r,t)$ take into account an inhomogeneous distribution of *l*-type mobile point defects. Due to the stoichiometric relation $j+a \longrightarrow ja$, the local alterations in the concentrations of complexes , intrinsic point defects and impurity atoms should be of the same order. (An increase in the complex concentration $\chi_{ja}C_{ja}(r,t)$ in (5.7) leads to an appropriate decrease in $C_{j}(r,t)$ in (5.5), which, in its turn, raises the probability of the complex formation, $\alpha_{ja}C_{j}C_{a}$, and hence, according to (5.6), produces an increase in $C_{a}(r,t)$ nearby the precipitate). So we have the relation

$$\Delta C_{ja}(r,t)\sim\Delta C_{j}(r,t)\sim\Delta C_{a}(r,t),$$

and in line with (5.52), the values $\Delta C_{a}(r,t)/C_{a}(t)$ and $\Delta C_{j}(r,t)/\hat{C}^{+}_{j}$ are small values of a higher order than $\Delta C_{ja}(r,t)/C_{ja}(t)$, i.e.

$$C_{j}(r,t)\simeq C^{+}_{j},\quad C_{a}(r,t)\simeq C_{a}(t) \qquad (5.53)$$

may be thought of a good degree. Substitute (5.53) into (5.7) and obtain a linear differential equation (5.28) for the complex concentration.

If the solid solution had been decomposed before the beginning of radiation, then, in keeping with (2.26) and (2.27), the distribution

$$C_{ja}(r,t_*)\simeq\left(1-\frac{R_P^*}{r}\right)C_{ja}^0\exp\left(-\frac{E_P^{ja}}{kT}\right), \qquad (5.54)$$

can be taken as an initial condition for the complex concentration to the instant of $\hat{C}_j(t)$ reaching their stationary values ($\hat{C}_j(t_*)=\hat{C}_j^+$). Here R_P^* is the precipitate radius at $t=t_*$, and C_{ja}^0 is determined from (5.46). If we have the separation between the forming precipitates $2L_P \gg R_P^*$, then it follows from (5.54) that the complex concentration is equal to $C_{ja}^0\exp\left(-E_P^{ja}/kT\right)$ in a large region of the decomposing solid solution apart from the spheres of the radius of several R_P^* which envelope the nuclei of a new phase. So we have, away from the precipitate ($r=L_P$),

$$C_{ja}(r,t_*)\simeq C_{ja}^0\exp\left(-\frac{E_P^{ja}}{kT}\right)=\tilde{C}_{ja}^0.$$

At the boundary of the precipitate, the complexes dissociate. The impurity atoms find a place for themselves near the precipitate following the stoichiometry of the formed structure. If the precipitate surface rearranges itself much more quickly than the diffusion of impurity atoms toward the precipitate takes place, then the complexes have not managed to gather nearby the precipitates, their concentration is equal at $r\simeq R_P(t)$ to its thermodynamically equilibrium value $\tilde{C}_{ja}^e=C_{ja}^e\exp\left[-E_P^{ja}/kT\right]$, which can be assumed zero on the conditions of our problem, i.e. (see (5.12))

$$\lim_{r\to R_P} C_{ja}(r,t)=C_{ja}(r,0)\simeq 0. \qquad (5.55)$$

By this means, the problem on inhomogeneous complex distribution in the volume of solid solution under irradiation is reduced to the diffusion equation (5.28) with the initial condition (5.54) and the boundary conditions (5.12b) and (5.55).

A more correct solution to such problems requires a joint search for the concentration profiles of impurity atoms or complexes and the dependency of the precipitate radius $R_P(t)$ on the time (the Stephan problem) [396, 397]. For this purpose, an additional boundary condition at the boundary of the growing precipitate is usually introduced (see, e.g., Refs. 184, 390, 398, 399):

$$\frac{dR_P(t)}{dt} = \frac{D_a}{C_a^P - C_a(r,t)} \left. \frac{\partial C_a(r,t)}{\partial r} \right|_{r=R_P(t)}, \tag{5.56}$$

where C_a^P is the homogeneous concentration of impurity atoms inside the precipitate.

We shall demonstrate that if the impurity concentration C_a^P in the precipitate is more than that in the solid solution, the Stephan problem is divided into two independent subproblems: a search for instantaneous concentration profiles $C_{ja}(r,t)$ at a fixed value of the precipitate radius $R_P(t)=\bar{R}_P$ and the calculation of the dependency $R_P(t)$ for a given complex concentration at the instant t. This is possible if the average precipitate sizes $\bar{R}_P(t)$ (and $R_R(t)$) change in time slower than the complex concentration $C_{ja}(t)$. Otherwise the condition

$$\left(\frac{L_P}{R_P}\right)^2 \frac{C_{ja}(t)}{C_a^P} \ll 1 \tag{5.57}$$

must be fulfilled. It is clear that for this purpose it is necessary that the diffusion shift rate of the impurity atoms in our case equal by the order of magnitude to D_{ja}/L_P, be much more than the precipitate growth rate dR_P/dt, i.e.

$$\frac{D_{ja}}{L_P} \gg \frac{dR_P}{dt}. \tag{5.58}$$

It follows from the condition of the impurity balance that the growth rate of the impurity quantity per the precipitate volume $(4/3)\pi(dR_P^3/dt^3)C_a^P$ is equal to the flux $J_P^a(R_P)$ of impurity atoms toward one precipitate from the whole volume $4\pi L_P^3/3$ of the solid solution, related to one precipitate. By the order of magnitude, $J_P^a(R_P)\simeq D_{ja}\nabla C_{ja}L_P^2/2$ and $\nabla C_{ja} \simeq C_{ja}/L_P$, so that

$$J_P^a(R_P)\simeq D_{ja}C_{ja}L_P\simeq C_a^P R_P^2\frac{dR_P}{dt}. \tag{5.59}$$

If the condition (5.58) is fulfilled, it is (5.57) that follows from (5.59). In our case $L_P/R_P<100$, and $C_{ja}/C_a^P\simeq 10^{-6}$- 10^{-5}, so that the inequality (5.57) is valid, i.e. the growth rate of the precipitate radius is less than the average diffusion shift rate of impurity atoms $v_a=D_{ja}/2L_P$ in the solid solution. Therefore, the condition of adiabatic precipitate growth has been fulfilled and the concentration profiles $C_{ja}(r,t)$ nearby the precipitates can be searched on the condition $R_P(t)\approx\tilde{R}_P$.

The differential equation (5.28) is a special case of the general linear differential equation (5.29). A solution to the equation(5.28) can be obtained from the found solution (5.33) of the general linear differential equation (5.29) assuming that in the expressions (5.29)-(5.32), in accord with (5.12b), (5.54), and (5.55),

$$X = 0,\ \ Y = b = 1,\ \ a = \chi_{ja} + D_{ja}k^2_{ja},$$

$$f(r,t) = f(t) = \alpha_{ja}C^{+}_{j}C_{a}(t), \tag{5.60}$$

$$\varphi^{01}(r,t)=C_{ja}(r,t),\ \ \varphi^{01}(r,0)=\left[1-\frac{R^*_P}{r}\right]\tilde{C}^0_{ja},\ \ \vartheta^{01}_{z_P}= 0.$$

takes place. Substituting (5.60) into (5.33), we find the complex concentration sought-for in the form

$$C_{ja}(r,t)=\tilde{C}^0_{ja}\sum_{n=1}^{\infty}\frac{\gamma_n(r)}{\lambda^{01}_n}\left\{\frac{\chi_{ja}+D_{ja}k^2_{ja}}{\chi_{ja}+D_{ja}k^2_{ja}+\lambda^{01}_n+t^{-1}_C}\times\right.$$

$$\left[\exp\left(-\frac{t-t_*}{t_C}\right)-\exp\left(-[\chi_{ja}+D_{ja}k^2_{ja}+\lambda^{01}_n][t-t_*]\right)\right]+ \tag{5.61}$$

$$\left.\left[1-\frac{R^*_P}{\bar{R}_P}(1-\cos\mu^{01}_n)\right]\exp\left(-[\chi_{ja}+D_{ja}k^2_{ja}+\lambda^{01}_n][t-t_*]\right)\right\}.$$

Here

$$\gamma_n(r)=\frac{2\lambda^{01}_n\bar{R}_P(L_P-\bar{R}_P)}{\mu^{01}_n r(L_P\sin^2\mu^{01}_n-\bar{R}_P)}\sin\left[\mu^{01}_n\frac{r-\bar{R}_P}{L_P-\bar{R}_P}\right], \tag{5.62}$$

$$\lambda^{01}_n=D_{ja}\left[\frac{\mu^{01}_n}{L_P-\bar{R}_P}\right]^2, \tag{5.63}$$

μ^{01}_n is determined from

$$\mathrm{tg}\ \mu^{01}_n=\mu^{01}_n\frac{L_P}{L_P-\bar{R}_P}, \tag{5.64}$$

and $\bar{R}_P$ is the average precipitate radius.

When deriving (5.61), we used the relation

$$\sum_{n=1}^{\infty} \frac{\gamma_n(r)}{\lambda_n^{01}} \cos\mu_n^{01} \simeq \left(1 - \frac{\bar{R}_P}{r}\right),$$

whose validity can be verified in the given limiting case $t_C > t_*$ by a numerical calculation.

The expression (5.61) is an exact solution of the appropriate balance equation (5.7) in the case of slow depletion of the solid solution under irradiation. However, it is difficult to employ it in further calculations to determine the $R_P(t)$-dependency of t because of its unwieldiness. Let us simplify it. It follows from the condition (5.57) that $D_{ja}k_{ja}^2 \leq \chi_{ja}$, and the left-hand side of (5.47) is small as compared with $\chi_{ja}C_{ja}$ if $\alpha_{ja}C_j^+C_a^0 \approx \chi_{ja}\tilde{C}_{ja}^0$. Now using (1.29), (1.29a) and (5.51), we have

$$D_{ja}k_{ja}^2 \gg \left(\frac{\tilde{C}_{ja}^0}{C_a^0}\right) \frac{\alpha_{ja}C_j^+C_a^0}{\chi_{ja}\tilde{C}_{ja}^0} D_{ja}k_{ja}^2 = t_C^{-1},$$

and hence, the inequality

$$\chi_{ja} + D_{ja}k_{ja}^2 + \lambda_n^{01} \gg t_C^{-1}. \tag{5.65}$$

is valid. With allowance for (5.65) the expression (5.61) can be presented in a more simple form

$$C_{ja}(r,t) = C_{ja}(t)\Phi(r), \tag{5.66}$$

where

$$\Phi(r)=\chi_{ja}\sum_{n=1}^{\infty}\frac{\gamma_n(r)}{\lambda_n^{01}(\chi_{ja}+D_{ja}k_{ja}^2+\lambda_n^{01})}, \tag{5.67}$$

and $C_{ja}(t)$ is the complex concentration in case of their homogeneous distribution in the volume of the solid solution (see (5.49)).

At $\bar{R}_P \sim L_P$ (i.e. $t \gg t_C$) when the influence of coalescence on the precipitate growth must be taken into account [186-188], the series (5.67) is converged very slowly, and further simplification of (5.66) is impossible. However, at the initial stage of precipitate formation, i.e. when $t \leq t_C$, or at a low segregation density, the following inequality is valid:

$$\bar{R}_P L_P^{-1} \ll 1. \tag{5.68}$$

And we need only keep the first summand in (5.67). In this case we do have for μ_1^{01} from (5.64)

$$\mu_1^{01} \simeq \sqrt{3\bar{R}_P L_P^{-1}}, \tag{5.69}$$

and hence,

$$\mu_1^{01} \ll \mu_2^{01} < \mu_3^{01} < \ldots < \mu_n^{01} < \ldots$$

Now, according to (5.62), (5.63), and (5.65),

$$\Phi(r) \approx \Phi_1(r) \approx \left[1-\frac{\bar{R}_P}{r}\right]\chi_{ja}(\chi_{ja}+D_{ja}k_{ja}^2+\lambda_1^{01})^{-1}. \tag{5.70}$$

We substitute (5.70) and (5.49) into (5.66), then the expression for the complex concentration can be presented in the form

$$C_{ja}(r,t) \simeq \left[1-\frac{\bar{R}_P}{r}\right] \frac{\chi_{ja}\tilde{C}^0_{ja}}{\chi_{ja}+D_{ja}k^2_{ja}+\lambda^{01}_1} \exp\left[-\frac{t-t_*}{t_C}\right]. \qquad (5.71)$$

It is seen from (5.71) that if $R_P \ll L_P$, the complex concentration $C_{ja}(r,t)$ in the most portion of the solid solution weakly depends on r and descends with time, as in the case of homogeneous distribution of mobile point defects in the material volume (see (5.49)).

5.4 Rapid Depletion of Solid Solutions

If the bound energy ε^b_{ja} of impurity atoms with the intrinsic point defects is sufficiently high, then the complexes dissociate seldom in their path toward the precipitate and the decomposition goes rapidly. The solid solution is depleted in the time $t_C < t_*$ and the condition (5.27) is not fulfilled. In this situation it is necessary to solve the nonstationary balance equations (5.5)-(5.8), as the concentrations of vacancies and interstitials do not manage to take their stationary values $\hat{C}^+_j$ in the characteristic time of the solid solution depletion t_C. Such a problem has been first considered analytically in Refs. 289, 290. According to the model offered in Ref. 289, the transfer of impurity atoms toward the growing segregate is carried out at the cost of the complexes vacancy - impurity atom diffusion (*va*) toward the segregate as follows. At the initial instant t=0, the most of vacancies are bound into complexes *va*.

The complexes are diffused toward a segregate boundary, where they dissociate into their constituents. The impurity atoms form a part of the

growing segregate, and the vacancies go into the solid solution. Due to this, the free vacancy concentration nearby the segregate becomes higher than that away from it. The vacancies are migrating back into the solid solution until they are tied up in new complexes which are rediffusing toward the segregate. Therein lies the mechanism of vacancy pump whereby a rapid growth of the segregates was explained which is not easily compatible with an ordinary impurity transfer [289, 290].
impurity transfer [289, 290].

Logically consistent analytical consideration of the kinetic balance equations (5.5)-(5.8) in the case of rapid solid solution depletion, $t_C<t_*$, was first performed in Refs. 91, 400, 401.

Consider the mechanism of impurity transfer toward a segregate through complexes when it is operating at its best, that is to say, the complexes forming in the solid solution will not be considered, as in Refs. 289, 290, to dissociate in this path toward the segregate. Let the condition of local equilibrium of mobile point defects in the material volume be rejected [289], but a more strong condition (5.10) be assumed.

The balance equations for the intrinsic point defect concentrations $C_{ja}(r,t)$, the impurity atom ones $C_a(r,t)$, and the complex ones $C_{ja0}(r,t)$ have the form (5.13)-(5.15) in thecase where, by the condition, (5.10) we have

$$D_{ja} \simeq D_j. \tag{5.72}$$

Summing (5.13) and (5.15), we obtain for the function $\Psi_0(r,t)=C_{j0}(r,t)+C_{ja0}(r,t)-gt$ a diffusion equation of the form (5.16) with the initial condition (see (5.11))

$$\Psi_0(r,0) = \tilde{C}^e_j + \tilde{C}^e_{ja} \tag{5.73}$$

and the boundary conditions

$$\lim_{r \to R_P} \nabla\Psi_0(r,t)=0, \quad \nabla\Psi_0(r,t)\Big|_{r=L_P} = 0, \tag{5.74}$$

which follow from the equality of the diffusion coefficients (see (5.72) and from the symmetry conditions of the problem (see (5.12)), respectively. The unique solution to the diffusion equation (5.16) with the initial and boundary conditions (5.73), (5.74) can be shown to be

$$\Psi_0(r,t)=\tilde{C}^e_j+\tilde{C}^e_{ja}$$

or

$$C_{j0}(r,t)+C_{ja0}(r,t)=gt+\tilde{C}^e_j+\tilde{C}^e_{ja}. \tag{5.75}$$

The boundary conditions (5.12) for the equation (5.13)-(5.15) are now formulated as follows:

$$\lim_{r \to R_P} C_{j0}(r,t)=\tilde{C}^e_j+gt; \quad \nabla C_{j0}(r,t)\Big|_{r=L_P}=0,$$

$$\lim_{r \to R_P} C_{ja0}(r,t)=\tilde{C}^e_{ja}; \quad \nabla C_{ja0}(r,t)\Big|_{r=L_P}=0,$$

because nearby the boundary the complex concentration remains constant and equal to its thermodynamically equilibrium value.

The relation (5.25) makes it possible instead of three equations (5.13)-(5.15), to solve only the first two. Having found C_{j0}(r,t), we obtain the complex concentration $C_{ja0}(r,t)$ from (5.75), and the impurity concentration from (5.74).

After the variable separation in equation (5.74) and integration, we obtain the relation between $C_a(r,t)$ and $C_{j0}(r,t)$ in the form

$$C_a(r,t)=C_a^0\exp[-y(r,t)], \tag{5.76}$$

where

$$y(r,t)=\alpha_{ja}\int_0^t C_{j0}(r,\tau)d\tau,$$

or

$$C_{j0}(r,t)=\alpha_{ja}^{-1}\partial y(r,t)/\partial t. \tag{5.77}$$

The initial and boundary conditions for the function $y(r,t)$, can be written as follows:

$$y(r,0)=0;\ \lim_{r\to R_P} y(r,t)=\alpha_{ja}t\left(\tilde{C}_j^e+\frac{gt}{2}\right);\ \nabla y(r,t)\Big|_{r=L_P}=0. \tag{5.78}$$

Substitute (5.76) and (5.77) into (5.13) and integrate over t. Using the conditions (5.78), we obtain the equation for $y(r,t)$ in the form

$$\frac{\partial y(r,t)}{\partial t}=D_j\Delta y(r,t)+\alpha_{ja}C_a^0\left[e^{-y(r,t)}-1\right]+\alpha_{ja}(\tilde{C}_j^e+gt). \tag{5.79}$$

The equation (5.79) can be solved analytically in two limiting cases: when $y(r,t)\gg1$ and $y(r,t)\ll1$. The condition

$$y(r,t_C)=1 \tag{5.80}$$

identifies the time t_C which establishes a line of demarcation between both cases, which further will be named the long- and short-times cases.

Consider the long-time case $y(r,t)\gg1$. Then neglect the term

$\exp[-y(r,t)]$ in (5.79) to obtain the equation

$$\frac{\partial y(r,t)}{\partial t}=D_j\Delta y(r,t)+\alpha_{ja}gt-\alpha_{ja}(C_a^0-\tilde{C}_j^e).$$

Let us make a substitution

$$y(r,t)=\varphi(r,t)-\alpha_{ja}t(C_a^0-\tilde{C}_j^e)+\frac{\alpha_{ja}gt^2}{2}, \tag{5.81}$$

and obtain for the function $\varphi(r,t)$ (see (5.17)) an ordinary diffusion equation (5.16) with the initial and boundary conditions (5.18).

A solution to the diffusion equation (5.16) can be obtained from the above found solution (5.33) to a general differential equation (5.29). Indeed, when

$$X = a = b = 0,\quad Y = 1,\quad \varphi^{01}(r,0) = 0,$$
$$\vartheta_{R_P}^{01}=\frac{\alpha_{ja}C_a^0 t}{\bar{R}_P},\quad \vartheta_{L_P}^{01}=0, \tag{5.82}$$

where $\bar{R}_P$ is the average dimension of segregates, the differential equation (5.29) is reduced to the equation (5.16) for the function $\varphi(r,t)\equiv\varphi^{01}(r,t)$ defined by equation (5.17). Substitute (5.82) into (5.33) and find a solution to the diffusion equation (5.16) with the initial and boundary conditions (5.18) in the form

$$\varphi(r,t)=\sum_{n=1}^{\infty}\gamma_n(r)e^{-\lambda_n^{01}t}\left[\bar{R}_P\int_0^t\vartheta_{R_P}^{01}(\tau)e^{\lambda_n^{01}\tau}d\tau+\frac{\varphi(r,0)}{\lambda_n^{01}}\right], \tag{5.83}$$

where $\gamma_n(r)$ and λ_n^{01} are determined from (5.62), (5.63). With a knowledge of $\varphi(r,t)$ from the relations (5.77) and (5.81), we obtain an expression for the intrinsic point defect concentration $C_{j0}(r,t)$ in the form

$$C_{j0}(r,t)=\tilde{C}^{e}_{j}+gt-C^{0}_{a}+\alpha^{-1}_{ja}\frac{\partial\varphi(r,t)}{\partial t}. \tag{5.84}$$

Using (5.75), the complex concentration $C_{ja0}(r,t)$ can be obtained.

We substitute an explicit form of the function $\vartheta^{01}_{R_P}$ into (5.83) and obtain

$$\varphi(r,t)=\alpha_{ja}C^{0}_{a}\left[t\sum_{n=1}^{\infty}\frac{\gamma_n(r)}{\lambda^{01}_{n}}-\sum_{n=1}^{\infty}\frac{\gamma_n(r)}{(\lambda^{01}_{n})^2}\left(1-e^{-\lambda^{01}_{n}t}\right)\right]. \tag{5.85}$$

It can be shown by numerical calculation that

$$\sum_{n=1}^{\infty}\frac{\gamma_n(r)}{\lambda^{01}_{n}}=1; \qquad \lim_{r\to R_P}\sum_{n=1}^{\infty}\frac{\gamma_n(r)}{(\lambda^{01}_{n})^2}=0. \tag{5.86}$$

With (5.86) we simplify the expression (5.85):

$$\varphi(r,t)=\alpha_{ja}C^{0}_{a}t-\alpha_{ja}C^{0}_{a}\sum_{n=1}^{\infty}\frac{\gamma_n(r)}{(\lambda^{01}_{n})^2}\left[1-e^{-\lambda^{01}_{n}t}\right]. \tag{5.87}$$

After substituting (5.87) into (5.84) with taking account of (5.75), we obtain the following expressions for the intrinsic point defects $C_{j0}(r,t)$ of the type j and the complexes $C_{ja0}(r,t)$ in the limiting long-time case:

$$C_{j0}(r,t)=\tilde{C}^{e}_{j}+gt-C^{0}_{a}\sum_{n=1}^{\infty}\frac{\gamma_n(r)}{\lambda^{01}_{n}}e^{-\lambda^{01}_{n}t}, \tag{5.88}$$

$$C_{ja0}(r,t)=\tilde{C}^{e}_{ja}+C^{0}_{a}\sum_{n=1}^{\infty}\frac{\gamma_n(r)}{\lambda^{01}_{n}}e^{-\lambda^{01}_{n}t}. \tag{5.89}$$

By virtue of the fact that analytical work with the sums (5.88) and (5.89) is always a problem, we consider the first approximation to there exact solution, substituting λ_1^{01} and $\gamma_1(r)$, according to (5.62) and (5.63),

$$C_{j0}(r,t)\simeq\tilde{C}_j^e+gt-C_a^0G_1(r)\exp\left[-\frac{D_j(\mu_1^{01})^2t}{(L_P-\bar{R}_P)^2}\right], \qquad (5.90)$$

$$C_{ja0}(r,t)\simeq\tilde{C}_{ja}^e+C_a^0G_1(r)\exp\left[-\frac{D_j(\mu_1^{01})^2t}{(L_P-\bar{R}_P)^2}\right], \qquad (5.91)$$

where

$$G_1(r)=\frac{\gamma_1(r)}{\lambda_1^{01}}=\frac{2\bar{R}_P(L_P-\bar{R}_P)}{\mu_1^{01}r(L_P\sin^2\mu_1^{01}-\bar{R}_P)}\sin\left[\mu_1^{01}\frac{r-\bar{R}_P}{L_P-\bar{R}_P}\right].$$

Now we obtain conclusively, in accord with (5.89) and (5.70),

$$C_{ja}(r,t)\simeq\tilde{C}_j^e+gt-C_a^0\left[1-\frac{\bar{R}_P}{r}\right]\exp\left[-\frac{t}{t_{R_P}}\right], \qquad (5.92)$$

$$C_{ja0}(r,t)\simeq\tilde{C}_{ja}^e+C_a^0\left[1-\frac{\bar{R}_P}{r}\right]\exp\left[-\frac{t}{t_{R_P}}\right], \qquad (5.93)$$

where the expression for the characteristic time t_{R_P} has the form

$$t_{R_P}=\frac{L_P(L_P-R_P)^2}{3D_j\bar{R}_P}. \qquad (5.94)$$

Now we consider the case of short times. If $y(r,t)\ll1$, then expanding $\exp[-y(r,t)]\simeq1-y(r,t)$, we have from (5.79)

$$\frac{\partial y(r,t)}{\partial t}=D_j\Delta y(r,t)+\alpha_{ja}C_a^0\left[\tilde{C}_j^e(C_a^0)^{-1}-y(r,t)\right]+\alpha_{ja}gt.$$

By the substitution

$$y(r,t)=\frac{\tilde{C}_j^e}{C_a^0}-\varphi(r,t)\exp\left[-\alpha_{ja}C_a^0t\right], \tag{5.95}$$

we reduce the problem to the following differential equation

$$\frac{\partial\varphi(r,t)}{\partial t}=D_j\Delta\varphi(r,t)-\alpha_{ja}gt\ \exp\left[\alpha_{ja}C_a^0t\right] \tag{5.96}$$

for the function $\varphi(r,t)$ with the initial and boundary conditions

$$\varphi(r,0)=\tilde{C}_j^e(C_a^0)^{-1};\quad \nabla\varphi(r,t)\Big|_{r=L_P}=0;$$

$$\lim_{r\to R_P}\varphi(r,t)=\left[\tilde{C}_j^e(C_a^0)^{-1}-\alpha_{ja}t\left(\tilde{C}_j^e+\frac{gt}{2}\right)\right]\exp\left[\alpha_{ja}C_a^0t\right].$$

A solution to the obtained differential equation (5.96) is found analogously to (5.97) and has the form

$$\varphi(r,t)=\int_0^t\left[\vartheta_{R_P}^{01}(\tau)-(\bar{R}_P)^{-1}\int_0^\tau f(\tau')d\tau'-\vartheta_{R_P}^{01}(0)\right]\Omega_{R_P}^{01}(r,t-\tau)d\tau+ \int_0^t f(\tau)d\tau+\varphi(r,0), \tag{5.97}$$

where

$$\vartheta_{R_P}^{01}(\tau)=\left[\frac{\tilde{C}_j^e}{C_a^0}-\alpha_{ja}t\left(\tilde{C}_j^e+\frac{gt}{2}\right)\right]\frac{\exp(\alpha_{ja}C_a^0t)}{\bar{R}_P},$$

$$f(t)=-\alpha_{ja}gt\exp\left[\alpha_{ja}C_a^0t\right],$$

and $\Omega_{R_P}^{01}(r,t)$ is defined by the relation obtained in Refs. 285, 393.

At the initial stage of solid solution irradiation when the concentrations of mobile point defects C_j do not manage markedly as compared with $\tilde{C}_j^e$, the expression (5.97) becomes more simple. Falling back on the relations (5.82), (5.77), and (5.95) we define $C_{j0}(r,t)$ and $C_{ja0}(r,t)$ in the form

$$C_{j0}(r,t\ll t_C)\simeq\tilde{C}_j^e-\tilde{C}_j^e\sum_{n=1}^{\infty}\frac{\gamma_n(r)}{\lambda_n^{01}}\frac{\alpha_{ja}C_a^0}{\lambda_n^{01}+\alpha_{ja}C_a^0}\times$$

$$\left\{1-\exp\left[-\left(\lambda_n^{01}+\alpha_{ja}C_a^0\right)t\right]\right\},$$

$$C_{ja0}(r,t\ll t_C)\simeq\tilde{C}_{ja}^e+\tilde{C}_j^e\sum_{n=1}^{\infty}\frac{\gamma_n(r)}{\lambda_n^{01}}\frac{\alpha_{ja}C_a^0}{\lambda_n^{01}+\alpha_{ja}C_a^0}\times$$

$$\left\{1-\exp\left[-\left(\lambda_n^{01}+\alpha_{ja}C_a^0\right)t\right]\right\}.$$

As a first approximation in the small parameter $\bar{R}_P/L_P$ we have analogously to (5.92), (5.93)

$$C_{j0}(r,t\ll t_C)\simeq\tilde{C}_j^e-\tilde{C}_j^e\left(1-\frac{\bar{R}_P}{r}\right)\left[1-\exp(-\alpha_{ja}C_a^0t)\right],$$

$$C_{ja0}(r,t\ll t_C)\simeq\tilde{C}_{ja}^e+\tilde{C}_j^e\left(1-\frac{\bar{R}_P}{r}\right)\left[1-\exp(-\alpha_{ja}C_a^0t)\right]. \quad (5.98)$$

Now we define from (5.80) the value of the critical time $t_C(r)$. So far as we shall be interested at a later time in the dependency of the segregate radius $R_P(t)$ on the time t (see Chapter 6), the value of $t_C\ (\bar{R})$ should be searched. Formula (5.80) with allowance for (5.78) gives

$$t_C(\bar{R}_P)=\frac{\tilde{C}^e_j}{g}\left[\sqrt{1+\frac{2g}{\alpha_{ja}(\tilde{C}^e_j)^2}}-1\right]. \qquad (5.99)$$

The times $t<t_C(\bar{R}_P)$ are not interesting from physical point of view, as the segregate actually does not grow when $t< t_C(\bar{R}_P)$. The times in which a sensible segregate growth is beginning (i.e. the depletion of the solid solution) fall within the interval $t>t_C(\bar{R}_P)$. On the other hand, in accord with (5.92) and (5.98), the above obtained exact solutions (5.88) and (5.89) to the balance equations (5.83)-(5.85) hold good if

$$t_C(\bar{R}_P)<t<t_{R_P}. \qquad (5.100)$$

It follows from the condition (5.100) that if (5.94) and (5.98) are taken into account, the parameters of the problem must satisfy the relation

$$\frac{L_P(L_P-\bar{R}_P)^2}{3D_j\bar{R}_P}>\frac{\tilde{C}^e_j}{g}\left[\sqrt{1+\frac{2g}{\alpha_{ja}(\tilde{C}^e_j)^2}}-1\right], \qquad (5.101)$$

or we have at large generation rates of intrinsic point defects ($g\gg\alpha_{ja}(\tilde{C}^e_j)^2$) and with allowance for (5.68) the condition

$$L_P^3(3D_j\bar{R}_P)^{-1}\gg\sqrt{2}(\alpha_{ja}g)^{-1/2} \qquad (5.102)$$

that defines the application range of the above obtained relations (5.88)-(5.93).

Listed in Table 5.3 are the values of the characteristic times t_* (see (1.27)), t_{R_P} (see (5.94)), and $t_C(\bar{R}_P)$ (see (5.99)) for various generation rates and the temperature T of Ni-based solid solution irradiation (see Table 5.1) in the case of vacancy mechanism of impurity transfer toward the segregates ($j=v$, $n=i$).

It is seen from the Table 5.3 that the expressions (5.88)- (5.93) obtained above for the concentration profiles of mobile point defects nearby the precipitates hold in a wide interval of the irradiation temperatures T and the generation rates g.

Table 5.3

Characteristic times $t_c(\bar{R}_P)$, t_{R_P} and t_* for a simulated alloy (see Table 5.1) at various temperatures T (K) and the generation rate of intrinsic point defects g (cm^{-3}s^{-1}) and

$$k_v^2=10^{10}\text{cm}^{-2},\ \bar{R}_P\simeq10^{-1}L_P\simeq10^2a_0$$

T	$t_c(\bar{R}_P)$ (s)			t_{R_P}	t_*
	$g=10^{15}$	$g=10^{16}$	$g=10^{17}$	(s)	(s)
500	$2.24\cdot10^4$	$7.07\cdot10^3$	$2.24\cdot10^3$	$1.30\cdot10^6$	$6.25\cdot10^6$
550	$4.86\cdot10^3$	$1.54\cdot10^3$	$4.86\cdot10^2$	$6.15\cdot10^4$	$2.95\cdot10^5$
600	$1.36\cdot10^3$	$4.30\cdot10^2$	$1.36\cdot10^2$	$4.82\cdot10^3$	$2.31\cdot10^4$
650	$4.63\cdot10^2$	$1.46\cdot10^2$	$4.63\cdot10^1$	$5.59\cdot10^2$	$2.69\cdot10^3$
700	$1.84\cdot10^2$	$5.82\cdot10^1$	$1.84\cdot10^1$	$8.83\cdot10^1$	$3.24\cdot10^2$

It should be remembered that the consideration of the recombination of intrinsic point defects does not change the characteristic times $t_C(\bar{R}_P)$ and t_{R_P}, the relations (5.99)- (5.102) remaining valid in the $\mu D_i>0$ case.

CHAPTER 6

RATES AND EFFICIENCIES OF POINT DEFECT ADSORPTION BY SINKS

6.1. General Remarks

Knowledge of the point defect distribution around sinks of various nature permits to calculate such important characteristics of the solid solutions decomposing under irradiation as the rates $I_q^j(R_P)$ of j defect adsorption on q sinks of R_q dimension (by (1.38), (1.39)) and the efficiencies $K_D^j(r_*)$ of j defect adsorption on edge dislocations $(q=D)$ by (1.40). An important contribution of the values $I_q^j(R_P)$ and $K_D^j(r_*)$ to a general procedure of self-consistent calculations for the changes in the material properties under irradiation is seen from Diagram in Fig. 2.7. Proper expressions for the rates and efficiecies are possible to obtain by (1.38)-(1.40) merely when the inhomogeneous defect distributions with whose help the fluxes $\mathbf{J}_q^j(\mathbf{r},t)$ of j point defects move towards q sinks by (1.31) and then the values $I_q^j(R_P)$, $K_D^j(r_*)$ can be calculated, are taken into account. It follows from balance equations (2.11)-(2.14) and conditions (2.15), (2.16) that the rates and efficiencies of point defect adsorption depend not only on the q-sink dimensions and nature, but on the sink strength sums k_j^2. Therefore, as pointed out in Chapter 2, Section 2.6, a proper calculation of the point defect adsorption rates and efficiencies, as well as the dependencies $R_P(t)$, requires to calculate the sink strength sums k_j^2 simultaneously (see Diagram in Fig. 2.7).

Analytical solution to such self-consistent a problem has proved to be difficult to find in view of its severity, in any event before the mere system of self-consistent equations is formulated (Chapter 2, Section 2.6, and the Diagram in Fig. 2.7) [92-94, 96]. Analogously to Refs. 93, 94, a part of the account for the self-consistency in the solution of the stationary problems for a thin film was attempted in Ref. 402. The self-consistency in the calculation of the diffusion rate in an ensemble has been considered in Ref. 403.

To perform an intricate procedure of self-consisting (see Chapter 8) one derives the functional dependencies $I^j_q(R_P)$ and $K^j_D(r_*)$ arbitrary k^2_j.

6.2. Rates of Intrinsic Point Defect Adsorption by Coherent Precipitates

The distributions of intrinsic point defects under irradiation both inside (4.26), and nearby (4.28), the coherent precipitates in an ambient decomposing solid solution have been derived in Chapter 4 (Section 4.2) [381]. With allowance for the transformation (4.17), the expression for the density (1.31) of j defect flux toward the precipitates becomes

$$\mathbf{J}_P^{jF}(r)=-D_j\chi_j^F\,\frac{d\varphi_j^F(r)}{dr}\,. \tag{6.1}$$

Use the expressions for $\varphi_j^{\mathrm{I}}(r)$ from Section 4.2 and for $C_j^{\mathrm{II}}(r)$ (see (4.28)) and substitute them into (6.1) taking account of (4.17), which yields an expression for the density of defect fluxes in regions I and II.

Substituting now (6.1) into (1.38), we get

$$I_P^{j\mathrm{I}}(R_P)=\frac{4}{3}\pi R_P^3 g_j^{\mathrm{I}}=g_j^{\mathrm{I}}V_P(R_P), \tag{6.2}$$

$$I_P^{j\,\mathrm{II}}(R_P)=\alpha_P^j R_P D_j(\tilde{C}_j^{+}+\kappa_P^{j\,\mathrm{II}}), \tag{6.3}$$

where

$$\tilde{C}_j^{+}=\hat{\tilde{C}}_j-\chi_j^{\mathrm{II}}C_j^{e\,\mathrm{II}}, \qquad \alpha_P^j=4\pi,$$

$$\kappa_P^{j\mathrm{II}}=\left[\chi_j^{\mathrm{II}}C_j^{e\mathrm{II}}+\frac{g_j^{\mathrm{II}}}{D_j k_j^2}\right]\left[1-\frac{k_jL_P}{\mathrm{sh}\,k_jL_P}+\frac{k_jR_P}{\mathrm{th}\,k_jL_P}\right]-\hat{\tilde{C}}_j^{+}\left[1-\frac{k_jL_P}{\mathrm{sh}\,k_jL_P}\right]. \tag{6.4}$$

In the case of compressed precipitate, the expression (6.2) is meaningful only for the interstitials, being the rate of interstitial emission from the precipitates (region I). The expression (6.3) is negligible for $j=i$, since $\chi_i^{\mathrm{II}}<1$ (see Section 4.2), and large for $j=v$ ($\chi_v^{\mathrm{II}}>1$). Addition of $\kappa_P^{j\,\mathrm{II}}$ to the precipitate $\tilde{C}_j^{+}$ in (6.3) is determined by the inhomogeneity in the j-defect distribution in region II nearby the forming coherent precipitate.

Because k_j^{-1} is the length of the diffusion path of a j defect before its adsorption on any one of the sinks, then $k_j^{-1}\ll L_P$. This means that $k_jL_P\gg 1$, and $k_jR_P\gg R_P/L_P\ll 1$. Now when $k_jL_P\gg 1$, we have the ratio $k_jL_P/\mathrm{sh}\,k_jL_P<1$, but $\mathrm{th}\,k_jL_P\approx 1$ and $k_jR_P/\mathrm{th}\,k_jL_P\ll 1$. Consequently, the expression (6.3) takes the form

$$I_P^{j\,\mathrm{II}}(R_P)\simeq\alpha_P^j R_P g_j^{\mathrm{I}} k_j^{-2} \tag{6.5}$$

in case when $\hat{\tilde{C}}_j^{+}$ is (1.29a).

6.3. Efficiency of Intrinsic Point Defect Adsorption by an Edge Dislocation in the Field of Coherent Precipitates

The concentration profiles $C^{\mathrm{I}}_{jD}(r,\vartheta)$ of the intrinsic point defects j nearby the edge dislocation ($F_D=1$) given some coherent secondary-phase precipitates (see (4.38)), have been found in Chapter 4. Substitute (4.38) into (1.31) and have

$$\mathbf{J}^j_D(r,\vartheta)=-D_j e^{-U^j_D(r,\vartheta)/2}\left[\nabla\Phi^{\mathrm{I}}_{jD}\left(\frac{R^j_0}{2r}\right)+\frac{1}{2}\Phi^{\mathrm{I}}_{jD}\left(\frac{R^j_0}{2r}\right)\nabla U^j_D(r,\vartheta)\right]. \tag{6.6}$$

The efficiency of point defect adsorption per unit dislocation length, $K^j_D(r_*)$, is related to $\mathbf{J}^j_D(r,\vartheta)$ by (1.40). Substituting (6.6) into (1.40), we have [383, 384]

$$K^j_D(r_*)=K^j_{D\mathrm{I}}(r_*)+K^j_{D\mathrm{II}}(r_*)+K^j_{D\mathrm{III}}(r_*)+K^j_{DL}(r_*), \tag{6.7}$$

where

$$K^j_{D\mathrm{I}}(r_*)=\chi^j_D(r_*)\,\frac{\alpha^{\mathrm{I}}_{0D}}{\Lambda_0}\left[P^{\mathrm{I}}_0(x_R)-\frac{x_R\Gamma_1}{k_{j\mathrm{II}}R\Gamma_2}\left.\frac{dP^{\mathrm{I}}_0(x)}{dx}\right|_{x=x_R}\right], \tag{6.8}$$

$$K^j_{D\mathrm{II}}(r_*)=\chi^j_D(r_*)\,\frac{\alpha^{\mathrm{II}}_{0D}\Gamma_3}{\Lambda_0\Gamma_2}, \tag{6.9}$$

$$K^j_{DIII}(r_*) = \chi^j_D(r_*)\sqrt{(L_D/R)}(\alpha^{III}_{0D}/\Lambda_0\Gamma_2)\left\{\sqrt{(r_D/L_D)}\times\right.$$

$$\left.\mathrm{ch}\left[k_{jIII}(L_D-r_D)\right]-1\right\}, \tag{6.10}$$

$$K^j_{DL}(r_*)=\chi^j_D(r_*)\sqrt{\frac{L_D}{R}}\,\frac{C^+_j(g^{III})}{\Lambda_0\Gamma_2}, \tag{6.11}$$

$$\chi^j_D(r_*)=\alpha^j_D(r_*)D_j=\frac{2\pi D_j I_0(R^j_0/2r_*)}{I_0\left[R^j_0/2r_*\right]K_0\left[R^j_0/2R\right]-I_0\left[R^j_0/2R\right]K_0\left[R^j_0/2r_*\right]} \tag{6.12}$$

Here the relations (6.8)-(6.10) correspond to the contributions into the intrinsic point defect adsorption efficiency by an edge dislocation from the regions F_D=I, II, III respectively, and the expression (6.11) to the contribution into $K^j_D(r_*)$ from the "outer" surface of $r=L_D$ radius.

When the point defects interact with the edge dislocation profoundly and (4.33) is taken into account, the expression (6.12) is essentially simplified and becomes

$$\chi^j_D(r_*) = \alpha^j_D D_j = \frac{2\pi D_j}{\ln(2R/R^j_0)}, \tag{6.13}$$

which, in essence, coincides with a known expression in Ref. 256.

For the solid solution involved (see assumptions (4.1)- (4.9) in Chapter 4), i.e. when $(r_D-R)k_{jII}\gg1$, $(L_D-r_D)k_{jIII}\gg1$, which means that $L_D>r_D>R$, the constants Γ_1, Γ_2, Γ_3 prove to be about the same and equal to

$$\Gamma_1 \simeq \Gamma_2 \simeq \Gamma_3 \simeq \frac{k_{jII}+k_{jIII}}{4k_{jIII}} \times$$

$$\exp\left[k_{jII}(r_D-R)+k_{jIII}(L_D-r_D)\right] \equiv \Gamma \gg 1, \tag{6.14}$$

and $\Lambda_0 \simeq 1$.

Now, in accord with (4.33)-(4.35), we derive from (6.8)-(6.11)

$$K^j_{DI}(r_*)=\alpha^j_D \frac{g^{\mathrm{I}}(R^j_0)^2}{4}\left[P^{\mathrm{I}}_0(x_R)-\frac{x_R}{k_{jII}R}\frac{dP^{\mathrm{I}}_0(x)}{dx}\bigg|_{x_R}\right], \tag{6.15}$$

$$K^j_{DII}(r_*)=\alpha^j_D D_j\left[\frac{g^{\mathrm{II}}}{D_j k^2_{jIII}}-\tilde{C}^e_j\right], \tag{6.16}$$

$$K^j_{DIII}(r_*)=\alpha^j_D\sqrt{\frac{r_D}{R}}\, D_j\left[\frac{g^{\mathrm{III}}}{D_j k^2_{jIII}}-\tilde{C}^e_j\right]\frac{e^{k_{jIII}(L_D-r_D)}}{2\Gamma}, \tag{6.17}$$

$$K^j_{DL}(r_*)=\alpha^j_D\sqrt{\frac{L_D}{R}}\, D_j\tilde{C}^+_j(g^{\mathrm{III}})\Gamma^{-1}. \tag{6.18}$$

When (6.15)-(6.18) are compared, we see that the main contribution into (6.7) is given by the summands $K^j_{DI}(r_*)$ and $K^j_{DII}(r_*)$, while $K^j_{DIII}(r_*)$ and $K^j_D(r_*)$ are exponentially small (see (6.14)). Thus we ultimately have

$$K^j_D(r_*) \simeq K^j_{DI}(r_*)+K^j_{DII}(r_*)=\alpha^j_D D_j(\tilde{C}^+_j+\kappa^j_D), \tag{6.19}$$

where

$$\kappa^j_D=\frac{g^{\mathrm{I}}(R^j_0)^2}{4D_j}\left[P^{\mathrm{I}}_0(x_R)-\frac{x_R}{k_{jII}R}\frac{dP^{\mathrm{I}}_0(x)}{dx}\bigg|_{x=x_R}\right]-\tilde{C}^e_j.$$

It is to be noted that as $K^j_{DIII}(r_*)$ is exponentially small as compared with $K^j_{DI}(r_*)$ and $K^j_{DII}(r_*)$, the distribution of vacancies and interstitials in the region F_D=III can be believed to be roughly homogeneous. The concentration of intrinsic point defects $\hat{\tilde{C}}^+_j$ set in the form (1.29) is worth to choose and permits to take into account the recombination of opposite point defects in the most part of the solid solution (in the region F_D=III).

In the limiting case of a pure crystal and in the absence of sinks between the dislocations (see (4.42)), the expression (6.7) is essentially simplified:

$$K^j_{D0}(r_*)\simeq\alpha^j_D D_j\left[\hat{C}^+_j - C^e_j + \frac{g^{\rm I}(R^j_0)^2}{4D_j} P^{\rm I}_0(x_L)\right], \tag{6.20}$$

where α^j_D is defined by the relation (1.35), and $x_L = R^j_0/2L_D$. In a special case $g^{\rm I}=0$, the formula (6.20) changes into a known relation (1.34) first arrived at in Ref. 204.

In line with (6.19) and the expression (2.27) for $\tilde{C}^e_j$, the field of elastic stresses of the precipitates can essentially change the adsorption efficiency of intrinsic point defects $K^j_D(r_*)$ depending on the type of the point defect and on the precipitate (i.e. on the signs of $\Delta\Omega_j$ and ε_P). Thus, if the relative deformation inside the precipitate is $\varepsilon_{\rm I}=\varepsilon_P<0$ (compression) and in the solid solution $\varepsilon_{\rm II}>0$, then as seen from (6.19) and (2.27), the efficiency $K^i_D(r_*)$ for the interstitials decreases with increasing a volume fraction of the precipitates.

As a case in point, we present a numerical evaluation of $K^j_D(r_*)$ by formula (6.19) for the metal parameters close to those of nickel (see Table 5.1). We select $r_* \simeq 1.5a$, $\rho_D = 10^{10}\text{cm}^{-2}$, and the generation rates of intrinsic point defects $g^{\text{I}}=g^{\text{II}}=g^{\text{III}}=10^{16}\text{cm}^{-3}\text{s}^{-1}$, thus finding $K^j_D(r_*)$ for various values of $k_{j\text{II}}$ and the precipitate volume fraction.

The estimates for T=600 K, v_P=0.1, and ε_{I}=-0.02 demon- strating a magnitude of the κ^j_D contribution to $K^j_D(r_*)$ as compared with that from the supersaturation, are given in Table 6.1.

Table 6.1

Comparison of the contributions to the adsorption efficiency $K^j_D(r_*)$ from k^j_D and the supersaturations $\tilde{C}^+$

	$\tilde{C}^+_j$ (cm^{-3})	κ^j_L(cm^{-3})	$\kappa^j_L/\tilde{C}^+_j$(%)
$j=i$	$3.6 \cdot 10^7$	$2.6 \cdot 10^7$	72.2
$j=v$	$3.2 \cdot 10^{14}$	$2.5 \cdot 10^{14}$	78.0

It is seen from Table 6.1 that one should not neglect the summands κ^j_D in $K^j_D(r_*)$, because they are comparable in their value to $\tilde{C}^+_j$. Consequently, the calculation of $K^j_D(r_*)$ must be performed necessarily with account for the inhomogeneous concentration distributions in the volume nearby the dislocations.

The results of calculating the efficiencies $K^j_D(r_*)$ by (6.20) for the temperatures T=600 and 900 K are shown in Figs. 6.1 and 6.2.

It is seen from Figs. 6.1 and 6.2 that in the case of a pure crystal (E^j_P=0) and in the absence of sinks between the dislocations ($k^2_{jF_D}$=0), the efficiencies of point defect adsorption by an edge dislocation, $K^j_{D0}(r_*)$, are always greater than the efficiencies $K^j_D(r_*,k^2_{j\,\mathrm{II}})$ given the sinks (Curves AB, $A_I B_I$), in this case the values $K^j_D(r_*)$ drop smoothly with increasing $k_{j\,\mathrm{II}}$. In the solid solution decomposing under irradiation, the efficiency of vacancy adsorption by an edge dislocation $K^v_D(r_*,k^2_{v\,\mathrm{II}})$ does not actually depend on the volume fraction v_P of the surface $A_1B_1C_1D_1$ by virtue of the fact that $\tilde{C}^e_v \ll \tilde{C}^+_v$. At the same time $K^i_D(r_*,k^2_i)$ decreases with increasing v_P (the surfaces ABCD), becoming equal to $K^v_D(r_*)$ at certain values of v_P (the curves *DG*). Further increase in the precipitate volume fraction leads theoretically to a preferential vacancy adsorption by the edge dislocation. This means physically that when the volume fraction v_P is growing, the number of interstitial sinks increases in the regions between the precipitates, so that less the interstitials than the vacanciesare going to the dislocations from region II.

Such changes in the efficiencies $K^j_D(r_*)$ of the material containing the coherent segregates can bring about, with increasing v_P, the changed sign of the difference $K^i_D(r_*)$- $K^v_D(r_*)$ that determines the dislocation contribution to the rate of radiation swelling of the materials (see Chapter 8).

6.4. Rate of Intrinsic Point Defect Adsorption by a Short-Range Dislocation Loop

In Chapter 4 the distributions $C^{\mathrm{I}}_{jL}(r)$ of the intrinsic point defects j

have been found in the neighborhood of a short-range dislocation loop (F_L=I) (see (4.57)). Substituting (4.57) into (4.51), one can find the densities of defect fluxes j toward the loop in the form

$$\mathbf{J}_L^j(r)=-\frac{D_j}{r}\left(\frac{B_L^j}{r}-\frac{g^{\mathrm{I}}r^2}{3D_j}\right)\mathbf{i}_r, \tag{6.21}$$

Then, substituting (6.21) into (1.38), we have, in accordance with (4.60) [385],

$$I_L^j(R_L)=4\pi\alpha_L^j R_L D_j\left[\left[\hat{C}_{j0}^{+}-\mathrm{V}_{1L}\tilde{C}_j^e\exp\left(-U_{L_C}^j\right)\right]+\frac{g^{\mathrm{I}}r_L^2}{6D_j}\mathrm{V}_{1L}(\mathrm{V}_{2L}-H_L^j)-\frac{g^{\mathrm{I}}R_L^2}{3D_j\alpha_L^j}\right]. \tag{6.22}$$

Since it is difficult to operate with the expression (6.21) analytically (see (4.64)-(4.66)), we consider its approximate representation. When the condition (4.67) is fulfilled, we have H_L^j<1 (see (4.70)) and this value can be neglected as compared with V_{2L} (see (4.68)). After the substitution of (4.68) into (6.22), we obtain an approximate value for the rate of intrinsic point defect adsorption by a dislocation loop in the form

$$I_L^j(R_L)\simeq 4\pi\alpha_L^j R_L D_j\left[\hat{C}_{j0}^{+}-\tilde{C}_j^e+\frac{g^{\mathrm{I}}r_L^2}{6D_j}\left(k_{j\mathrm{II}}r_L+3\right)\left(k_{j\mathrm{II}}r_L+1\right)^{-1}-\frac{g^{\mathrm{I}}R_L^2}{3D_j\alpha_L^j}\right]. \tag{6.23}$$

If the sink L attracts the intrinsic point defects j (i.e. $E_L^j(r)$<0),

$\alpha_L^j > 1$ takes place and is determined by the expression (4.69). In this case we rewrite (6.23) as

$$I_L^j(R_L) \simeq \alpha_L^j I_{L0}^j(R_L) > I_{L0}^j(R_L), \tag{6.24}$$

where

$$I_{L0}^j(R_L) \simeq 4\pi R_L D_j \left[\hat{C}_{j0}^+ - \tilde{C}_j^e + \frac{g^{\mathrm{I}} r_L^2}{6D_j} \, \frac{k_{j\mathrm{II}} r_L + 3}{k_{j\mathrm{II}} r_L + 1} \right] - \tag{6.25}$$

is the rate of j point defect adsorption by the sink L when $E_L^j(r)=0$. It follows from the relation (6.24) that the adsorption rate $I_L^j(R_L)$ increases at the cost of an attraction stress field of the sink L.

The factor $\left(k_{j\mathrm{II}} r_L + 3\right)\left(k_{j\mathrm{II}} r_L + 1\right)^{-1}$ in the expressions (6.23) and (6.25) has occurred due to the fact that the inhomogeneous distribution of j defects was considered in the region F_L=II (see (4.58)). However, the value of this factor is variable from 1 to 3 depending on the value $k_{j\mathrm{II}} r_L$. Therefore, taking it equal to 1, one can consider a homogeneous distribution of intrinsic point defects $\hat{C}_j^+$ in the region F_L=II with an error being no more than a few per cents. The substitution of $\hat{C}_{j0}^+$ for $\hat{C}_j^+$ in the expressions (6.23) and (6.25) permits to take into account the recombination of opposite point defect in the region F_L=II. Now we obtain from (6.24), (6.25) the expressions for the rates of j-defect adsorption by a dislocation loop (q=L) with allowance for a mutual recombination of vacancies and interstitials in the region F_L=II in the following form:

$$I_L^j(R_L) \simeq 4\pi R_L D_j z_L^j \left[\hat{C}_j^+ - \tilde{C}_j^e + \kappa_L^j\right]. \tag{6.26}$$

Here $\hat{C}_j^+$ is determined from (1.29), and the quantities z_L^j, by (4.69), take the values $z_L^v \simeq 1$, $z_L^i \simeq 1.14 \cdot \sqrt[3]{|E_L^i(R_L)|/kT}$ and

$$\alpha_L^j = 4\pi z_L^j, \quad \kappa_L^j = \frac{g^I r_L^2}{6D_j}. \tag{6.27}$$

For a material of *Ni*-type (the parameters see in Table 5.1, Chapter 5) at the temperature T=600 K, one can evaluate z_L^j =1.22, z_L^v=1. Correspondingly, at $g=10^6 \text{cm}^{-3}\text{s}^{-1}$ and $k_i^2 = k_v^2$, one can compare the supersaturations $\tilde{C}_j^+$ and κ_L^j in (6.26) and (6.27) (see Table 6.2).

Table 6.2

Comparison of the contributions of κ_L^j and the precipitates $\tilde{C}_j^I$ into the adsorption rate $I_L^j(R_L)$

	$\tilde{C}_j^+$ (cm^{-3})	κ_L^j(cm^{-3})	$\kappa_L^j/\tilde{C}_j^+$ (%)
$j=i$	$1.03 \cdot 10^8$	$3.74 \cdot 10^7$	36.3
$j=v$	$2.71 \cdot 10^{18}$	$9.87 \cdot 10^{17}$	36.4

From Table 6.2 well seen is an essential role of the summands κ_L^j in (6.26) as well as for the dislocations.

When z_L^j=1, the expression (6.26) determines the rate of intrinsic point defect adsorption by the sink whose interaction energy with vacancies and interstitials is equal to zero (see (4.69)). Such a sink may be, for example, a spherical void. In this case, suffices it to substitute the index L in (6.26) by the index V. If one make tend the dimensions of the region $F_V=1$ to zero, i.e. take $r_V \equiv R_V$ in the expression (4.72), then (6.26) is transformed to a known form for the voids

$$I_V^j(R_V) \simeq 4\pi R_V D_j(\hat{C}_j^+ - \tilde{C}_j^e). \tag{6.28}$$

In the subsequent discussion we shall be interested in the kinetics of intrinsic point defects nearby the prismatic interstitial loops, as they are always present in the volume of the material under irradiation. The interaction energy of such loops with an interstitial atom is presentable in the form [232]

$$E_L^i(r,\vartheta) = \frac{1+\nu}{2(1-\nu)} \frac{Ga_0 \Delta\Omega_i R_{L0}^2}{r^3} (1-3\cos^2\vartheta), \tag{6.29}$$

where R_{L0} is the dimension of a prismatic interstitial loop that is equivalent in the quantity of adsorbed point defects to a spherical aggregate of the intrinsic point R_L-radius defects ($R_L \geq R_{L0}$); G is the elastic modulus of the material; ν is the Poisson coefficient; and $\Delta\Omega_i$ is the relaxation of the volume about the interstitial. Averaging (6.29) over the angle ϑ, we obtain

$$E_L^i(r) \simeq -\frac{1}{4}\left(\frac{1+\nu}{1-\nu}\right) \frac{Ga_0 \Delta\Omega_i R_{L0}^2}{r^3}, \quad r \geq R_L. \tag{6.30}$$

The interaction energy of a prismatic interstitial loop with a vacancy intricately depends on the elastic constants of the material and decreases with the separation as r^{-6} [232] (i.e. $E_L^v(r) \sim r^{-6}$). Since $E_L^v(r)$ decreases with increasing r much more steeply than $E_L^i(r)$, and what is more $|\Delta\Omega_v|<\Delta\Omega_i$ we define R_L as the radius of a surface at which

$$E_L^v(R_L) \ll kT, \tag{6.31}$$

$$E_L^i(R_L) \simeq kT. \tag{6.32}$$

Now, in accordance with (6.24) and (6.25),

$$I_L^i(R_L) \simeq z_L^i I_{L0}^i(R_L), \tag{6.33}$$

$$I_L^v(R_L) \simeq z_L^v I_{L0}(R_L) = I_{L0}^v(R_L), \tag{6.34}$$

where z_L^i and z_L^v are given by the expressions (4.69) (see the explanations after (4.70)).

The formulas (6.33), (6.34) determine the rates of vacancy and interstitial adsorption by the R_L-radius surface at whose center the prismatic loop is located.

6.5. Collection of Formulas for the Rates and Efficiencies of Point Defect Adsorption

From comparison of the obtained analytical expressions (6.3), (6.19), (6.26) it will be obvious that they may be written in the case of a stationary defect distributions uniformly:

$$I_q^j(R_q)=\alpha_q^j R_q D_j(\tilde{C}_j^{+}+\kappa_q^j), \tag{6.35}$$

$$K_D^j(r_*)=\alpha_D^j D_j(\tilde{C}_j^+ + \kappa_D^j), \tag{6.36}$$

where

$$\alpha_P^j=4\pi, \qquad \kappa_P^j=\left[\tilde{C}_j^e+\frac{g_j^{\mathrm{II}}}{D_j k_j^2}\right]\left[1-\frac{k_j L_P}{\mathrm{sh}\, k_j L_P}+\frac{k_j R_P}{\mathrm{th}\, k_j L_P}\right]-$$

$$\tilde{C}_j^+\left[1-\frac{k_j L_P}{\mathrm{sh}\, k_j L_P}\right],$$

$$\alpha_L^j=4\pi z_L^j, \quad z_L^v\simeq 1, \quad z_L^i\simeq 1.14\cdot\sqrt[3]{|E_L^i(R_L)|/kT}; \quad \kappa_L^j\simeq g^{\mathrm{I}} r_L^2/6D_j,$$

$$\alpha_D^j=2\pi/\ln(2R/R_0^j); \quad \kappa_D^j=\frac{g^{\mathrm{I}}\,(R_0^j)^2}{4D_j}\left[P_0^{\mathrm{I}}(x_R)-\frac{x_R}{k_{j\,\mathrm{III}}R}\left.\frac{dP_0^{\mathrm{I}}(x)}{dx}\right|_{x=x_R}\right]-\tilde{C}_j^e.$$

The values α_q^j are the preferential factors, and κ_q^j are changes in the stationary point defect concentrations, concerned with their inhomogeneous distribution about the sink q and depending on the superposition of the fields of sink elastic stresses and on the coherent secondary-phase precipitates, as well as on the rates of defect generation around the sinks.

Some appraisals show that the contributions of κ_q^j to the expression (6.35) and (6.36) are comparable with the supersaturations $\tilde{C}_j^+$, making an essential contribution to the values $I_q^j(R_q)$ and $K_D^j(r_*)$. Hence, one needs to perform the calculation of the rates and efficiencies of j-defect adsorption on q sinks with the use of spatial distributions of j defects around the sinks q.

CHAPTER 7
EVOLUTION OF SECONDARY-PHASE SEGREGATES UNDER IRRADIATION

7.1. Changes in the Sinks of Metals under Irradiation

The diffusion transfers of the point defects under irradiation could give rise to the evolution in the material structure, i.e. to the formation of both intrinsic point defect and impurity clusters. As noted above, possible for solid solutions is a decomposition with the secondary-phase precipitation. This stage can be called the stage of sink evolution (see Figs. 1.1 and 2.7, stage III).

The segregate growth in the solid solutions could be determined by both diffusion of single impurity atoms (*a*) and their transfer in the form of complexes (*ja*).

With a knowledge of the space-time distribution of impurity atoms in keeping with the condition (5.1) and the equations (5.2), (5.3), one can, invoking (1.31) and the mass balance [184, 185], find an important characteristic of the solid solution decomposition: the law of growing the precipitate radius $R_P(t)$ growth with the time t. Thus, for a binary solid solution (see, e.g., Ref. 183), the following dependency of the spherical segregate radius on the time t takes place (see Fig. 7.1):

$$R_P(t) \simeq 2\sqrt{D_a t} + R_{P0}, \qquad (7.1)$$

where R_{P0} is the initial precipitate dimension.

In actual conditions, the segregate "is fed" from a limited volume, and when $C_a(\mathbf{r},t)$ is approaching its thermodynamically equilibrium value C_a^e, the precipitate growth is cut off. An analogous type of the problems was considered in Refs. 387, 388 for the arbitrary-shaped particles. In this analysis, a number of simplifying assumptions was made. The particles were believed to be distributed in the matrix of the solid solution periodically, their initial dimensions R_{P0} being negligible against the average separation $2L_P$ between the segregates which are growing with no change in their shape. In this situation, the average concentration of impurity atoms $\bar{C}_a(t)$ in the solid solution has the form

$$\bar{C}_a(t) = C_a^0 \exp(-\xi_a t^{3/2}),$$

where ξ_a is the constant depending on the diffusion mobility of impurity atoms, on the mutual arrangement of the segregates and on their dimensions. Taking into account the matter- conservation condition, one can show (see Fig. 7.1) that

$$R_P^3(t) - R_{P0}^3 \sim C_a^0 \left[1 - \exp(-\xi_a t^{3/2})\right]. \tag{7.2}$$

One of the first theoretical works devoted to the segre- gate growth in a radiation-accelerated decomposition is Ref. 389, where the method of Refs. 387, 388 is extended to the case of homogeneous irradiation of a solid solution. It was shown in Ref. 389 that the segregates are growing by the linear law at the initial stage of irradiation when $D_j k_j^2 t \ll 1$. If $D_j k_j^2 t \gg 1$, the rate of precipitate growth is essentially decreased.

In Refs. 289, 290, an analytical expression for the concentration profiles vacancy - impurity atom around the segregate has been obtained, and the dependency of the spherical precipitate radius $R_p(t)$ on the time t determined. For the initial stage of solid solution decomposition when $D_{va}R_p(t)t \ll L_p^3$, the expression

$$R_p(t) \simeq \sqrt{R_{p0}^2 + 2\Omega C_{va}^e D_{va} t} \, . \qquad (7.3)$$

has been obtained. The dependency (7.3) is in agreement with the experimental data on the initial stage of ageing the aluminum alloy with 3.4 and 5.3 at. % *Zn* at the room temperature [289]. However, in Refs. 289, 290, it was assumed to simplify the calculations, that the concentrations of impurity atoms, vacancies, and complexes are in a local equilibrium. Such a condition runs counter to the assumption (5.4) and based on it that the complexes are decomposing only on the segregate surface. This contradiction was noted in Ref. 404, but the goal to obtain a solution to the problem without this condition was not sought.

7.2. Mass Balance on the Segregate Surface

The expressions for the point defect concentration profiles about the isolated secondary-phase segregates under irradiation which were found in Chapter 5, can be used to obtain the equations determining $R_p(t)$.

It should be noted that the calculations of the point defects about the forming segregates, listed in Chapter 5, have been performed as an adiabatic assumption (see condition (5.57)): the segregate dimension changes essentially more slowly than the point defect migration toward the segregate. This leads, in finding the concentration-time dependency, to the necessity of assuming the dimension R_p and the segregate concentration C_p on the average constant and hence, to that of considering all the coefficients of the appropriate equations for the concentration as independent of t and r. However, as one passes to the calculation of the rate of changing the segregate dR_p/dt, it should be remembered that the t-dependency of this rate obtained with the formulas (5.31) or (5.93) will be valid only on a small time interval dt, inasmuch as the average radius $\bar{R}_p$ (over the time) that enters the coefficients, is also changes with time (see Fig. 7.2). Therefore, the obtained t-dependency of dR_p/dt cannot be extendible to any time interval without reserve. For this purpose, $\bar{R}_p$ must be given properly. The averaging

$$\bar{R}_p(t)=\frac{1}{\Delta t}\int_t^{t+\Delta t} R_p(t')dt'$$

can be shown to lead to the following results. When $R_p(t)$ are changed weakly, we have

$$\bar{R}_p=\frac{1}{\Delta t}\int_t^{t+\Delta t} R_p(t')dt' \simeq R_p(t)\left[1+\frac{\Delta R_p}{2R_p(t)}\right]\simeq R_p(t),$$

where $\Delta R_p=R_p(t+\Delta t)-R_p(t)$. When $R_p(t)$ changes rapidly (i.e. when $\Delta t \longrightarrow 0$, we obtain

$$\bar{R}_p=\lim_{\Delta t\to 0}\frac{1}{\Delta t}\int_t^{t+\Delta t} R_p(t')dt'=$$

$$\lim_{\Delta t\to 0}\frac{1}{\Delta t}\left\{R_p(t)\Delta t+\frac{1}{2}\left[R_p(t+\Delta t)-R_p(t)\right]\Delta t\right\}\longrightarrow R_p(t).$$

Consequently, the substitution $\bar{R}_p \longrightarrow R_p(t)$ can be performed in the expression for the appropriate concentrations when one need to calculate the rate of changing the segregate radius $dR_p(t)/dt$.

Let us calculate the amount of impurity atoms landing in a forming segregate as free atoms $l=a$ or in the form of complexes $l=ja$ and offering a change in the segregate volume

$$\frac{dV_p(R_p)}{dt}=\Omega(1+\xi)I_p^l(R_p).$$

Here ξ describes the stoichiometric composition of the precipitate (i.e. when ξ of matrix atoms fell on one impurity atom), the parameter ξ taking into account the difference between the specific volumes in the solid solution and in the segregate, as well as possible additions of the intrinsic point defects. Passing from the volume $V_p(R_p)=4\pi R_p^3/3$ to $R_p(t)$, one obtains an equation to determine $R_p(t)$ in the form

$$\frac{dR_p(t)}{dt}=\frac{\Omega(1+\xi)}{4\pi}\,\frac{I_p^l(R_p)}{R_p^2}. \tag{7.4}$$

Consider various chances depending on a specific problem on finding the l-type defect concentration C_l.

Inhomogeneous stationary concentrations. Make use of the expression for the adsorption rate $I_p^j(R_p)$ in the form (6.35):

$$I_p^j(R_p)=\alpha_p^j R_p D_j(\tilde{C}_j^e+\kappa_p^j). \tag{6.35}$$

The substitution (6.35) into (7.4) yields the equation

$$\frac{dR_P}{dt}=B_j\,\frac{\tilde{C}_j^{+}+\kappa_P^j}{R_P},\quad B_j=a(1+\xi)D_j, \tag{7.5}$$

which causes the following time dependency of the segregate radius:

$$R_P(t)=\sqrt{R_{P0}^2+2B_j(\tilde{C}_j^{+}+\kappa_P^j)(t-t_0)}\,, \tag{7.6}$$

where $R_{P0}=R_P(t_0)$ is the initial segregate radius.

Inhomogeneous nonstationary concentrations. In this case, one should use the definition (1.31) for the density $\mathbf{J}_P^j(\mathbf{r},t)$ of j defect flux toward the segregate with account for (1.38), which produces the following expression for the adsorption rate $I_P^j(R_P)$:

$$I_P^j(R_P)=\alpha_P^j R_P^2\left|\mathbf{J}_P^j(\mathbf{r},t)\right|_{r=R_P}=\alpha_P^j D_j R_P^2 \nabla C_j(r,t)\Big|_{r=R_P}. \tag{7.7}$$

The substitution of (7.7) into (7.4) leads to the equation

$$\frac{dR_P}{dt}=B_j\nabla C_j(r,t)\Big|_{r=R_P}. \tag{7.8}$$

Depending on a specific form of $\nabla C_j(r,t)$ as a function of t and R_P, one obtains the appropriate equations to determine $R_P(t)$.

Homogeneous nonstationary concentrations. Proceeding from physical meaning of the rate of j-defect adsorption by the segregates, one can write the following expression if he recalls the form of sink terms in the equations (2.12) and (2.13) and the definition of sink strength sums (1.26),

$$I_P^j(R_P,t)=D_j S_P^j(t)C_j(t)C_P^{-1}, \tag{7.9}$$

where

$$C_P(t)=\int f_P(R_P,t)dR_P \tag{7.10}$$

is the segregate concentration in the material volume in the time instant t. Based on (2.19), the sink strength $S_P^j(t)$ has the form

$$S_P^j(t)=\left[D_jC_j(t)\right]^{-1}\int I_P^j(R_P,t)f_P(R_P,t)dR_P. \tag{7.11}$$

Then, substituting (7.11) into (7.9), we obtain

$$I_P^j(R_P,t)=\frac{\int I_P^j(R_P,t)f_P(R_P,t)dR_P}{\int f_P(R_P,t)\,dR_P}\equiv\overline{I_P^j(t)},$$

i.e. the quantity inserted in (7.9) is an average adsorption rate for the segregates of an arbitrary dimension R_P, and naturally does not depend on R_P. Based on the adiabatic assumption (see the condition (5.57) used in the calculation of the concentrations in Chapter 5), we have $S_P^j(t)$=const, i.e. is this quantity is a parameter for the problem involved. Now, substituting (7.9) into (7.4), we obtain the equation for $R_P(t)$ in the form

$$\frac{dR_P(t)}{dt}=B_jS_P^jC_j(t)(4\pi R_P^2C_P)^{-1},$$

which give the following dependency:

$$R_P(t)=\sqrt[3]{R_P^3+3\frac{B_jS_P^j}{4\pi C_P}\int_{t_0}^{t}C_j(t')dt'}. \tag{7.12}$$

7.3. Time-Dependency of the Segregate Radius in a Slow Depletion of a Solid Solution

When the mobile point defects are volume-distributed homogeneously, the *ja*-complex concentration has the form (5.49). Substitute (5.49) into (7.12) and integrate, then we have if $\chi_{ja} > D_{ja}k^2_{ja}$ [285] (the complex dissociation goes more rapidly than their exit toward the sinks)

$$R_P(t)=\sqrt[3]{R^3_{P*}+3\,\frac{B_{ja}C^0_a}{4\pi D_{ja}C_P}\,\frac{S^{ja}_P}{k^2_{ja}}\left[1-\exp\left(-\frac{t-t_*}{t_C}\right)\right]}\ , \qquad (7.13)$$

where R_{P*} is the segregate radius in the time instant $t\text{-}t_*$. It is seen from (7.13) that in the case of negligibly small initial segregate dimensions R_{P*} the volume $4\pi R^3_P(t)/3$ of the forming precipitate is proportional to the initial impurity concentration C^0_a, the volume of the solid solution $4\pi L^3_P/3=C^{-1}_P$ falling on one precipitate, and also the efficiency of the impurity atom capture by the segregates, $S^{ja}_P(\bar{R}_P)k^{-2}_{ja}$.

It should be, however, noted that the value $S^{ja}_P(\bar{R}_P)$ in fact depends on the instantaneous value $\hat{R}_P(t)$ rather than the average segregate radius $\bar{R}_P$. The characteristic time t_C also depends on t through the sink strength sums $k^2_l(t)$ (see (5.51)). Therefore, if one takes into account the $\bar{R}_P$ growth with time, then $R_P(t)$ increases, and (7.13) is a lower evaluation of the segregate radius.

After the completion of the solid solution decomposition when the impurity concentration reaches its thermodynamically equilibrium value C^e_a, one can obtain

$$R_p(\infty)=\lim_{t\to\infty} R_p(t)\simeq L_p\sqrt[3]{\Omega(1+\xi)C_a^0 S_p^{ja}(R_p)k_{ja}^{-2}}\ . \tag{7.14}$$

The allowance for the inhomogeneous *ja*-complex dis- tribution close by a spherical coherent segregate (see expression (5.66)) permits more precisely to determine the dependency of the precipitate radius $R_p(t)$ on the time t at an initial stage of segregate formation ($t_*<t<t_C$) when $R_p \ll L_p$.

Substituting the simplified expression (5.71) for $C_{ja}(r,t)$ into the differential equation (7.8) and taking into account (5.69) and the condition $\chi_{ja} \gg D_{ja}k_{ja}^2$, we obtain

$$\frac{dR_p(t)}{dt}=B_{ja}\frac{\tilde{C}_{ja}^0}{R_p(t)}\exp\left[-\frac{t-t_*}{t_C}\right]. \tag{7.15}$$

A solution to the equation (7.15) can be represented in the form

$$R_p(t)\simeq\sqrt[3]{R_{p*}^2+2\Omega(1+\xi)\frac{C_a^0}{k_{ja}^2}\left[1-\exp\left(-\frac{t-t_*}{t_C}\right)\right]}. \tag{7.16}$$

The formulas (7.13), (7.16) obtained determine the law of growing the segregate radius with the time t in two limiting cases. At the initial decomposition stage ($t_*<t<t_C$) when the segregate dimensions are much less than the average separation between them ($R_p \ll L_p$), the precipitate radius increases in line with (7.16). At later decomposition stages ($t>t_C$) when the solid solution is depleted, i.e. when the segregates have reached some substantial dimensions ($R_p \sim L_p$), the formula (7.16) is invalid. The decomposition goes more slowly and the precipitates increase following the law (7.13). One has not succeeded in finding an analytical form for the law of $R_P(t)$ growing with the time t when the time is intermediate, $t \sim t_C$. The dependencies $R_P(t)$ are demonstrated in Fig. 7.3 in these two limiting cases.

The investigation of the decomposing *Cu-Re* solid solutions irradiated by high-energy electrons (E=1 MeV) and *Cu* ions (E=300 keV) has been carried out in the experimental Refs. 405-408. The authors of these works have explained the growth of the secondary-phase segregates by the interstitial mechanism of impurity-atom transport toward the precipitates ($j=i$, $n=v$, $a=Be$).

In Ref. 406, the bound and migration energies have been determined for some mixed dumbbells ($\varepsilon^b_{ia} \simeq 1.45$ eV, $\varepsilon^m_{ia} \simeq 0.65$ eV). Since the bound energy is high, the complexes do not dissociate in their path toward the segregates. This should lead to a very rapid precipitate growth which does not agree with the experimentally observed data (see Ref. 405). The authors of Ref. 406 offered a model explaining a more slow diffusion of the impurity atoms toward the segregates. According to this model, the characteristic lifetime of the mixed dumbbells in the material volume is limited by the traps available in the solid solution (e.g. interstitial loops) rather than the complex dissociation. Such traps confine *Be* atoms for a while, effectively increasing their migration energy or decreasing the bound energy of the complex. In the growth mechanism suggested by us (see the balance equations (5.5)-(5.8), such traps can be effectively taken into account in the kinetic coefficient χ_{ja} by renormalizing ε^b_{ja}. In so doing the lifetime of the complex *ia*, $\tau_{ia}=\chi_{ia}^{-1}$, will be determined by the relation

$$\tau_{ia}=\frac{1}{4\pi}\frac{D_v}{D_{ia}D_i}\left(\frac{aL_q}{\bar{R}_q}\right)^2\frac{C_v^+}{C_i^+}, \tag{7.17}$$

where $2L_q$ is the separation between the traps, and $\bar{R}_q$ is their average dimension. Considering that $\bar{R}_q \sim 10^{-1}L_q$, we determine τ_{ia} with the use of the experimental data of Refs. 405, 406: ε_v^m=0.74 eV, ε_i^m=0.12 eV, ε_{ia}^m=0.65 eV, ε_{ia}^b=1.45 eV, T=583 K, the relative trap concentration is $C_q \simeq 10^{-7}$, the electron flux is $\Phi=2.9\cdot10^{18}$cm^{-2}s^{-1}, and $k_v^2=k_i^2$ is assumed. Substitute the data listed into (7.17) and obtain $\tau_{ia}\simeq 5\cdot10^{-7}$ s. Now, in accord with (2.10), the effective complex bound energy is $\varepsilon_{ia}^{b*}\simeq 0.9$ eV.

The results of a numerical calculation by (7.18) and (7.16) of the $R_p(t)$-dependencies and the experimental data [405] are shown in Fig. 7.4 from where one can see a good accordance of the theoretical segregate growth model de- veloped with the experimental data on the decomposition of the solid solution *Cu* - 0.18 weight % *Be* [405].

7.4. The Dependency of Segregate Radius on the Time in the Rapid- Depletion of a Solid Solution

If the complexes *ja* do not dissociate in their path toward the segregate, then at $t>t_C(\bar{R}_q)$ the concentration profiles of mobile point defects close by the segregate are presentable in the form (5.92), (5.93). Substituting the expression (5.93) for $C_{ja0}(r,t)$ into (7.8), we obtain a differential equation for $R_{P0}(t)$ in the form [285,400]

$$\frac{dR_{P0}(t)}{dt}=\Omega(1+\xi)D_{ja}\frac{C_a^0}{R_{P0}(t)}\exp\left[\frac{t}{t_{R_P}}\right], \tag{7.18}$$

where t_{R_P} is determined by the relation (5.94). In the case of small segregate dimensions ($R_P \ll L_P$) and $t\cdot t_{R_P}^{-1}<1$, we have an approximate solution to this equation,

$$R_{P0}(t)\simeq\sqrt{R_{PC}^2+2\Omega(1+\xi)D_{ja}C_a^0[t-t_C(\bar{R}_P)]}. \tag{7.19}$$

Here R_{PC} is the precipitate radius in the time instant $t_C(\bar{R}_P)$. Using this line of reasoning in the $t<t_C(\bar{R}_P)$ case, we obtain, in accord with (5.98), the expression

$$R_{P0}(t)\simeq\sqrt{R_{PC}^2+2\Omega(1+\xi)D_{ja}\tilde{C}_j^e\{t-[1-\exp(-\alpha C_a^0)](\alpha C_a^0)^{-1}\}}. \tag{7.20}$$

where R_{PC} is the initial segregate radius ($t=0$).

Note that the times $t<t_C(\bar{R}_P)$ are of no physical interest, since it may be shown by numerical calculation (see Fig. 7.5) that, according to (7.20), the segregates do not virtually grow at the time $t_C(\bar{R}_P)$. The times in which a detectable segregate growth is beginning find themselves in the interval $t>t_C(\bar{R}_P)$. On the other side, it must be remembered that the expansion of the exponent accepted in (7.18) is valid on the condition $t<t_{R_P}$. Therefore the expression (7.19) holds within the interval limited on either side (see (5.100)). For the times $t>t_{R_P}$, the differential equation (7.18) should be solved by numerical methods.

A consideration of the complex dissociation in the solid solution volume does not change the time-dependency $C_{l0}(r,t)$, but causes only a renormalization of the preexponential factor in the concentration profiles of the mobile point defects. Making use of the relations (5.75) and (7.8), we have analogously to (7.19)

$$R_{P0}(t)\simeq\sqrt{R^2_{PC}+2\Omega(1+\xi)D_{ja}[C^0_a-\chi_{ja}\tilde{C}^e_{ja}t_C(\bar{R}_P)][t-t_C(\bar{R}_P)]}.$$

In Ref. 91 it is shown that the recombination of opposite intrinsic point defects inessentially influences the dependency of the segregate radius $R_{P0}(t)$ on the time t, and a decrease in $R_{P0}(t)$ for a given t may reach some percents in reasonably deciding on the problem parameters.

Let us consider the results of Refs. 289, 290 on the decomposition of the *Al* alloy with 3.4 and 5.3 at. % *Zn* to compare the expressions (7.19) with the experiment. In this system the impurity-atom transfer toward the segregate is performed by the vacancy - impurity atom complexes (i.e. $j=v$). The energies of vacancy formation and migration are equal: $\varepsilon^f_v\simeq\varepsilon^m_v\simeq0.6$ eV (see Ref. 409). If the initial vacancy concentration C^0_v corresponds to the equilibrium one equal to $2.4\cdot10^{17}$ cm^{-3} at T=600 K, the average segregate dimensions are $\bar{R}_P\simeq10^{-1}$, and the distance between the precipitates is equal to 10^2a, then in the absence of irradiation we have by (5.94), (5.99) for the characteristic times of solid solution depletion: $t_C(\bar{R}_P)=2.7\cdot10^{-4}$s, $t_{R_P}=5.3\cdot10^{-4}$s. The process goes rapidly, and at t=$4\cdot10^{-4}$s, the radius $R_{P0}(t)$, in line with (7.19), reaches $18a$. At the room temperature and the same initial vacancy concentration C^0_v (produced by quenching), we have

$t_C(\bar{R}_P)$=44 s, t_{R_P}=90 s, and in the time t=67 s the segregate radius is building up to the same value $18a$. Thus, as one would expect, the rate of segregate growth decreases with the temperature. Comparing the expression (7.19) with the analogous formula for the segregate radius [299], we find that the rate of precipitate growth obtained from (7.19) is $\beta=\sqrt{D_v C_a^0 / D_{va} C_{va}^0}$ times as great as that in Refs. 289, 290. When $C_a^0/C_{va}^0 \simeq 10^4$ and the temperature is room one, $\beta \simeq 1$, and at T=600K K_a-$\beta \simeq 10$. This means that at the room temperatures, the relation (7.19) is in a good accordance with the data on the initial stage of ageing of the *Al* alloys with 3.4 and 5.3 at. % *Zn* listed in Ref. 289 (see Fig. 7.6), therefore the incorrectness of the local equilibrium approximation in Refs. 289, 290 has seemed to escape notice, while the abandonment of the local equilibrium condition gives rise, at elevated temperatures, to a more rapid precipitate growth, as far as the complexes do not actually dissociate in their path toward the segregate.

CHAPTER 8

SOLID SOLUTION DECOMPOSITION AND RADIATION SWELLING

8.1. Rate of Radiation Swelling in a Heterogeneous Material

The vacancies and interstitials forming in the materials under irradiation settle on sinks of *q* type to change them, i.e. finish building or diminishing the appropriate planes of atoms in the crystal lattice. The processes of disappearing the mobile point defects can be divided into three basic groups [91, 92, 410].

1. Outgoing to "far" structural sinks: (a) outside surface, (b) dislocations, (c) boundaries of grains, fragments, blocks, and so on.

2. Outgoing to "neighboring" sinks: (a) intrinsic clusters (the formation of clusters, dislocation loops, voids), (b) impurities (point defect capture, heterogeneous nucleation of intrinsic clusters and secondary phases, (c) the secondary-phase segregates (various stages from the nucleation to the decoherency) producing stresses in the matrix.

3. The recombination, i.e. mutual annihilation of opposite radiation defects: (a) annihilation of migrating interstitials and vacancies at their meeting, i.e. ordinary recombination; (b) annihilation of point defects on the opposite-defect clusters (on the intrinsic-defect sinks), the recombination on the bound states; (c) annihilation of the migrating point defects both at the meeting and on additional opposite-defect clusters due to flux changes and the redistribution of the vacancy and interstitial

concentrations in the stress fields, i.e. an extra (anomalous) - recombination.

The radiation swelling of materials is due to arriving the interstitials knocked-on by the radiation, at the "far" and "neighboring" sinks which increase the material volume by ΔV. The radiation void formation is developing in the material when there is an excess concentration of noncompensated point defects, for the interstitials, due to their great mobility, are going toward the sinks more rapidly than the vacancies.

By the radiation swelling of the material is meant the change in the volume resulting from the formation and evaluation of structural sink-type defects which produces surplus or deficient volume in relation to the initial structure. The relative swelling is evaluated as

$$\gamma = \frac{\Delta V}{V_0} = \sum_q \gamma_q, \tag{8.1}$$

where V_0 is the initial volume, γ_q is the contribution of the q-type sinks to the swelling. One therewith assumes that the structural defects do not interact between each other. In the subsequent discussion we assume that V_0 is a unit volume of the material. The swelling rate

$$\dot{\gamma} = \sum_q \dot{\gamma}_q. \tag{8.2}$$

is more conveniently calculated rather than the swelling γ.

If the interstitials and vacancies forming in the irradiated material completely annihilated both at the meeting and on the opposite defect clusters, there would not take place the radiation changes other than the accumulated products of nuclear reactions, the radiation buffeting phenomenon [146, 147] and their consequences. Therefore, the formation of the conditions favorable for the recombination processes must occur with decreasing the radiation swelling of the material. Hence to surplus on depress the swelling, the material must be prepared, so that either the additional "neighboring" sinks be formed onto which the interstitials landed to disappear there or some "obstacles" to interstitial runoff to the "far" sinks be built up [91, 93, 94, 410].

As the dislocations and the grain boundaries are excellent structural sinks for the interstitials and the operating conditions (operating temperature, fluency of the irradiating particles) are commonly set for a given material, the interstitial runoff to these "dangerous" ("far") sinks can be controlled if only a specialized alloyage which builds up a definite structure in the space between the "far" sinks. The processes of solid solution decay accompanied by a secondary-phase precipitation (most of the structural alloys being subject to such decays) are a means that permits to build up such a specialized structure (see Refs. 8, 295-300).

Consider the contributions to the radiation swelling (8.1) from various structural q defects both available in the material and occurring in response to irradiation or other attacks [92-94, 410, 411]. Let there be, in the material, dislocations ($q=D$), innerside interfaces ($q=\Lambda$). Let the secondary-phase segregates be formed as well which dominantly adsorb either interstitials ($q=P_i$) or vacancies ($q=P_v$), and some clusters of intrinsic defects be set up: interstitials being clusters small in size ($q=Q$) and interstitial dislocation loops ($q=L_i$); vacancies being voids ($q=V$) and vacancy dislocation loops ($q=L_i$).

Let v_{P_j} be a volume fraction of the forming segregates of P_j type, j be the type of the defect adsorbed by the segregate, and $\dot{\xi}_g^{jF}$ be the number of the j defects generated in a unit time in a unit volume of the phases $F=P_v$, P_i, and M (M is the matrix). Now, signifying as $\dot{\hat{\xi}}_q^{jF}$ a total number of the j-type defects thrown onto all q-type sinks in a unit time in a unit volume of the phase F, we write for the matrix ($F=M$)

$$\dot{\xi}_g^{jM}=\dot{\hat{\xi}}_j+\sum_{q_j=1}^{n_j}\dot{\hat{\xi}}_{q_j}^{j}, \qquad (8.3)$$

where $q_j=D$, Λ, P_j, Q, L_j, V; n_j is the number of type for the j-type defect sinks, and the value $\dot{\hat{\xi}}_j$ is the number of the free j-type defects generated in the matrix in a unit volume in a unit time. An analogous expression for the segregate $F=P_j$ will take the form

$$\dot{\xi}_{jg}^{P_j} = g_j^{P_j}, \qquad (8.4)$$

where $g_j^{P_j}$ is the rate of j-defect generation in the segregate P_j. The balance of particles - interstitials and vacancies which occur in equal amounts when all the material volume is irradiated - yields

$$\nu_{P_i} g_v^{P_i} + \nu_{P_v} g_v^{P_v} + \left(1-\nu_{P_i}-\nu_{P_v}\right)\left(\sum_{q_v} \dot{\hat{\xi}}^v_{q_v} + \dot{\hat{\xi}}_v\right) =$$
$$\nu_{P_i} g_i^{P_i} + \nu_{P_v} g_i^{P_v} \left(1-\nu_{P_i}-\nu_{P_v}\right)\left(\dot{\hat{\xi}}_i + \sum_{q_i} \dot{\hat{\xi}}^i_{q_i}\right). \tag{8.5}$$

Since the same sink may be settled by both interstitials and vacancies, the expressions

$$\dot{\xi}^j_q = \dot{\hat{\xi}}^j_q - \dot{\hat{\xi}}^k_q, \tag{8.6}$$

are more conveniently operated, where $j \neq k$ and $j,k=i,v$ and $\dot{\xi}^j_q$ is the resulting number of the j defects having settled on the q sinks in a unit volume in a unit time. Using (8.6), we rewrite the balance condition (8.5) as

$$\nu_{P_i} g_v^{P_i} + \nu_{P_v} g_v^{P_v} + \left(1-\nu_{P_i}-\nu_{P_v}\right)\left(\dot{\xi}_v + \dot{\xi}^v_V + \dot{\xi}^v_{L_v} + \dot{\xi}^v_{P_v}\right) =$$
$$\nu_{P_i} g_i^{P_i} + \nu_{P_i} g_i^{P_v} + \left(1-\nu_{P_i}-\nu_{P_v}\right)\left(\dot{\xi}^i_D + \dot{\xi}^i_\Lambda + \dot{\xi}^i_Q + \dot{\xi}^i_{L_i} + \dot{\xi}^i_{P_i}\right). \tag{8.7}$$

Here account has been taken of the fact that the number of free interstitials in the material is small, i.e. $\dot{\xi}_i=0$, and $\dot{\xi}_v=\dot{\hat{\xi}}_v$. Moreover, taken into consideration was a preference in the j-defect adsorption by various q sinks (see, e.g., Refs. 204, 206, 247, 248, 251, 412). The swelling rate for a heterogeneous material may be presented in the form

$$\dot{\gamma} = \nu_{P_v} \dot{\gamma}_{P_v} + \nu_{P_i} \dot{\gamma}_{P_i} + \left(1-\nu_{P_i}-\nu_{P_v}\right)\dot{\gamma}_M. \tag{8.8}$$

If one assumes that the j defects do not form clusters in the forming secondary-phase segregates, then written may be

$$\nu_{P_v}\dot{\gamma}_{P_v}=-\Delta\Omega_v^{P_v}\left[\nu_{P_v}g_v^{P_v}+\left(1-\nu_{P_i}-\nu_{P_v}\right)\dot{\xi}^v_{P_v}\right]+\left(\Omega+\Delta\Omega_i^{P_v}\right)\nu_{P_v}g_i^{P_v},$$
$$\nu_{P_i}\gamma_{P_i}=\left(\Omega+\Delta\Omega_i^{P_i}\right)\left[\nu_{P_i}g_i^{P_i}+\left(1-\nu_{P_i}-\nu_{P_v}\right)\dot{\xi}^i_{P_i}\right]-\Delta\Omega_v^{P_i}\nu_{P_i}g_v^{P_i}, \tag{8.9}$$

where $\Delta\Omega_j^{Pj}$ and $\Delta\Omega_k^{P_j}$ are the volume changes of the defects of j or k types in the segregate P_j. We shall refer to the definition (8.2) for the value $\dot{\gamma}_M$.

Consider various contributions $\dot{\gamma}_{Mq}$ into the value $\dot{\gamma}_M$. When the interstitials and vacancies are adsorbed by dislocations, there takes place a volume change as follows:

$$\gamma_{MD}=\Omega\xi^i_D \quad \text{or} \quad \dot{\gamma}_{MD}=\Omega\dot{\xi}^i_D, \tag{8.10}$$

and one can write from (8.6)

$$\dot{\xi}^i_D=\rho_D(K^i_D-K^v_D). \tag{8.11}$$

For the innerside surfaces Λ, the expressions $\dot{\gamma}_{M\Lambda}$ will have the form analogous to (8.10).

If one considers a q_i sink of R_{q_i} radius, forming from the interstitials (e.g. Q, L_i, P_i), for which $\Omega_i^{q_i}(R_{q_i})=\Omega+\Delta\Omega_i^{q_i}(R_{q\theta_i})$ is valid, then each new arriving interstitial increases it by the value $\Omega_i^{q_i}(R_{q_i})$. Each occurring vacancy will lead to the one particle decrease in this q_i sink. In view of the fact that the q_i-type sinks can be of various sizes and distributed in the sizes with the function $f_{q_i}(R_{q_i})$, as well as the value $I^j_{q_i}(R_{q_i})$ of the adsorption rate, one can write the contribution of q_i sinks into the swelling rate of the matrix in the form

$$\dot{\gamma}_{Mq_i} = \int \Omega_i^{q_i}(R_{q_i}) \left[I^i_{q_i}(R_{q_i}) - I^v_{q_i}(R_{q_i}) \right] f_{q_i}(R_{q_i}) dR_{q_i} \,. \tag{8.12}$$

For the R_{q_v}-dimension q_v sinks formed from the vacancies, e.g., this may be voids, the vacancy dislocation loops segregate adsorbing vacancies, the volume relaxation is combined from the volume relaxation of the constituent vacancies, $\Delta\Omega_v^{q_v}(R_{q_v})$. Hence, the occurring interstitials decrease this relaxation. Consequently, if one uses the distribution function for such sinks in their dimensions, $f_{q_v}(R_{q_v})$, and the expression (6.35) for the rate of defect adsorption $I^j_{q_v}$, the contribution of such sinks into the swelling rate can be written as

$$\dot{\gamma}_{Mq_v} = -\int \Delta\Omega_v^{q_v}(R_{q_v}) \left[I^v_{q_v}(R_{q_v}) - I^i_{q_v}(R_{q_v}) \right] f_{q_v}(R_{q_v}) dR_{q_v} \,. \tag{8.13}$$

If one assumes that $\Omega_i^{q_i}(R_{q_i})$ and $\Delta\Omega_v^{q_v}(R_{q_i})$ do not depend on the R_{q_j}-sink sizes, one then obtains instead of (8.12) and (8.13), using (8.6), that

$$\dot{\gamma}_{Mq_i} = \Omega_i^{q_i} \int \left[I^i_{q_i}(R_{q_i}) - I^v_{q_i}(R_{q_i}) \right] f_{q_i}(R_{q_i}) dR_{q_i} = \Omega_i^{q_i} \dot{\xi}^i_{q_i} \,,$$

$$\dot{\gamma}_{Mq_v} = -\Delta\Omega_v^{q_v} \int \left[I^v_{q_v}(R_{q_v}) - I^i_{q_v}(R_{q_v}) \right] f_{q_v}(R_{q_v}) dR_{q_v} = -\Delta\Omega_v^{q_v} \dot{\xi}^v_{q_v} \,, \tag{8.14}$$

where

$$\dot{\xi}^j_{q_j} = \int \left[I^j_{q_j}(R_{q_j}) - I^k_{q_j}(R_{q_j}) \right] f_{q_j}(R_{q_j}) dR_{q_j} \,, \quad j \neq k. \tag{8.15}$$

The free vacancy contribution, from (8.15), has the form

$$\dot{\gamma}_{Mv} = -\Delta\Omega_v \dot{\xi}_v \,. \tag{8.16}$$

When assembling the contributions of various phases (8.8) and sinks (8.10), (8.14)-(8.16) into the expression (8.8), in view of the fact that (8.9) takes into account the j defects landing into the forming P_j segregates from the matrix, one obtains

$$\dot{\gamma}=\nu_{P_v}\left\{-\Delta\Omega_v^{P_v}\left[g_v^{P_v}+(1-\nu_{P_i}-\nu_{P_v})\nu_{P_v}^{-1}\dot{\xi}_{P_v}^{v}\right]+\left[\Omega+\Delta\Omega_i^{P_v}\right]g_i^{P_v}\right\}+$$

$$\nu_{P_i}\left\{\left[(\Omega+\Delta\Omega_i^{P_i}\right]\left[g_i^{P_i}+(1-\nu_{P_i}-\nu_{P_v})\nu_{P_i}^{-1}\dot{\xi}_{P_i}^{i}\right]-\Delta\Omega_v^{P_i}g_v^{P_i}\right\}+ \tag{8.17}$$

$$(1-\nu_{P_i}-\nu_{P_v})\left[\Omega\left(\dot{\xi}_D^i+\dot{\xi}_\Lambda^i\right)+\left[\Omega+\Delta\Omega_i^{Q}\right]\dot{\xi}_Q^i+\left[\Omega+\Delta\Omega_i^{L_i}\right]\xi_{L_i}^i-\Delta\Omega_v\dot{\xi}_v-\right.$$

$$\left.\Delta\Omega_v^{\mathrm{V}}\dot{\xi}_{\mathrm{V}}^v-\Delta\Omega_v^{L_v}\dot{\xi}_{L_v}^v\right].$$

Make use of the balance relation (8.7) and obtain the expression

$$\dot{\xi}_v=\dot{\xi}_D^i+\dot{\xi}_\Lambda^i+\dot{\xi}_\vartheta^i+\dot{\xi}_{L_i}^i+\dot{\xi}_{P_i}^i-\dot{\xi}_{\mathrm{V}}^v-\dot{\xi}_{L_v}^v-\dot{\xi}_{P_v}^v-$$

$$\frac{\nu_{P_v}\left(g_v^{P_v}-g_i^{P_v}\right)+\nu_{P_i}\left(g_v^{P_i}-g_i^{P_i}\right)}{1-\nu_{P_i}-\nu_{P_v}}. \tag{8.18}$$

Substituting (8.18) into (8.17), one obtains the following expression for the swelling rate of a heterogeneous material:

$$\dot{\gamma}=(1-\nu_{P_i}-\nu_{P_v})\left\{(\Omega-\Delta\Omega_v)\left[\dot{\xi}_D^i+\dot{\xi}_\Lambda^i\right]+\left[\Omega+\Delta\Omega_i^{Q}-\Delta\Omega_v\right]\dot{\xi}_Q^i+\right.$$

$$\left[\Omega+\Delta\Omega_i^{L_i}-\Delta\Omega_v\right]\dot{\xi}_{L_i}^i+\left[\Omega+\Delta\Omega_i^{P_i}-\Delta\Omega_v\right]\dot{\xi}_{P_i}^i-\left[\Delta\Omega_v^{\mathrm{V}}-\Delta\Omega_v\right]\dot{\xi}_{\mathrm{V}}^v-$$

$$\left(\Delta\Omega_v^{L_v} - \Delta\Omega_v\right)\dot{\xi}_{L_v}^v - \left(\Delta\Omega_v^{P_v} - \Delta\Omega_v\right)\dot{\xi}_{P_v}^v - (1-\nu_{P_i}-\nu_{P_v})^{-1}\nu_{P_v}\times \tag{8.19}$$

$$\left[g_v^{P_v}\left(\Delta\Omega_v^{P_v} - \Delta\Omega_v\right) - g_i^{P_v}\left(\Omega - \Delta\Omega_v + \Delta\Omega_i^{P_v}\right)\right] + \nu_{P_i}(1-\nu_{P_i}-\nu_{P_v})^{-1}\times$$

$$\left[g_i^{P_i}\left(\Omega + \Delta\Omega_i^{P_i} - \Delta\Omega_v\right) - g_v^{P_i}\left(\Delta\Omega_v^{P_i} - \Delta\Omega_v\right)\right].$$

In case if formed in the material are only coherent vacancy-adsorbing precipitates ($P_v = P$, $\nu_{P_i} = 0$, $\nu_{P_v} = \nu_P$, $g_i^{P_v} = g_v^{P_v} = g^P$), i.e. if the segregates P are compressed, the expression (8.19) takes the form

$$\dot{\gamma} = (1-\nu_P)\left\{(\Omega - \Delta\Omega_v)\left(\dot{\xi}_D^i + \dot{\xi}_\Lambda^i\right) + \left(\Omega + \Delta\Omega_i^Q - \Delta\Omega_v\right)\dot{\xi}_Q^i + \right.$$

$$\left(\Omega + \Delta\Omega_i^{L_i} - \Delta\Omega_v\right)\dot{\xi}_{L_i}^i - \left(\Delta\Omega_v^{V} - \Delta\Omega_v\right)\dot{\xi}_V^v - \left(\Delta\Omega_v^{L_v} - \Delta\Omega_v\right)\dot{\xi}_{L_v}^v - \tag{8.20}$$

$$\left.\left(\Delta\Omega_v^P - \Delta\Omega_v\right)\left(\frac{\nu_P}{1-\nu_P}g^P + \dot{\xi}_P^v\right) + \frac{\nu_P}{1-\nu_P}g^P\left(\Omega + \Delta\Omega_i^P - \Delta\Omega_v\right)\right\}.$$

The last summand in (8.20) describes the effect of outgoing the interstitials from the compressed precipitate P into the solid solution.

If, in addition: 1) one ignores the relaxations $\Delta\Omega_j^a$ as compared to the atomic volume Ω, and 2) one assumes equal the relaxation of vacancy volumes in the voids, loops and in the free state $\Delta\Omega_v^V \simeq \Delta\Omega_v^{L_v} \simeq \Delta\Omega_v$, then the following will be valid:

$$\dot{\gamma} \simeq (1-\nu_P)\Omega\left(\dot{\xi}_D^i + \dot{\xi}_\Lambda^i + \dot{\xi}_Q^i + \dot{\xi}_{L_i}^i + \frac{\nu_P}{1-\nu_P}g^P\right) - \tag{8.21}$$

$$\left(\Delta\Omega_v^P - \Delta\Omega_v\right)\left[\nu_P g^P + (1-\nu_P)\dot{\xi}_P^v\right].$$

Substituting the expressions (8.11) and (8.15) into (8.21) and leaving out of account the role of $q=\Lambda$ sinks (surfaces), one obtains

$$\dot{\gamma} = (1-v_P)\Omega\left\{\rho_D(K_D^i-K_D^v)+\int\left[I_Q^i(R_Q)-I_Q^v(R_Q)\right]f_Q(R_Q)dR_Q + \right.$$

$$\left.\int\left[I_{L_i}^i(R_{L_i})-I_{L_i}^v(R_{L_i})\right]f_{L_i}(R_{L_i})dR_{L_i}\right\} + \qquad (8.22)$$

$$v_P g^P\Omega-\left[\Delta\Omega_v^P-\Delta\Omega_v\right]\left[v_P g^P+(1-v_P)\int I_P^v(R_P)f_P(R_P)dR_P\right].$$

It is seen from the formulas (8.20)-(8.22): (1) the influence of the secondary-phase segregates on the swelling rate of the material, (2) the relation of the swelling process to such sink characteristics as their density ρ_D, the size-distri- butions $f_q(R_q)$, the rates $I_q^j(R_q)$, and the efficiencies K_D^j of the point defect adsorption.

In particular, the preferential factors α_D^j (see (6.13)) or α_{D0}^j (for the material without precipitates, (1.35)) vary in inverse proportion to $\ln(2R/R_0^j)$. In this case, $R_0^j \sim |\Delta\Omega_j|$, i.g. to the volume relaxations of j defects (see (1.35)). The denser the atomic stacking in the crystal lattice is the more will be the values $|\Delta\Omega_j|$. Therefore, a decrease in $\Delta\Omega_j$ will give rise to a decrease in R_0^j and α_D^j , and hence, in accordance with the expressions (8.22) and (6.36), to a decrease in both swelling rate $\dot{\gamma}$ and the value γ. As we are known (see, e.g., Refs. 1-7) that the materials with a body-centered lattice (more loose ones) swells to a lesser extent than those with a face-centered lattice (more dense ones), i.e. $\Delta\Omega_j^{fcc}>\Delta\Omega_j^{bcc}$, then a correct account for the defect relaxations $\delta\Omega_j$ in these lattices is likely to give more exact estimates for the values of their swelling.

Besides, the formula (8.22) in the combination with the expressions (1.38)-(1.40) shows a close relation of the calculations to those for the values K_D^j, $I^j(R_q)$, $f_q(R_q)$, and $R_q(t)$. A self-consistent procedure of such a calculation is persecuted in Diagram of Fig. 2.7 (Level III), where by the block "property" is meant j in the form (8.22).

Thus, the expressions (8.20)-(8.22) have been obtained to calculate the radiation swelling rate of a heterogeneous material, which can be employed in case when pre-evaluated have been the values of the adsorption rates and efficiencies of the defect adsorption by the sinks, those of the sink radii and the functions describing their distribution in sizes.

8.2. Influence of Coherent Precipitates on the Radiation Swelling of Decomposing Solid Solutions

Chapter 2 (Section 2.5) presents the experimental results which show the influence of the coherent precipitates on decreasing the radiation swelling in a homogeneous continuous decomposition of solid solutions as a result of continuous solid solution decomposition due to the anomalous recombination of opposite point defects. Formulate a model of the anomalous recombination and evaluate the rate of radiation swelling basing on the above mentioned formula (8.22).

Considering a solid solution decomposition under irradiation where the coherent secondary-phase precipitates of R_P radius with the deformation $\varepsilon_I>0$ drop out in the volume between the "far" sinks [92-95, 410, 413-418]. The decay goes with a positive volume effect, since the impurity atomic radius r_a is more than that of a matrix atom. Since, according to Refs. 295-300, the largest effects take place at the stage of forming the precipitates, the best plan to be followed is to correlate r_a and r_0 rather than the ratios of specific volumes of the matrix and the segregates with various crystal lattices.

Neglect the elastic anisotropy (and that of segregate shape) and assume, as above (Chapter 2, Section 2.6) that the internal stresses in the compressed precipitates σ_P^{I} are constant (see (2.22)), and for an extended matrix, we have $\mathrm{Sp}\sigma_P^{II}<0$, for σ_P the equilibrium (2.24) (see Fig. 2.9) being fulfilled on average over the volume. Until the coherency loss has taken place on the precipitate surface in the process of solid solution decay, there are, around them, essential tangent stresses. The role of internal the stresses in the mechanism of anomalous recombination is characterized by the following features:

(a) the local elastic fields σ_P the more, the more is the difference between the matrix atomic radii r_0 and those of the precipitated impurity r_a (dimensional incompatibility);

(b) the stresses σ_P are the greatest ones at the stage of precipitate formation, i.e. the effect manifests itself more essentially in alloys with a long incubation period and increases when the number of precipitate nucleation loci increase;

(c) the influence on the point defect fluxes toward the dislocations and other "far" sinks the more drastically, the more homogeneous are distributed the segregates over the matrix volume, i.e. the more homogeneous is the decay.

The opposite irradiation-generated point defects, as shown in various calculated approximations of Chapter 4, behave differently in the compressed (regions I in Fig. 2.9) and expanded matrix (regions II). The precipitate are dominantly gone out by the interstitials, and the matrix by the vacancies which are going into the forming precipitates, but with a lower high velocity, as the vacancies are less mobile. In this process of matrix irradiation, the supersaturation over the interstitials proves higher than that over the vacancies. In this case the diffusion fluxes of interstitials toward the "far" (structural) sinks (dislocations, grain boundaries, surface) are lowered (see, e.g., (6.19) and Figs. 6.1 and 6.2) as a result of an occasional rearrangement of opposite point defect concentrations, which favors the occurrence of additional small-sized interstitial-type clusters in the matrix volume and their ensuing recombination with the slowly moving vacancies. A change in the supersaturation conditions for the opposite defects in the presence of precepitates against their absence should be exerted on the sink-distribution functions $f_q(R_q)$.

Due to the elevated concentration of the interstitials (both free and in the makeup of clusters) in a solid solution (against the solid solution without precipitates), the probability of their recombination with the vacancies increases. It is the internal precipitate-generated stresses of a wanted sign available in the decomposing solid solution that leads to the anomalous recombination of the opposite defects. This effect is immediately manifests itself in the suppression of radiation swelling, i.e. the decreased interstitial outgoing to the structural ("far") matrix sinks: the dislocations, grain boundaries, surface.

Let in the precipitateless case (v_P=0) there be, in a solid solution, the structural ("far") sinks in the form of dislocations (q=D), and the "neighboring" sinks - voids (q=V) - be formed under irradiation, as well as, the interstitial dislocation loops (q=L_i), i.e. we have for the sink-strength sum k^2_{j0}

$$k^2_{j0}=S^j_{D0}+S^j_{V0}+S^j_{L_i0} \tag{8.23}$$

(the index "0" means the case v_P=0).

By reference to the definitions (2.19), (2.20) of the sink strengths S^j_q and the expressions (6.35), (6.36) obtained for the rates $I^j_q(R_q)$ and the efficiencies $K^j_D(r_*)$ of j-point defect adsorption, we rewrite (8.23) in the form

$$k^2_{j0}=\alpha^j_{D0}\rho_D\left[1+\frac{\kappa^j_{D0}}{C^+_{j0}}\right]+\alpha^j_{V0}\bar{R}^0_V C^0_V\left[1+\frac{\kappa^j_{V0}}{C^+_{j0}}\right]+$$

$$\alpha^j_{L_i0}R^0_{L_i}C^0_{L_i}\left[1+\frac{\kappa^j_{L_i0}}{C^+_{j0}}\right], \tag{8.24}$$

where $\bar{R}^0_q$ and C^0_q are the average radius and the concentration of sinks in the v_P=0 material.

Using the expression (4.24) for the stationary supersaturations $C^+_j=\hat{C}^+_j-C^e_j$, the relations (6.35), (6.36), (8.24) and taking into account the absence of Q and P sinks, we obtain, from (8.23), a known expression for the swelling rate in a pure metal:

$$\dot{\gamma}_0=\Omega\left\{\rho_D(K^i_D-K^v_D)+\int\left[I^i_{L_i}(R_{L_i})-I^v_{L_i}(R_{L_i})\right]f_{L_i}(R_{L_i})dR_{L_i}\right\}=$$

$$\Omega\bar{R}^0_V C^0_V\left[\alpha^v_{V0}D_v C^+_{v0}\left(1+\frac{\kappa^v_{V0}}{C^+_{v0}}\right)-\alpha^i_{V0}D_i C^+_{i0}\left(1+\frac{\kappa^i_{V0}}{C^+_{i0}}\right)\right]. \qquad (8.25)$$

Consequently, the particle balance (see (8.5)) offers, in such simplified a scheme, an ordinary formula for $\dot{\gamma}_0$ which allows to evaluate the swelling from the number of the uncompensated vacancies assembled together as vacancy voids. This is precisely a balance which has formed the basis of the integration of such terms as "vacancy swelling" and "vacancy transformation" into the united "vacancy swelling" term. As seen from (8.20) and (8.22), the expression (8.25) is actually valid in only special cases.

Now we consider the case of a solid solution of interest to us into where the coherent precipitates with $\nu_P\neq0$ fall out.

Let there be the following sinks in the material: dislocations ($q=D$), precipitates ($q=P$), interstitial dislocation loops ($q=L_i$), interstitial clusters ($q=Q$), voids (q=V). We have thus for the sink-strength sums

$$k^2_i=S^i_D+S^i_{L_i}+S^i_Q+S^i_V;\quad k^2_v=S^v_D+S^v_{L_i}+S^v_Q+S^v_V+S^v_P. \qquad (8.26)$$

By reference to the expressions (2.19), (2.20), (6.35), (6.36), we rewrite

(8.26) in the form

$$k_i^2=\alpha_D^i\rho_D\left[1+\frac{\kappa_D^i}{\tilde{C}_i^+}\right]+\alpha_{L_i}^i\bar{R}_{L_i}C_{L_i}\left[1+\frac{\kappa_{L_i}^i}{\tilde{C}_i^+}\right]+$$

$$\alpha_Q^i\bar{R}_QC_Q\left[1+\frac{\kappa_Q^i}{\tilde{C}_i^+}\right]+\alpha_V^i\bar{R}_VC_V\left[1+\frac{\kappa_V^i}{\tilde{C}_i^+}\right]; \tag{8.27}$$

$$k_v^2=\alpha_D^v\rho_D\left[1+\frac{\kappa_D^v}{\tilde{C}_v^+}\right]+\alpha_{L_i}^v\bar{R}_{L_i}C_{L_i}\left[1+\frac{\kappa_{L_i}^v}{\tilde{C}_v^+}\right]+$$

$$\alpha_Q^v\bar{R}_QC_Q\left[1+\frac{\kappa_Q^v}{\tilde{C}_v^+}\right]+\alpha_V^v\bar{R}_VC_V\left[1+\frac{\kappa_V^v}{\tilde{C}_v^+}\right]+\alpha_P^v\bar{R}_PC_P\left[1+\frac{\kappa_P^v}{\tilde{C}_v^+}\right]. \tag{8.28}$$

Substituting (6.35), (6.36) into (8.22) and using (8.27), (8.28), we obtain for the swelling rate of a solid solution with falling-out precipitates as follows:

$$\dot{\gamma} = (1-v_P)\Omega\Big\{g_i-g_v-\frac{\Delta\Omega_v^P-\Delta\Omega_v}{\Omega}\,\frac{v_P}{1-v_P}g^P+$$

$$\alpha_P^v\bar{R}_PC_PD_v\tilde{C}_v^+\left[1+\frac{\kappa_P^v}{\tilde{C}_v^+}\right]\left[1-\frac{\Delta\Omega_v^P-\Delta\Omega_v}{\Omega}\right]+ \tag{8.29}$$

$$\bar{R}_VC_V\left[\alpha_V^vD_v\tilde{C}_v^+\left[1+\frac{\kappa_V^v}{\tilde{C}_v^+}\right]-\alpha_V^iD_i\tilde{C}_i^+\left[1+\frac{\kappa_V^i}{\tilde{C}_i^+}\right]\right],$$

where $g_i-g_v=v_Pg^P(1-v_P)^{-1}$.

Comparing $\dot{\gamma}$ by (8.29) and $\dot{\gamma}_0$ by (8.25), we see that the deterioration (or suppression) in swelling can take place in the model involved merely by way of decreasing the number of vacancies in voids. In its turn, this can be reached by raising the vacancy recombination with the interstitials both free and collected into clusters (e.g., when $q=Q$).

8.3. Estimation of the Radiation Swelling Rate in a Solid Solution with Coherent Precipitates

To calculate the ratio $\dot{\gamma}/\dot{\gamma}_0$ it is necessary in both cases ($v_P=0$ and $v_P\neq0$) to accomplish a self-consistent calculation procedure (see Fig. 2.7) for the purpose of excluding the dependencies on k^2_{j0} and k^2_j from C^+_{j0} and $\tilde{C}^+_j$. To this end the functional relations $\bar{R}^0_q C^0_q$ and $\bar{R}_q C_q$ should be found to the supersaturations C^+_{j0} and $\tilde{C}^+_j$. This procedure is rather complicated to be performed analytically.

As seen from Fig. 2.7, one needs to have for such a calculation some expressions for the dimensions and the q-sink size-distribution functions. The equations for the sink dimensions R_q have the form (7.5). Now, e.g. for the voids, we obtain the following equation:

$$\frac{dR_V}{dt}=\frac{h_V}{R_V}; \quad h_V=(\Omega-\Delta\Omega^I_{vV})\left[D_v(\tilde{C}^+_v+\kappa^v_V)-D_i(\tilde{C}^+_i+\kappa^i_V)\right], \tag{8.30}$$

which gives (see (7.6))

$$R_V(t)=\sqrt{R^2_{V0}+2h_V(t-t_0)}. \tag{8.31}$$

Here R_{V0} is the initial radius of void nuclei in the instant t_0 corresponding to the beginning of the irradiation. For the rates of interstitial dislocation loop growth, the expression

$$\frac{dR_L}{dt}=Q_L-\frac{h_L}{R_L}, \qquad (8.32)$$

was obtained in Ref. 231, where the values Q_L and h_L can be improved against those in Ref. 231 using (6.35) in the form

$$Q_L=\frac{\Omega_{iL}}{b}\left[\alpha_L^i D_i(\tilde{C}_i^+ +\kappa_L^i)-\alpha_L^v D_v(\tilde{C}_v^+ +\kappa_L^v)\right],$$

$$h_L=\frac{G\Omega_v^2 D_v \tilde{C}_v^e}{2(1-\nu)kT}\,\frac{\ln\frac{R_L}{r_*}+\zeta'}{\ln\frac{R_L}{R_L^*}},$$

where ζ' takes into account the energy of the dislocation loop [231], $R_L^*=(3\pi kT)^{-1}\Delta\Omega_v^I Gb(1+\nu)(1-\nu)^{-1}$. If the condition

$$\frac{h_L}{Q_L(R_L-R_{L0})}\ln\frac{Q_L R_L-h_L}{Q_L R_{L0}-h_L}<1,$$

is fulfilled, then, integrating (8.32), we obtain

$$R_L(t)=R_{L0}+Q_L(t-t_0). \qquad (8.33)$$

where R_{L0} is the initial radius of the loops in the instant t_0.

For the coherent precipitates, one can use the dependencies obtained in Chapter 7 selecting the expressions (2.27) as thermodynamically equilibrium defect concentrations. Now with the proviso that $2R_P/L_P\ll 1$ we have: for small times, the expression for the radius R_P has the form (7.16), and for long times the form(7.13).

It can be observed from the formulas (8.22) and (8.25) for the swelling rate that the distribution functions of sinks $f_q(R_q)$ appear in the expressions for $\dot{\gamma}$ (or $\dot{\gamma}_0$) in the integral expression with respect to dR_q in the form of products with the adsorption rates $I^j_q(R_q)$. Write one of such integrals, using the expression (6.35):

$$\int f_q(R_q)I^j_q(R_q)dR_q=\alpha^j_q D_j(C^+_j+\kappa^j_q)\int f_q(R_q)R_q dR_q =$$
$$\alpha^j_q D_j(C^+_j+\kappa^j_q)\bar{R}_q C_q. \tag{8.34}$$

Thus, to calculate the swelling rate, it is necessary to know the products $\bar{R}_q C_q$.

We have the equation (2.18) to determine the distribution functions $f_q(R_q)$. Its solution can be written in the form

$$f_q(R_q)=\exp\left[-\int_{t_0}^{t} P_q(t')dt'\right]\left[f_q(R_q,t_0)+\right.$$
$$\left.\int_{t_0} W_q(t')\exp\left[\int_{t_0}^{t'} P_q(\tau)d\tau\right]dt'\right], \tag{8.35}$$

where $P_q(t)=d(dR_q/dt)/dR_q$. The most intricate question is to preset the rates W_q of occurring the q sinks. Assume for our estimates that the voids, loops and precipitates are nucleated of only one dimension R_{q0} each by a heterogeneous way on the nucleation centers whose concentration is C_{3q}. The rates of sink occurrence can now be presented as follows:

$$W_q=\left[I^{j+}_q(R_q)-I^{j-}_q(R_q)\right]\frac{C_{3q}}{R_q}\,\delta\left(\frac{R_q}{R_{q0}}-1\right), \tag{8.36}$$

where, as in (2.17), the indexes j^+ and j^- signify the j-type defects which lead to a growth of q sink or its decrease, respectively. With a knowledge of the growth rates dR_q/dt of the R_q-sized sinks one can find the expressions for the sink-distribution functions, and then, using (8.34), express the sought-for products $\bar{R}_q C_q$, which yields

$$\bar{R}_V C_V = \alpha_V C_{3V} R_{V0} \frac{R^3_{Vm} - R^3_{V0}}{\Omega - \Delta\Omega^v_V} \equiv \Psi_V, \tag{8.37}$$

$$\bar{R}_L C_L = \Psi_L \left[1 + \frac{k^2_i k^2_v}{\varphi_L}\right] \left[\alpha^i_L k^2_v (1+\beta_i) - \alpha^v_L k^2_i\right], \tag{8.38}$$

where R_{qm} is a maximum value of the q-sink radius, and

$$\beta_i = \frac{\delta_i}{g}; \qquad \Psi_L = bC_{3L} R_{L0} (R^2_{Lm} - R^2_{L0})(2\Omega^i_L)^{-1};$$

$$\varphi_L = \Omega^i_L g (R_{lm} - R_{L0})(2h_L b)^{-1}.$$

For the compressed precipitates, one can obtain

$$\bar{R}_P C_P \simeq \frac{5\nu_P}{4\pi R^2_{Pm}} = \nu_P \Psi_P. \tag{8.39}$$

When used is the effective medium method, the sink strengths S^j_q (see Chapter 2) are determined by the expressions (2.19), (2.20). It is to be noted that such a definition of sink strengths means that a R_q-sized sink of q type at an arbitrary point of the crystal is adequate to an identical sink at any other point of the crystal. Such a relation takes place merely in the case of homogeneous sink distribution over the crystal volume, which, in its turn, is a consequence of the homogeneity in the intrinsic point defect distribution.

To derive the equation determining k_j^2, the definitions (2.19), (2.20) need to be used into whose right-hand side the expression for K_D^j by (6.36) and $I_q^j(R_q)$ by (6.35) must be substituted with allowance for (8.34). Then we derive

$$S_D^j = \alpha_D^j \rho_D \left[1 + \frac{\kappa_D^j}{\tilde{C}_j^+} \right], \tag{8.40}$$

$$S_q^j = \alpha_q^j \bar{R}_q C_q \left[1 + \frac{\kappa_q^j}{\tilde{C}_j^+} \right]. \tag{8.41}$$

Using the definition (1.26) for the sink-strength sum or the expression (2.21), we obtain a system of equations for k_j^2 in the form

$$k_j^2 = \alpha_D^j \rho_D \left[1 + \frac{\kappa_D^j}{\tilde{C}_j^+(k_j^2)} \right] + \sum_{q=1}^{n} \alpha_q^j \bar{R}_q C_q \left[1 + \frac{\kappa_q^j}{\tilde{C}_j^+(k_j^2)} \right], \quad q \neq D. \tag{8.42}$$

The unknown k_j^2 appears on the right-hand side of equation (8.42): first, in $\tilde{C}_j^+$ (see, e.g., (4.24)), and secondly, in $\bar{R}_q C_q$ (see (8.37)-(8.39)) through $\tilde{C}_j^+$. Considering that $\tilde{C}_j^+$ are large as compared with κ_q^j and using the expressions (8.37)- (8.39), we simplify (8.42), which gives

$$k_j^2 = k_{j*}^2 + \alpha_L^j \bar{\Psi}_L, \tag{8.43}$$

$$\bar{\Psi}_L = \Psi_L \varphi_L \frac{k_i^2 k_v^2}{\alpha_L^i k_v^2 - \alpha_L^v k_i^2 + \beta_i \alpha_L^i k_v^2}, \tag{8.44}$$

where

$$k_{i*}^2 = \alpha_D^i \rho_D + \alpha_L^i \Psi_L + \alpha_V \Psi_V; \quad k_{v*}^2 = \alpha_D^v \rho_D + \alpha_L^v \Psi_L + \alpha_V \Psi_V + \alpha_P \Psi_P. \tag{8.45}$$

Transform (8.43)-(8.45) and derive the equations

$$k_i^2 = k_{i*}^2 + \frac{\alpha_L^i}{\alpha_L^v}(k_v^2 - k_{v*}^2), \tag{8.46}$$

$$y = y_* - H\frac{y(1-y)}{1+\beta_i y}, \tag{8.47}$$

where

$$y = \alpha_L^i k_v^2 (\alpha_L^i k_{i*}^2 - \alpha_L^v k_{v*}^2), \qquad H = \Psi_L \varphi_L^{-1},$$

$$y_* = y(k_v^2 \to k_{v*}^2).$$

Solving (8.46) and (8.47) simultaneously, we have

$$k_v^2 = k_{v*}^2 \left[1 + 0.25(H + 3\beta_i)\right],$$

$$k_i^2 = k_{i*}^2 \left[1 + 0.25(H + 3\beta_i)\frac{\alpha_L^v k_{i*}^2}{\alpha_L^i k_{v*}^2}\right]. \tag{8.48}$$

In the case without precipitations ($v_P=0$, $\beta_i=0$), the following substitutions should be made: $k_j^2 \to k_{j0}^2$; $k_{j*}^2 \to k_{j*0}^2$, $H=H_0$, where

$$k_{j*0}^2 = \alpha_{D0}^j \rho_D + \alpha_L^v \Psi_{L0} + \alpha_V \Psi_V \tag{8.49}$$

and Ψ_{q0} and α_{D0}^j are the values Ψ_q and α_D^j when leaving out of account are the precipitate. Assuming that the dislocations are screened by the precipitates, and

$$\Psi_L \simeq \Psi_{L0}, \quad \Psi_V \simeq \Psi_{V0}, \quad H \simeq H_0,$$

one obtains the following approximate expressions for the sink-strength sums from (8.48), (8.49):

$$k_v^2 = k_{v0}^2 \left[1 + \nu_P \left(\frac{3g^I}{4g} + \frac{\alpha_P \Psi_P}{k_{v*0}^2} \right) \right],$$

$$k_i^2 \simeq k_{i0}^2 \left[1 - \nu_P \left(\frac{\alpha_{D0}^i \rho_D}{k_{i*0}^2} - \frac{3g^I}{g} \frac{\alpha_L^i k_{v*0}^2}{\alpha_L^v k_{i*0}^2} \right) \right]. \tag{8.50}$$

Now we substitute (8.50) into (4.24) and have

$$\tilde{C}_i^+ \simeq C_{i0}^+ \left\{ 1 + \nu_P \left[\frac{\alpha_{D0}^j \rho_D}{k_{i*0}^2} + \frac{g^I}{g} \left(1 - \frac{3\alpha_L^i k_{v*0}^2}{4\alpha_L^v k_{i*0}^2} \right) \right] \right\},$$

$$\tilde{C}_v^+ \simeq C_{v0}^+ \left[1 - \nu_P \left(\frac{3g^I}{4g} + \frac{\alpha_P \Psi_P}{k_{i*0}^2} \right) \right], \tag{8.51}$$

where $C_{j0}^+ = g(D_j k_{j0}^2)^{-1}$ (see (1.29a)).

With a knowledge of the expressions (8.51) for the j-defect concentrations, one can now write a specific form of the efficiencies K_D^j and the rates I_q^j of j-defect adsorption, substitute them into the expression (8.22), isolating the expression for the swelling rate $\dot{\gamma}_0$ in a material without precipitate (in the form (8.25)), and obtain

$$
\begin{aligned}
\frac{\dot{\gamma}}{\dot{\gamma}_0} = (1-\nu_P)\Bigg\{ 1 - & \frac{\Omega \alpha^i_{D0} \rho_D D_i C^+_{i0}}{\dot{\gamma}_0} \left[1 - \nu_P \frac{\alpha^v_{D0} D_v C^+_{v0}}{\alpha^i_{D0} D_i C^+_{i0}} \left(\frac{3g^{\mathrm{I}}}{g} + \frac{\alpha_P \Psi_P}{k^2_{v*0}} \right) \right] + \\
& \frac{\Omega \alpha^i_L \bar{R}^0_L C^0_L D_i C^+_{i0}}{\dot{\gamma}_0} \Bigg[\frac{\alpha^i_{D0} \rho_D}{k^2_{i*0}} + \\
& \nu_P \left[\frac{\alpha^v_L D_v C^+_{v0} \alpha_P \Psi_P}{\alpha^i_L D_i C^+_{i0} k^2_{v*0}} + \frac{g^{\mathrm{I}}}{g} - \frac{3 g^{\mathrm{I}} \alpha^i_L k^2_{v*0}}{4 g \alpha^v_L k^2_{i*0}} + \frac{3}{4} \frac{g^{\mathrm{I}} \alpha^v_L D_v C^+_{v0}}{g \alpha^i_L D_i C^+_{i0}} \right] \Bigg] - \qquad (8.52) \\
& \nu_P \frac{\Delta\Omega^P_v g^{\mathrm{I}}}{\dot{\gamma}_0} \Bigg[\frac{1 - \dfrac{\Delta\Omega_v}{\Delta\Omega^P_v}}{1 - \nu_P} + \\
& \left(1 - \frac{\Delta\Omega_v}{\nu_P \Delta\Omega^P_v} \right) \frac{\alpha_P \Psi_P D_v C^+_{v0}}{g^{\mathrm{I}}} \left(1 - \frac{3}{4} \nu_P \frac{g^{\mathrm{I}}}{g} - \nu_P \frac{\alpha_P \Psi_P}{k^2_{v*0}} \right) \Bigg] \Bigg\}.
\end{aligned}
$$

one can see from the expression (8.52) the role of both dominant sinks being radiation-swelling dangerous dislocations and dislocation loops of interstitial type and the secondary- phase precipitate effect. The value of the ratio $\dot{\gamma}/\dot{\gamma}_0$ should fall with increasing the precipitate volume fraction, as well as with decreasing the interstitial fluxes toward the dislocations as a consequence of rearrangement in the precipitate elastic-stress fields, which is evident from (8.52).

Now we evaluate the value of the swelling-rate ratio $\dot{\gamma}/\dot{\gamma}_0$. The formula (8.52) is rather cumbersome for this purpose. Suppose, though, that the dislocation loops are more effective sinks than the voids and dislocations, i.e. $\alpha^j_L\Psi_{L0} > \alpha_V\Psi_{V0} > \alpha^j_{D0}\rho_D$. Such supposition may be warranted if the dislocation density remains the same for both material without the precipitates and the material with a coherent precipitate, i.e. in these cases the diffusion processes are developing on the same dislocation background. Then $k^2_{j*0} \simeq \alpha^j_L\Psi_{L0}$. It can be shown that $H_0 \to H \to 0$. Assuming that the dislocations are almost completely shielded from the interstitials by the segregates, one obtains a more simple expression

$$\frac{\dot{\gamma}}{\dot{\gamma}_0} \simeq (1-\nu_P)\left[\frac{\nu_P}{(1-\nu_P)}\,\frac{\Delta\Omega^0_v g^I \alpha^i_L\Psi_{L0}}{\Omega g \alpha^i_{D0}\rho_D} - \left(1+\frac{\Delta\Omega^i_L}{\Omega-\Delta\Omega_v}\right)\right] \tag{8.53}$$

instead of (8.52). Here it was taken into account that $\alpha_P\Psi_P \ll \alpha^v_L\Psi_{L0}$ and $\alpha^v_{D0} \ll \alpha^i_{D0}$. One can obtain the following estimate for $\alpha^i_L\Psi_{L0}$ from the condition $\dot{\gamma}/\dot{\gamma}_0 < 1$ if use is made of (8.53):

$$\frac{1-\nu_P}{\nu_P}\,\frac{\Omega}{\Delta\Omega^P_v}\,\frac{1-\Delta\Omega_v/\Omega}{1-\Delta\Omega_v/\Omega^P_v}\,\frac{g}{g^I}\,\alpha^i_{D0}\rho_D < \alpha^i_L\Psi_{L0} <$$

$$\frac{1-\nu_P}{\nu_P}\,\frac{\Omega}{\Delta\Omega^P_v}\,\frac{2(1-\Delta\Omega_v/\Omega)}{1-\Delta\Omega_v/\Delta\Omega^P_v}\,\frac{g}{g^I}\,\alpha^i_{D0}\rho_D. \tag{8.54}$$

In order to evaluate the value of the ratio $\dot{\gamma}/\dot{\gamma}_0$ by (8.53), one selects a material with *Ni* parameters (see Chapter 5, Table 5.1). Now when $\alpha^i_L\Psi_{L0} = 1.5\cdot10^{12}\,cm^{-2}$ and satisfies the condition (8.54) and when $\nu_P = 10^{-1}$, one obtains by (8.53) the following value for the ratio involved: $\dot{\gamma}/\dot{\gamma}_0 \simeq 0.18$. The value $\dot{\gamma}/\dot{\gamma}_0$ as a function of ν_P is given in Fig. 8.1.

Therefore, it has been shown that the swelling rate of the decomposing

If the processes of sequential self-consistent expression of the sink strengths $\bar{R}_q C_q$ through $\tilde{C}^+_j$ is not carried out, one can, as in many works (e.g. Refs. 29, 203, 204, 206, 207, 215, 218, 240, 246), evaluate the sink strength sums (8.26) proceeding from the experimental data on the concentrations and the average sink dimensions in the form of $k_i^2 \simeq 8.5 \cdot 10^{11} \text{cm}^{-1}$, $k_v^2 \simeq 7 \cdot 10^{11} \text{cm}^{-2}$. Use of these values at T=600 K and $g \simeq 10^{16} \text{cm}^{-3} \text{s}^{-1}$ for the material with Ni parameters (see above) gives for the swelling rate $\dot{\gamma}$ evaluated by formula (8.22) that the main contribution is given by the first dislocation summand. So, the swelling rate will change depending on how the coherent precipitates influence the efficiencies K_D^j of the point defect adsorption by the dislocations . The expression (6.36) describes the change K_D^j when the volume precipitate fraction is such as demonstrated in Fig. 6.1 and 6.2 of Chapter 6 for the temperatures T=600 and 900 K. The value $\dot{\gamma}_0/\dot{\gamma} \simeq 5$ is also obtained when use is made of the dependency (6.36) for v_P=0.1-0.14.

One may note, by reference to the available experimental data on studying the structure and suppression of the radiation swelling in dispersion-hardening alloys with a developed homogeneous decay (see Chapter 2, Section 2.5 and Refs. 295-300), that, as a rule, neither there are voids observed nor their concentration is small. This correlates with the model developed for the mechanism of opposite radiation defect anomalous recombination. Thus, if there takes place an intense recombination with the vacancies on the small-sized interstitial clusters Q formed in the matrix when decay is in progress, there must be no free vacancies for the voids to be formed, these clusters Q being recombinators. Due to this, we suppose, as a result of Refs. 295-300, that in a decomposing solid solution with $\nu_P \neq 0$: (a) the voids are missing (q=V), (b) the interstitial clusters Q formed are recombinators, i.e. the number of the opposite point j defects precipitated on these sinks Q in a unit time in a unit volume, are the same, hence (see (8.15))

$$\dot{\hat{\xi}}^i_Q = \dot{\hat{\xi}}^v_Q \quad \text{or} \quad \dot{\hat{\xi}}^i_Q = 0. \tag{8.55}$$

Now due to the assumption (a), one should take $\bar{R}_V C_V = 0$ in the formula (8.29), which yields

$$\dot{\gamma} = (1-\nu_P)\Omega\left(1-\frac{\Delta\Omega^P_v - \Delta\Omega_v}{\Omega}\right)\left[\frac{\nu_P}{1-\nu_P} g^P + \alpha^v_P \bar{R}_P C_P D_v \tilde{C}^+_v \left(1+\frac{\kappa^v_P}{\tilde{C}^+_v}\right)\right]. \tag{8.56}$$

To compare the values $\dot{\gamma}_0$ numerically by (8.25) and $\dot{\gamma}$ by (8.56), one assumes that $\Delta\Omega^P_v \simeq \Delta\Omega w_v$, and $\kappa^j_{q0}/C^+_{j0} \simeq \kappa^v_P/C^+_j \to 0$. Now one obtains

$$\frac{\dot{\gamma}_0}{\dot{\gamma}} = \frac{\bar{R}^0_V C^0_V}{\bar{R}_P C_P}\left(1-\frac{k^2_{v0}}{k^2_{i0}}\right)\frac{C^+_{v0}}{C^+_v}\left[1+\nu_P\left(\frac{k^2_v}{\alpha^v_P \bar{R}_P C_P}-1\right)\right]^{-1}. \tag{8.57}$$

The ratio between the sink strength sums k^2_{v0}/k^2_{i0} in the material without segregates is always less than 1, $C^+_{v0}/C^+_v \simeq 1$, and one may show that $k^2_v(\alpha^v_P \bar{R}_P C_P)^{-1} - 1 \simeq 1$. Now we obtain

$$\frac{\dot{\gamma}_0}{\dot{\gamma}} \simeq (1+\nu_P)^{-1}\frac{\bar{R}^0_V C^0_V}{\bar{R}_P C_P}. \tag{8.58}$$

By reference of the data for the material with $\nu_P \neq 0$ from Ref. 247 ($\bar{R}_P \simeq 2\text{-}3\cdot 10^{-7}$cm, $C_P \simeq 10^{16}$cm^{-3}) and for $\bar{R}^0_V C^0_V$, $\bar{R}^0_V \simeq 8\cdot 10^{-7}$ cm, $C^0_V \simeq 2\cdot 10^{16}$cm^{-3}, respectively, as in Refs. 207, 215, 218, 240, we also obtain that $\dot{\gamma}_0/\dot{\gamma}\simeq5$.

The decrease in the swelling derived by the theoretical estimates agrees with the experimental data of Refs. 295-300: observed in dispersion-hardening alloys of various compositions marked by a developed homogeneous precipitation of γ' and β phases of the Ni_3Ti type with the density up to $10^{16}cm^{-3}$, is a high resistance to the radiation swelling ($\gamma\simeq1$-3 %), and in the alloys of 18-8 and 15-15, where the selective phase precipitation takes place and the body volume remains free from the precipitates, a large swelling being seen up to $\gamma_0\simeq20$-50 %.

8.4. Small-Sized Interstitial Clusters as Opposite-Defect Recombinators in Decomposing Solid Solutions

The enhancement of opposite defect recombination can be realized in the materials while some additional recombinators are forming in the processes involved. This is the situation, for example, with dislocation dipoles [419] at deformation loads or with semi-interstitials and semi-vacancies on incoherent boundaries [420]. The most likely in our case are the small-sized interstitial clusters which have not yet turned into the dislocation loops [93,94,421-423]. Let us evaluate the possibility of such recombinators working.

Consider in more detail the assumption (8.55) on the availability of the recombinators $q=Q$. For this purpose, substitute into (8.15) the expressions (6.35) for the adsorption rates $I^j_Q(R_Q)$ of j defects on Q sinks and fulfill the condition (8.55), the result being

$$\alpha^{i}_{Q} D_{i} \tilde{C}^{+}_{i}\left[1+\frac{\kappa^{i}_{Q}}{\tilde{C}^{+}_{i}}\right]\bar{R}_{Q}C_{Q}=\alpha^{v}_{Q} D_{v} \tilde{C}^{+}_{v}\left[1+\frac{\kappa^{v}_{Q}}{\tilde{C}^{+}_{v}}\right]\bar{R}_{Q}C_{Q}. \tag{8.59}$$

Using (4.25), we obtain the condition

$$k^{2}_{v} = \frac{\alpha^{v}_{Q} g_{v}}{\alpha^{i}_{Q} g_{i}} \frac{k^{2}_{i}}{1+\frac{k^{2}_{i}}{\alpha^{i}_{Q} g_{i}}(\kappa^{i}_{Q}\alpha^{i}_{Q}D_{i}-\kappa^{v}_{Q}\alpha^{v}_{Q}D_{v})}. \tag{8.60}$$

If κ^{j}_{Q} is ignored and $\kappa^{j}_{Q}/\tilde{C}^{+}_{j} \ll 1$ assumed, we have

$$k^{2}_{v} \simeq \frac{\alpha^{v}_{Q} g_{v}}{\alpha^{i}_{Q} g_{i}} k^{2}_{i}. \tag{8.60a}$$

With no regard in (8.20) and (8.28) for the values $\kappa^{j}_{q}/\tilde{C}^{+}_{j}$, we assume that the interstitials in the decomposing solid solution do not manage to form interstitial dislocation loops, forming merely the clusters Q [1]. And then we obtain from (8.60a), substituting (8.57) and (8.28), the following relation

$$\bar{R}_{Q}C_{Q}=\frac{g_{i}}{g_{i}-g_{v}}\left[\rho_{D}\frac{g_{v}\alpha^{i}_{D}}{g_{i}\alpha^{i}_{Q}}\left(1-\frac{g_{i}\alpha^{i}_{Q}\alpha^{v}_{D}}{g_{v}\alpha^{v}_{Q}\alpha^{i}_{D}}\right)-\frac{\alpha^{v}_{P}}{\alpha^{v}_{Q}}\bar{R}_{P}C_{P}\right]. \tag{8.61}$$

The effective rates of vacancy and interstitial generation in the solid solution with coherent precipitates are just the ones which are different

---

[1]This assumption does not reflect on the evaluation of $\dot{\gamma}$ by (8.29) and (8.56), is not physically discrepant and allows to simplify the estimates for the clusters Q.

in the values of free defect yield. In our case, the interstitials come out from the compressed precipitates (see (8.28)) into the matrix. This difference δ_i can be had with the use of (4.23):

$$\delta_i = (1-\nu_P)^{-1} \int I_P^{II}(R_P) f_P(R_P) dR_P = \frac{g^P \bar{V}_P C_P}{1-\nu_P} = g^P \frac{\nu_P}{1-\nu_P}. \tag{8.62}$$

The number of interstitials N_Q^i containing in a unit volume of the cluster Q can be evaluated as

$$N_Q^i = C_Q \pi \bar{R}_Q^2 a_0^{-2},$$

the clusters being assumed plane. Then one may put into use the upper boundary for the total interstitial number in a volume unit when the irradiation lasts the time t, in the form $\hat{N}_i = g_i t$. Because of this, the condition $N_Q^i < \hat{N}^i$ should be always fulfilled, which gives

$$\bar{R}_Q < g_i t a_0^2 (\pi \bar{R}_Q C_Q)^{-1}. \tag{8.63}$$

Substitute the relations (8.61) for $\bar{R}_Q C_Q$ into (8.63) and derive with allowance for (8.62)

$$\bar{R}_Q < \frac{a_0^2 (g_i - g_v) t}{\pi \left[\rho_D \frac{g_v \alpha_D^i}{g_i \alpha_D^v} \left(1 - \frac{g_i \alpha_Q^i \alpha_D^v}{g_v \alpha_Q^v \alpha_D^i} \right) - \frac{\alpha_P^v}{\alpha_Q^v} \bar{R}_P C_P \right]}. \tag{8.64}$$

Choosing the same parameters for the material and its defect structure as above, we obtain

$$\bar{R}_Q < \frac{\alpha_Q^i}{\alpha_D^i} \frac{3\cdot 10^{-10} v_P (1-v_P)^{-1} t}{1-(1-v_P)^{-1} \dfrac{\alpha_Q^i \alpha_P^v}{\alpha_Q^v \alpha_D^i} - \dfrac{\alpha_Q^i \alpha_P^v}{5(1-v_P)\alpha_Q^v \alpha_P^i}} . \qquad (8.65)$$

As the clusters Q are recombinators, their preferential factors must fulfill the relation $\alpha_Q^i<\alpha_Q^v$. Such a relation follows, in particular, from the fact of existing the critical size R_Q^C for the cluster nucleus at which size the free cluster energy reaches its maximum value. When $R_Q<R_Q^C$, the free material energy increases with growing the nucleus [183]. The small-sized clusters possess a relatively large surface energy and their long life in the material is energetically useless. If one increases the dissolution rate of such clusters, say, at the cost of an extra force acting in our case on the vacancies on such interstitial cluster side ($\alpha_Q^v>\alpha_Q^i$), then the clusters Q will not reach their critical sizes R_Q^C, and hence, will not be able to metamorphose into steady and energetically useful clusters of large-sized interstitials, i.e. into the dislocation interstitial loops. The arriving vacancies will annihilate of them, and the clusters vanish, thus improving the efficiency of opposite defect recombination. It follows herefrom that the dimensions $\bar{R}_Q$ of such clusters must be not so large, they must be no more than several interstitials. By the way (see Ref. 325, v. 3), the conclusion on the formation of double and triple interstitials was made from the measurements of diffusion X-ray scattering, Huang scattering, and the mechanical relaxation. We drop, in (8.65), the terms in the denominator by virtue of the relations $\alpha_Q^i<\alpha_Q^v$, $\alpha_D^v<\alpha_D^i$, $\alpha_P^v<\alpha_D^i$ and estimate the average dimensions of the interstitial cluster as follows:

$$\bar{R}_Q < \alpha^i_Q(\alpha^v_Q)^{-1}v_P(1-v_P)^{-1}t\cdot 3\cdot 10^{-2}(\AA).$$

Hence, when v_P=0.1 and t=10^3s, we have $\bar{R}_Q$<3 .

Therefore, the matrix precipitation over the interstitials in the presence of the compressed coherent precipitates must favour the formation of minite bound interstitional states which, on the one hand, do not make it possible the itnterstitials to go out to the "far" sinks, thus making their contribution into the material swelling, and on the other hand, are the centers of additional vacancy recombination, which leads to disappearance of the Q clusters themselves.

Two approaching interstitials interact between each other through the field of atomic displacements. The interaction energy can be clculated with the Kansaki's forces stimulating the displacement fields resulting from the defects (see Ref. 325, v. 3). The first-order energy of the dimensional interaction falls with increasing the separation as r^{-3}. It depends on the direction of the line connecting the defects, to the elastic anisotropy axes of the crystal. That is why the interaction of such point defects can be, generally speaking, attractive and repulsive. It makes sense to study the formation of small interstitial clusters with computers (see, e.g., Refs. 38, 145, 424). The computations for fcc-crystals made it know that for all types of small interstitial clusters, the bound energy ε^b_{ni} is higher than the energy for migration activation ε^m_{ni}. Therefore, the dissociation of interstitional clusters can be ignored when kinetics is considered.

The stable bi-interstitials (2*i*) fall into two parallel dumbbells at the nearest neighboring position inclined to a small angle (<10°) in the planes {110}. The stable tri-interstitials (3*i*) are formed from the mutually orthogonal dumbbells at the nearest-neighboring position (see Ref. 325).

Consider, as above (see Chapters 1 and 4), the average solid solution where coherent precipitates are formed, becomes 5 times lower against the material that is not precipitate-containing. supersaturations over the interstitials, vacancies, bi- and tri-interstitials in two cases: without account for the coherent precipitates (see Chapter 1), C_{j0}, and with account for the precipitates (see Chapter 4), $\tilde{C}_j(t)$, where $j=i$, v, $2i$, $3i$. Let us investigate the possibility for the vacancies to recombine with bi- and tri-interstitials in both cases and compare this with an ordinary volume recombination. For this purpose we write a system of equations, analogous to (4.1), (4.3) for the average precipitations $\tilde{C}_j$ taking into account the creation and recombination reactions, as well as those of outcoming of mobile interstitials and vacancies onto the sinks:

$$\frac{d\tilde{C}_i}{dt} = g_i - \mu D_i \tilde{C}_i \tilde{C}_v - \alpha_2 D_i \tilde{C}_i^2 - D_i C_i k_i^2 + \beta_2 D_v \tilde{C}_v \tilde{C}_{2i} - \gamma_i D_i \tilde{C}_i \tilde{C}_{2i}, \qquad (8.66)$$

$$\frac{d\tilde{C}_v}{dt} = g_v - \mu D_i \tilde{C}_i \tilde{C}_v - D_v \tilde{C}_v k_v^2 - \beta_2 D_v \tilde{C}_v \tilde{C}_{2i} - \beta_3 D_d \tilde{C}_v \tilde{C}_{3i}, \qquad (8.67)$$

$$\frac{d\tilde{C}_{2i}}{dt} = \frac{\alpha_2}{2} D_i \tilde{C}_i^2 - \beta_2 D_v \tilde{C}_v \tilde{C}_{2i} - \gamma_i D_i \tilde{C}_i \tilde{C}_{2i} + \beta_3 D_v \tilde{C}_v \tilde{C}_{3i}, \tag{8.68}$$

$$\frac{d\tilde{C}_{3i}}{dt} = \gamma_i D_i \tilde{C}_i \tilde{C}_{2i} - \beta_3 D_v \tilde{C}_v \tilde{C}_{3i}. \tag{8.69}$$

Here α_2, β_2, β_3, γ_i are the kinetic coefficients of the corresponding reactions having significance of the separation of capturing one defect by another, multiplied by 4π.

Let the defect accumulation processes be stationarized, i.e. $d\tilde{C}_j/dt=0$. The equations for the precipitations C_{j0} in the absence of precipitates have the same form as (8.66)-(8.69). A main difference is in the values k_j^2 and k_{j0}^2. We consider in both cases that the structural sinks are represented merely by the dislocations ($q=D$) of the density ρ_D and that added in the second case are more vacancy-adsorbing precipitates ($q=P_v=P$). Now we have (see (8.24), (8.27), (8.28))

$$k_{j0}^2=\alpha_{D0}^j \rho_D \left[1+\frac{\kappa_D^j}{C_{j0}}\right], \tag{8.70}$$

$$k_i^2=\alpha_D^i \rho_D \left[1+\frac{\tilde{\kappa}_D^i}{\tilde{C}_i}\right], \tag{8.71}$$

$$k_v^2=\alpha_D^v \rho_D \left[1+\frac{\tilde{\kappa}_D^v}{\tilde{C}_v}\right]+\alpha_P^v \bar{R}_P C_P \left[1+\frac{\tilde{\kappa}_P^v}{\tilde{C}_v}\right], \tag{8.72}$$

where α_{D0}^j, α_D^j, α_P^v are determined in (1.35), (6.13), (6.4), respectively. To simplify our estimates, we neglect, as above, the contributions of κ^j_D and $\tilde{\kappa}^j_q$ into (8.70)-(8.72), thus underestimating (see Chapter 6) the efficiency of precipitate action in changing the dislocation prevalence.

Write from (8.66)-(8.69), given the precipitates, the ratio of the vacancy numbers recombining in a time unit with the bi- and tri-interstitials to that of the vacancies recombining with the single interstitials:

$$\tilde{\Theta}=\left[\beta_2 D_v \tilde{C}_v \tilde{C}_{2i}+\beta_3 D_v \tilde{C}_v \tilde{C}_{3i}\right]\left[\mu D_i \tilde{C}_i \tilde{C}_v\right]^{-1}. \tag{8.73}$$

There is an analogous expression for Θ_0 in case without the precipitates as well. Substitute the stationary expressions (8.68) and (8.69) into the difference of equations (8.66) and (8.67) and obtain the relation

$$D_v \tilde{C}_v = D_i \tilde{C}_i \frac{k_i^2}{k_v^2}\left[1-\frac{g_i-g_v}{D_i \tilde{C}_i k_i^2}\right]. \tag{8.74}$$

Then we have for $\tilde{\Theta}$ by (8.73)

$$\tilde{\Theta}=\frac{\alpha_2 D_v k_v^2}{2\mu D_i k_i^2}\left[1-\frac{g_i-g_v}{D_i \tilde{C}_i k_i^2}\right]^{-1}\left[1+\frac{\gamma_i k_v^2}{\beta_2 k_i^2}\left[1-\frac{g_i-g_v}{D_i \tilde{C}_i k_i^2}\right]^{-1}\right]. \tag{8.75}$$

Neglect the terms $(g_i-g_v)/D_i \tilde{C}_i k_i^2$, which again underestimates the efficiency of recombination on the bound interstitials, and have instead of (8.75)

$$\tilde{\Theta}\approx\frac{\alpha_2 D_v k_v^2}{2\mu D_i k_i^2}\left[1+\frac{\gamma_i k_v^2}{\beta_2 k_i^2}\right]. \tag{8.75a}$$

Now we write the ratio $\tilde{\Theta}/\Theta_0$ which characterizes the efficiency of recombination on the bound interstitial states ($2i$ and $3i$) in the precipitate case against that of $v_P=0$. Using (8.75a), we obtain

$$\frac{\tilde{\Theta}}{\Theta_0} \simeq \frac{k_v^2}{k_i^2}\frac{k_{i0}^2}{k_{v0}^2}\left(1+\frac{\gamma_i k_v^2}{\beta_2 k_i^2}\right)\left(1+\frac{\gamma_i k_{v0}^2}{\beta_2 k_{i0}^2}\right)^{-1}. \tag{8.76}$$

Substituting the expressions (8.70)-(8.72) into (8.76), we now have

$$\frac{\tilde{\Theta}}{\Theta_0} = \frac{\alpha_D^v}{\alpha_D^i}\left(1+\frac{\alpha_P^v \bar{R}_P C_P}{\alpha_D^v \rho_D}\right)\frac{\alpha_{D0}^i}{\alpha_{D0}^v}\left[1+\frac{\gamma_i \alpha_D^v}{\beta_2 \alpha_D^i}\left(1+\frac{\alpha_P^v \bar{R}_P C_P}{\alpha_D^v \rho_D}\right)\right]\times$$
$$\left(1+\frac{\gamma_i \alpha_{D0}^v}{\beta_2 \alpha_{d0}^i}\right)^{-1}. \tag{8.77}$$

For a numerical evaluation of the ratio $\tilde{\Theta}/\Theta_0$, we select, as above, the material with the parameters listed in Table 5.1 (see Chapter 5). Now, after the substitution of the numerical data into expression (8.77), we obtain $\tilde{\Theta}/\Theta_0 \simeq 20$ provided that $\gamma_i=\beta_2$ and $\tilde{\Theta}/\Theta_0 \simeq 16$ when $\gamma_i=\beta_2/2$.

By virtue of the fact that, as followed from the experiments [295-300], the radiation swelling in the materials with coherent precipitates becomes lower, then the recombinators forming as this takes place, remain at a lesser rate. Alternatively, the swelling should increase . So we can write the condition

$$\frac{C_{2i0}}{\tilde{C}_{2i}} > 1, \quad \frac{C_{3i0}}{\tilde{C}_{3i}} > 1. \tag{8.78}$$

We derive from the equations (8.66)-(8.69) in the stationary regime

$$\tilde{C}_{3i} = \tilde{C}_{2i}\,\frac{\gamma_i k_v^2}{\beta_3 k_i^2}, \tag{8.79}$$

$$\tilde{C}_{2i} = \tilde{C}_i\,\frac{\gamma_2 k_v^2}{2\beta_2 k_i^2}, \tag{8.80}$$

$$\tilde{C}_i = \frac{D_v k_v^2 k_i^2}{\mathcal{B}}\left[\sqrt{1+\frac{2g_i}{k_i^2}\frac{\mathcal{B}}{D_v k_v^2 D_i k_i^2}} - 1\right], \tag{8.81}$$

where

$$\mathcal{B} = 2\mu D_i k_i^2 + \alpha_2 D_v k_v^2 + (\alpha_2\gamma_i/\beta_2)(k_v^2/k_i^2)D_v k_v^2.$$

Substituting the expressions (8.79)-(8.81) into the conditions (8.78), we have with allowance for (8.75a)

$$\frac{C_{2i}}{\tilde{C}_{2i}} = \frac{C_i k_{v0}^2 k_i^2}{\tilde{C}_i k_v^2 k_{i0}^2} = \sqrt{\frac{g}{g_i}\left(\frac{k_{v0}^2}{k_v^2}\right)^3\left(\frac{k_i^2}{k_{i0}^2}\right)^3\frac{1+\tilde{\Theta}}{1+\Theta_0}} > 1,$$

$$\frac{C_{3i}}{\tilde{C}_{3i}} = \frac{C_{2i}}{\tilde{C}_{2i}}\left(\frac{k_{v0}^2}{k_v^2}\right)\left(\frac{k_i^2}{k_{i0}^2}\right)^2 = \tag{8.82}$$

$$\sqrt{\frac{g}{g_i}\left(\frac{k_{v0}^2}{k_v^2}\right)^5\left(\frac{k_i^2}{k_{i0}^2}\right)^5\frac{1+\tilde{\Theta}}{1+\Theta_0}} > 1$$

The numerical values of $\tilde{\Theta}$ and Θ_0 will be evaluated by (8.75a), as above, for the material with nickel parameters (see Table 5.1, Chapter 5).

When $\gamma_i=\beta_2/2$, we have $\tilde{\Theta}=250$, $\Theta_0=15$. If the sink strength sums change proportionally in going from the $v_p=0$ case to that of $v_p\neq0$ and $g\simeq g_i$, then with the use of the $\tilde{\Theta}$ and Θ_0 estimates, we can write that

$$\frac{C_{2i}}{\tilde{C}_{2i}}=\sqrt{\tilde{\Theta}/\Theta_0}>1,\quad \frac{C_{3i}}{\tilde{C}_{3i}}\simeq\sqrt{\tilde{\Theta}/\Theta_0}>1. \tag{8.82a}$$

It is seen from the given estimates that the recombination prevalent in the decomposing solid solutions against the ordinary volume recombination is that on the bound interstitials ($\mu D_i\tilde{C}_i\tilde{C}_v$). The results obtained correspond to the volume fraction $v_p=0.04\text{-}0.1$. When the average spherical precipitate radii $\bar{R}_p$ are lower, for example $\bar{R}_p=5\cdot10^7$ and 10^{-7}cm, we obtain for $\tilde{\Theta}/\Theta_0$ by (8.77) 3.4 and 1.8, respectively.

8.5. Physical Aspects of Reducing the Radiation Swelling in Materials

The theoretical investigations of physical reasons of lowering the radiation swelling in the decomposing solid solutions have permitted to understand some physical mechanisms of the anomalous recombination of opposite radiation defects with both mobile opposite defects and their clusters when a developed homogeneous continuous decay takes place in the solid solutions under irradiation as a result of rearranging the fluxes of vacancies and interstitials in the internal stress fields of the precipitates formed.

The theoretical and experimental investigations of the conditions and the mechanism of opposite defect anomalous recombination and of the related changes in the properties of solid solution under irradiation, can favor the solution of important applied problems in other branches of nuclear and thermonuclear energetics, space technics and radiation technology.

The expressions (8.19) and (8.22) for the radiation swelling rate of heterogeneous materials allow to establish the possibilities to lower the radiation swelling not due to lowering the total void number but for a variety of other physical mechanisms. The analysis of expressions (8.19) and (8.22) permits to note the following features.

1. The dislocations and interstitial dislocation loops present in crystals always make a positive contribution to the swelling due to the dislocations adsorbing the interstitials preferentially over the vacancies. This can be decreased merely by decreasing the number of the interstitials precipitated on the dislocations. There appears to exist two basic ways: (a) to decorate the dislocations by the interstitial so as they would not adsorb the interstitials or would adsorb them at a lesser rate [308]; (b) to decrease the number of readily mobile interstitials due to the anomalous recombination in the special-composition alloys.

2. For the most part, the experimental data (e.g. in Refs. 239, 309, 310, 425, 426) show that the swelling occurs with a vacancy void formation. This means that it is energetically more beneficial for the vacancies to unite into voids than to exist in the form of free vacancies or vacancy dislocation loops, and this tendency for void formation manifests itself at low temperatures too. In this case the role of a thermal void formation

under irradiation increases. It must be remembered that the volume relaxation due to the voids is lower than that due to the single vacancies or vacancy dislocation loops [427]. Two ways can keep from the void formation: (a) to enhance the recombination of opposite defects, thus decreasing the number of free vacancies; (b) to hinder the vacancy integration into voids, e.g. by way of producing the stress fields blocking the diffusion of both interstitial and vacancies.

3. The impurities, specially gaseous and particularly helium, gain in the rate of void nucleation. The amount of helium may be decreased by the introduction of various impurity traps.

4. A definite alloying provides the developed-in-time homogeneous continuous decomposition of a solid solution with the well pronounced incubation period when the volume dilatation at the interface "the forming secondary phase - matrix" has a definite value, as well as the decomposition of ordering type, the *K* state, the layering of the solid solution and so on, which leads to suppressing the swelling due to anomalous recombination of opposite radiation defects in the elastic stress fields and that of the falling-out coherent secondary-phase precipitates [295-300].

CHAPTER 9

SOME APPLICATIONS OF THE THEORY OF RADIATION PROCESSES IN ALLOYS

9.1. Influence of Thermal Spike Diffusion on the Resolution of Ionic Etching Methods for Analyzing the Impurity-Depth Distribution

The depth profiles of impurity concentration are determined by various methods which employ the ionic etching with a subsequent (or simultaneous) analysis of the surface by the SIMS methods, the electronic Auger spectroscopy, and so on. However, there can take place, in the sample, a mass transfer, which results in the violation of the initial impurity distribution profile and the deterioration of the resolution of the method. The most apparent mechanisms of mass transfer are the ballistic mixing and the radiation-quickened diffusion. The latter mechanism is realized at sufficiently high temperatures when the defects are mobile. It was customary to believe that at low temperatures (e.g. at the room temperature for silicon, germanium, gallium arsenide), the mass transfer goes due to the ballistic processes only. However, as shown in Refs. 428, 429, the calculations based on the ballistic model by pure are in a manifest conflict with the experimental results. To explain this effect, we must invoke the thermal-spike diffusion model [430]. According to this model, being in the zone of localized overheating produced by the motion of high-energy particles (thermal spike), the diffusate could perform some jumps in the time of these spike existence.

The analytical expressions for the rate of thermoactivated thermal-spike processes were derived in Ref. 431. In Refs. 432, 433, the computerized modeling of the cascade region relaxation was pursued at the thermal-spike stage. These works have shown that the number of atomic displacements occurring at this stage is in excess of that of ballistically displaced atoms. The authors of Refs. 428, 429 have considered the mixing going on at the sacrifice in thermal spikes while the energies of primary atoms are high and their path is long against both the characteristic dimension of a thermal spike and the width of the material layer atomized. In this case, the number of atomic displacements increases linearly with the irradiation dose.

Consider the mixing going on the thermal spikes when the ionic etching is used [434]. Then the primary ion energy E_0 is, as a rule, equal to 0.5-10 keV, and the thermal spike size is close to the ion path projected and equal to several lattice parameters. Usually this value is much lesser than the width of the layer etched away in the investigation. In proportion with etching, the primarily rested impurity atom (diffusate) approaches the surface atomized and reaches the region where it occasionally exposed to the thermal spikes. Let $\bar{n}$ signify an average number of diffusion atomic jumps under the action of these spikes in the time from the beginning of etching to the diffusate exit to the surface. The diffusion path L is related to $\bar{n}$ by the following relation:

$$L^2 = a^2\bar{n}, \quad a \sim \Omega^{1/3}, \tag{9.1}$$

where Ω is the atomic volume, and a is the interatomic spacing. It is the diffusion path (a measure of atomizing of a thin impurity layer) that is characterized the mass transfer influence on the resolution of the above mentioned methods. The value $\bar{n}$ is determined by the number of thermal spikes attacking the diffusate and by the number of diffusate jumps in these thermal spikes. Denote the distance between the diffusate to the surface etched by x, and the "impact parameter" of a primary ion by R (see Fig. 9.1), the average number of diffusate jumps under the thermal spike action by $P(x,R)$. Then the frequency of diffusate jumps at the depth x under the action of the ion flux striking the surface of the density j will be

$$\omega(x)=2\pi j\int_0^\infty RP(x,R)dR.$$

Suppose that $L \ll x_0$ (x_0 is the initial diffusate depth). Then

$$\bar{n}=\int_0^{t_0}\omega[x(t)]dt=\int_0^{x_0}\omega(x)\,\frac{dx}{v}\,, \qquad (9.2)$$

where v is the etching rate which is supposed to be constant and can be expressed through the atomizing coefficient Y as follows:

$$v = jY\Omega.$$

If at the initial instant the diffusate was sufficiently far from the surface and did not experience the thermal spikes, then the upper limit of interaction in expression (9.2) can be replaced by the infinity. Then we have

$$\bar{n}=\frac{2\pi}{\Omega Y}\int_0^\infty RdR\int_0^\infty dxP(R,x)=\frac{1}{\Omega Y}\int_V P(R,x)dV.$$

Here the integral is taken over the whole volume V of the sample.

Consider the energy relaxation in some thermal spike taken separately. Suppose that the thermal spike can be described by the space-time temperature distribution $T(x,R,\tau)$, where τ is the time measured from the moment of thermal spike initiation. The frequency of diffusate jumps can be described by the Arrenius equation

$$\nu(x,R,\tau)=\nu_0\exp\left[-\frac{Q}{kT(x,r,\tau)}\right].$$

Here Q is the energy of diffusion activation, k is the Boltzmann constant, ν_0 is the constant which is assumed to be close to the Debye frequency ν_D. Then for the thermal spike taken separately, the average number of diffusate jumps has the form

$$P(R,x)=\int_0^\infty \nu_0\exp\left[-\frac{Q}{kT(x,R,\tau)}\right].$$

In order to derive the expression for $T(x,R,\tau)$ in an explicit form, we consider a model of thermal spike relaxation, basing on the following assumptions.

1. The ion flux is not so large, so that a vicinity of the diffusate has not managed to "cool down" in the time between two thermal spikes, that is why their overlapping may be neglected.

2. At the instant τ=0, the heating source is distributed along the ion

path direction, and in the one perpendicular to the path, the-heat distribution has a zero dispersion.

3. The heat distribution along the path direction is determined by the average losses ε in the ion energy E per unit length:

$$\varepsilon(E)=-\frac{\partial E}{\partial \eta},$$

where η is the coordinate along the ion path direction.

4. The path is a broken line, but the thermal fields from some sections of it do not overlap.

With allowance for these assumptions, the space-time heat can be written in the coordinates η and s (s is the distance from the path) in the form [430]

$$T(\eta,s,\tau)=\frac{\varepsilon[E(\eta)]}{4\pi\kappa\tau}\exp\left(-\frac{cs^2}{4\kappa\tau}\right). \tag{9.3}$$

Here c is the heat capacity of a unit volume, κ is the lattice heat conductivity of a target sample. The energy density in a thermal spike is the value of the order of some eV's per atom. When the temperatures are so high, the heat conductivity can be assumed independent of the temperature (see Ref. 435),

$$\kappa\simeq(1/3)cu\Omega^{1/3}, \tag{9.4}$$

where u is the sound velocity in the sample. The expression (9.4) is a consequence of the fact that at such temperatures the phonon path length decreases down to the values of the order of the lattice parameter. Furthermore, when the temperatures are rather high,

$$c \approx 3k/\Omega. \tag{9.5}$$

According to Ref. 431, the average number of jumps per the unit length of the path in a thermal field (9.3) has the form

$$\frac{\partial \bar{n}}{\partial \eta} = \frac{\nu_0 \varepsilon^2 [E(\eta)]}{8\pi\kappa c (Q/k)^2}. \tag{9.6}$$

To find $\bar{n}$, we integrate (9.6) over the path length:

$$\bar{n} = \int_0^{\eta_{max}} \frac{\nu_0 \varepsilon^2 [E(\eta)]}{Y\Omega 8\pi\kappa c (Q/k)^2} d\eta = \int_0^{E_0} \frac{\nu_0 \varepsilon(E)\, dE}{Y\Omega 8\pi\kappa c (Q/k)^2}. \tag{9.7}$$

The value ε involves two loss components: a nuclear one ε_n and an electronic one ε_e. At the energies of the order of 1 keV (see Ref. 436), we have

$$\varepsilon_e \ll \varepsilon_n, \quad \varepsilon \approx \varepsilon_n \approx AE^B, \quad B \sim 0.5. \tag{9.8}$$

Since, as a rule, the energy fraction E_0 of the primary ion, which goes to the defect formation, is small, we suppose that the whole energy E_0 goes into the heat. Then, using (9.7), (9.8), we obtain

$$\bar{n} = \frac{\nu_0 A E_0^{B+1}}{(B+1)Y\Omega \cdot 8\pi\kappa c (Q/k)^2} = \frac{\nu_0 \varepsilon(E_0) E_0}{(B+1)Y\Omega \cdot 8\pi\kappa c (Q/k)^2}. \tag{9.9}$$

Considering that $\nu_0 \sim \nu_D$ and $u \sim \nu_D \Omega^{1/3}$, we obtain with account for (9.1), (9.4)

$$\bar{n} = \frac{\varepsilon(E_0) E_0 \Omega^{1/3}}{24\pi(B+1)YQ^2}, \tag{9.10}$$

$$L^2 = \frac{\varepsilon(E_0)E_0\Omega}{24\pi(B+1)YQ^2}. \tag{9.11}$$

Give a numerical evaluation to the values $\bar{n}$ and L by (9.10), (9.11) for the targets of *Si*, *Ge* and *GaAs* types. We have for these materials $\Omega \sim 20$ Å^3, and the activation energy Q is equal to several eV for the diffusion processes. When such targets are bombarded by the heavy 1 keV-energy ions, $Y{\sim}1$ and ε is of the order of tens of eV/atom [436, 437]. Then, taking Q=3 eV and ε = 20 eV/, we obtain $\bar{n}$=50 jumps, L=20 .

As seen, the diffusion in thermal spikes can substantively influence the resolution of the ion etching-used methods. The value L can reach the values of the order of the primary ion path, which is more than the depth resolution of the methods of analyzing the surface, for example the electron Auger spectroscopy [438]. Note that the sufficiently strong assumptions made in deriving the expressions (9.9)-(9.11) determine them qualitatively. It is necessary to use computer for a more precise treatment of the problem.

The thermal diffusion in thermal spikes makes possible the Gibbs adsorption in the low-temperature atomization of *GaAs*, which was observed in Ref. 439.

9.2. Impurity Concentrations with Secondary-Phase Precipitate Initiation

The question whether a volume-homogeneous precipitate initiation is possible and in what relation it is to the selective nucleation on the sinks, has been considered in Ref. 279 for the binary systems. However, the temperature dependence of the average impurity concentration to be necessary for the homogeneous nucleation was not considered. Furthermore, there are Refs. 280, 281 where the nucleation on the dislocations is compared with that in volume. Consequently, if one considers the possibilities of nucleating the precipitates in the material volume (homogeneous), various types of heterogeneous nucleation, for example on the dislocations, impurities, etc., must be taken into account.

Evaluate a dominant type of nucleation with the proviso that the rates of precipitate nucleation in the volume J_V and at the grain boundaries J_G are stationary [282]. Then one can write for the rate of nucleation in the volume J_V as follows:

$$J_V = J_0 + \sum_k J_k,$$

where J_0 and J_k are the rates of homogeneous and heterogeneous types of k-nucleation, respectively. Considered below for the k-type are the dislocations only, i.e. $\sum_k J_k = J_D$. In fact, one can assume that $J_V = J_m$ $(m=0,D)$ with m being the dominant type of volume nucleation. In Ref. 440 the following relation is offered:

$$J_k/J_e = (N_k/N_e)\exp\{(F_e - F_k)/kT\}, \qquad (9.12)$$

where F_j when $j=e,k$, is the energy for nucleation activation, and N_j is the number of sites for the j-type nucleation per unit volume.

It is seen from (9.12) that the condition

$$J_V/J_G>1. \tag{9.13}$$

must be fulfilled for the preferential nucleation.

We assume from the start that $J_V \approx J_0$. In the classical nucleation theory, the free energy of a n-particle cluster has the form

$$F(n)=-\Delta\varphi n+\gamma\chi n^{2/3}. \tag{9.14}$$

Here γ is the specific surface energy of the cluster (or precipitate) formed; χ is a geometrical factor;

$$\Delta\varphi=\Delta\mu(C_m)-\varphi_\delta, \tag{9.15}$$

$$\Delta\mu(C_m)=\mu(C_m)-\mu(C_a). \tag{9.16}$$

The difference in the chemical potentials (9.16) describes the thermodynamic driving force for the formation of a nucleus of the relative impurity concentration C_a if the relative impurity concentration in the matrix is C_m. The value φ_δ is the deformation energy which depends on the change in the lattice parameters on nucleation. The value φ_δ is often ignored ($\varphi_\delta \approx 0$).

The factor χ_0 for the homogeneous nucleation is determined from the condition that $\chi_0 n^{2/3}$ is equal to the nucleus area. When the selective nucleation takes place on some sink, for example on the grain boundaries, the value χ_G must depend on the specific surface energy γ_G of the grain boundary, i.e. $\chi_G=\chi_G(\gamma_G)$. Now suppose that $\gamma_G \leq 2\gamma$, since otherwise the precipitates formed will be located in the grain boundary.

A critical value of the number of atoms in the cluster, n_c, that is, such an atom number higher than that the clusters become stable, is determined from the condition $dF(n)/dn\Big|_{n=n_c}=0$ and is concerned to the energy value $F(n_c)$ which has the form

$$F\equiv F(n_c)=\frac{4}{27}\,\frac{(\gamma\chi)^3}{\Delta\varphi^2}\,, \qquad (9.17)$$

where χ takes two values, χ_0 and χ_G.

Using expressions (9.12), (9.15)-(9.17), the condition (9.13) can be written as follows:

$$\mu(C_m)-\mu(C_G)\geq\sqrt{\frac{4}{27}\,\frac{\gamma^3\chi_0^3[1-(\chi_G/\chi_0)^3]}{kT\,\ln(N_0/N_G)}}\,. \qquad (9.18)$$

The expression for the chemical potential difference has the form

$$\mu(C_m)-\mu(C_a)=kT\,\ln\left[\frac{\kappa(C_m)}{\kappa(C_a)}\cdot\frac{C_m}{C_a}\right], \qquad (9.19)$$

where κ is the coefficient of chemical activity of the element [441]; T is the temperature; and k is the Boltzmann constant.

We shall use the following simplifying assumptions.

1. The impurity concentration C_m in the matrix does not change with the nucleus formation and is equal to the average impurity concentration C.

2. The impurity concentration C_a in the precipitate nucleus is an equilibrium concentration C_2 in the precipitate, thus we obtain $\mu(C_a)\approx\mu(C_2)=\mu(C_1)$, the solubility limit being

$$C_1=\exp(-E/kT), \qquad (9.20)$$

where E is the solubility free energy.

3. The activation coefficients $\kappa(C)=\kappa(C_1)$, i.e. their changes with concentration, are small.

Then the expression (9.19) takes the form

$$\mu(C_m)-\mu(C_k)\approx\mu(C)-\mu(C_1)=kT\ \ln(C/C_1). \tag{9.21}$$

Assuming as well that only the interaction of the nearest neighboring atoms is taken into account at the nucleus-matrix boundary, we have for the specific surface energy

$$\gamma=\xi\varepsilon(C_m-C_a)^2a^{-2}, \tag{9.22}$$

where a is the lattice parameter,

$$\varepsilon=Z\left[\varepsilon_{ij}-\frac{1}{2}(\varepsilon_{ii}+\varepsilon_{jj})\right] \tag{9.23}$$

is a characteristic interaction energy; ε_{ij} is the interaction energy of the atom i with the atom j; Z is the coordination number.

The parameter ξ in (9.22) depends on the value of the area accounted for by an atom in the given crystal lattice and on the ratio Z'/Z, where Z' is the number of the nearest neighbors for a surface nucleus atom in the matrix, and $Z'<Z$. For a fcc-lattice, the estimates of the ξ value provide $\xi=0.6$, and for a bcc-lattice, $\xi=0.4$. Using the above mentioned assumptions, as well as the fact that $C\ll C_2$, we obtain

$$\gamma\approx\xi\varepsilon C_2^2a^{-2}. \tag{9.24}$$

In case when $C_1\ll 1$, the solubility limit can be calculated with the model of a regular solution [442], the solubility free energy E remaining constant on some temperature interval. Then we have $E\approx\varepsilon$.

Table 9.1

The dependence of the concentration of *Sc* in *Ni* for the volume-homogeneous nucleation of the secondary-phase at various temperatures

T, K	C_0, at. %	C_0, at. %	
		$N_0/N_G=10^4$	$N_0/N_G=10^5$
700	$4.8\cdot10^{-3}$	0.6	0.4
800	$1.7\cdot10^{-2}$	0.9	0.6
900	$1.4\cdot10^{-2}$	1.2	0.9
1000	$9.4\cdot10^{-2}$	1.6	1.2

In such a model $C_2=1-C_1$, and hence $C_2\approx1$. Substitute (9.20), (9.21), and (9.24) into the expression (9.18), we determine the value of the average impurity concentration higher than that the nucleation is anticipated homogeneous by volume, i.e.

$$C\geq C_0=\exp\left\{-\frac{\varepsilon}{kT}\left[1-\sqrt{\frac{4}{27}\left(\frac{\xi\chi_0}{a^2}\right)^3\frac{\varepsilon}{kT}\frac{1-(\chi_G/\chi_0)^3}{\ln(N_0/N_G)}}\right]\right\}. \tag{9.25}$$

If the needles on homogeneous decomposition are spherical, one can obtain $\chi_0=a^2\sqrt[3]{9\pi/4}$ for a fcc-lattice, and $\chi_0=a^2\sqrt[3]{9\pi}$ for a bcc-lattice. On a selective decomposition the nucleus located at the grain boundary can be shaped like a symmetric biconvex lens (see Ref. 442) with a characteristic contact angle ϑ, for which one can write as follows [442]:

$$2\cos\vartheta=\gamma_G/\gamma\equiv M.$$

Then the expression

$$(\chi_G/\chi_0)^3=1-\frac{3M}{4}+\frac{M^3}{16} \tag{9.26}$$

does not depend on the lattice type. In this case (9.26) takes the form

$$C\geq C_0=\exp\left\{-\frac{\varepsilon}{kT}\left[1-\beta\sqrt{\frac{\varepsilon}{kT\ \ln(N_0/N_G)}\ \frac{M}{4}\left[3-\frac{M^2}{4}\right]}\right]\right\}, \tag{9.27}$$

with β=0.48 for the fcc-crystals and β=0.52 for the bcc ones.

The expression (9.26) changes when the nuclei at the grain boundariesare shaped differently, however, as shown by the evaluations, the concentration C_0 is slightly dependent on this factor.

Therefore, if ε and M are known, and one assume as to the ratio N_0/N_G that it depends on the average grain dimension $\bar{R}$ as $N_0/N_G=\bar{R}/3a$, i.e. is of the order of 10^4-10^5, then, by (9.27), one can determine the value of the impurity con-centration for a homogeneous nucleation. For example, for the case of *Ni-Sc* alloy, we have ε=0.6 eV from the experimental evidence on *Sc* solubility in *Ni* at the temperatures from the interval $1073<T<1413$ K [443]. In addition, the specific surface energy at the grain boundary is equal to 0.8 Joule/m^2 [444], which gives $M\approx1.7$ for the scandium nickelide nucleation in nickel. With the use of above listed expressions, one can obtain for the concentration C_0 the results outlined in Table 9.1.

The concentration C_0 determines the impurity amount when there takes place a transition from the grain-boundary nucleation to the volume nucleation provided that the relation $J_0/J_D>1$ is fulfilled. The number of sites N_D for the nucleation on a dislocation is related to the dislocation density ρ_D through the relation $N_D=\nu a^2\rho_D N_0$, where $\nu=\pi(r_*/a)^2$, r_* being the radius of the dislocation core. For example, when $\nu=\pi/2$ and $\rho_D=10^7\text{-}10^{11}\text{cm}^{-2}$, we obtain $N_D/N_0=10^{-8}\text{-}10^{-4}$. As opposed to (9.14), there appears, in the expression for the free energy of the nucleus on the dislocation, some additional terms. They are related to the energy of dislocation core E_D and the stress field occurred around the dislocation interstitial. Let us assume, in accordance with Ref. 281, that the nucleus on the dislocation line has an ellipsoidal shape extended along this line, the radius of the section perpendicular to the dislocation line being denoted by r. Then we have for the nucleus free energy per the unit length of the dislocation

$$F_D/e = \begin{cases} -\dfrac{\Delta\mu}{\Omega}\pi r^2 + 2\pi\gamma r - \dfrac{E_D}{\Omega}\pi r_*^2 - B\ln\dfrac{r}{r_*}, & r > r_*, \\ -\dfrac{\Delta\mu}{\Omega}\pi r^2 + 2\pi\gamma r - \dfrac{E_D}{\Omega}\pi r^2, & r \leq r_*, \end{cases}$$

where Ω is the atomic volume; $E_D=B\Omega/\pi r_*^2$; $B=Gb^2/4\pi$, G is the shear modulus; b is the Burgers vector.

When $C\geq C_1$, then we have $r \gg r_*$, and one way fall back on the results of Ref. 280 without regard to the dislocation core energy. The activation energy F_D is now described by a something like the following expression:

$$F_D \approx (1-\alpha^{0.6})F_0 \tag{9.28}$$

with

$$C \ll C^{+} = \exp\left[-\frac{\varepsilon}{kT}\left(1 - \frac{\pi}{2}\xi^2\Omega\frac{\varepsilon}{a^4 B}\right)\right], \qquad \alpha = 2B\Delta\mu(C)/\pi\gamma^2\Omega.$$

It is shown in Ref. 280 that $\lim\limits_{C \to C_1}(F_D/F_0)=1$, and when $\lim\limits_{C \to C_1}(F_D/F_0)<1$ and $C \geq C_1$, the relation $J_G>J_D>J_0$ is fulfilled. When the supersaturation is reasonably high, $r \approx r_0$ and the elastic energy of the dislocation field is inessential. Then the nucleation process differs from the homogeneous one only by the existence of the dislocation core energy that now is functioning as a complementary thermodynamic force. Therefore, F_D has the following form:

$$F_D = \left(\frac{\Delta\mu}{\Delta\mu + E_0}\right)^2 F_0.$$

The role of the nucleation on dislocations is thus seen to be evaluated as follows. For example, in the *Ni-Sc* alloy ($G=8 \cdot 10^{10}$ N$\cdot$m^{-2}, $r_*=b=a/\sqrt{2}$) the relation $r \approx r_*$ is fulfilled if $C<C_0$ Then the nucleation on dislocations may be ignored due to the fact that, as a rule, $J_G>J_0>J_D$ for $C \leq C_0$ approximately. This stems from $F_g/F_0=0.03$ ($M=1.7$) and $F_D/F_0 \geq 0.2$. From this it is inferred that in the systems based on iron and nickel, a marked nucleation on dislocations is not expected when $M \geq 1.5$ approximately. It is possible only for the coarse-grained materials with the high dislocation density ($\rho_D \approx 10^{12}$cm^{-2}) close by C_0 that $J_G \approx J_D > J_0$. A detailed investigation of the value J_D is thus necessary merely in the cases of reasonably low M.

When a high impurity segregation arises on the dislocations, the nucleation rate essentially changes, and the expression (9.28) describes F_D

over all interval of changing the concentration $C_1<C<C^+$, and for $C \geq C^+$, one has $F_D=0$, as has been shown in Ref. 280. Now the relation $J_D=J_G>J_0$ takes place at the concentration value $C_D<C^+$ that is a transition one between the nucleation at the grain boundaries and that in the volume when $N_D>N_G$. However, the value C_D takes place in the case $N_G>N_D>N_G \exp[-F_G(C^+)/kT]$. In so doing, there exists the concentration C_* which is determined from $N_D=N_G \times \exp[-F_G(C_*)/kT]$, and hence once again we have $J_D=J_G$. If $C_*<C_0$ on the interval $C_*<C<C_0$, then the coarse-grained nucleation is dominant. For the *Ni-Sc* alloy, there is no evidence for such a segregation. And hence, the value C_0 describes the transition in this binary system between the nucleation at the grain boundaries and that in the volume.

It is essential to consider the results obtained only as a preliminary evaluation of the boundary concentration C_0, since it strongly depends on ε and γ. However, these values are not known well enough. Thus, we shall restrict ourselves to some general conclusions.

When the temperature becomes lower, the value C_0 decreases despite the fact that the ratio C_0/C_1 increases. However, the solubility C_1 decreases in this case more strongly than C_0. The supersaturations by impurity used in our model have the values nearby the solubility limit, the spinoidal decay being observed on a different temperature interval. Because of this, it is not improbable that attained in experiment is the volume-homogeneous classical nucleation. Furthermore, as seen from the above cited evidences, the boundary concentration C_0 depends on the grain dimensions, and the volume-homogeneous nucleation must be better realized in the coarse-grained materials.

9.3. Energy Parameters of Mobile Point Defects

For a correct description of the physical processes taking place in the system of crystal structure defects, it is necessary to know some energy characteristics of the point defects *j* such as the activation energies for migration ε^m_j, the bound energies ε^b_{ja} of the complexes *ja* and so on. Some of the energies of point defects in metals are determined rather reliably and given in various references. Part of the information on the point defect energies has been collected in the tabulated form in Ref. 181. However, the determination of such energies as the bound energy of the intrinsic defect-impurity atom complexes or their migration energy offers great experimental and theoretical problems. Therefore, it is very urgent for this problem to obtain some reliable results basing on the experimental data.

In connection with developing the procedure of a direct observation of the individual point defects with autoionic microscope [37, 351] that permits to evaluate the concentrations, the volume distribution, and the defect mobility at various temperatures most advantageously, it was attempted to use this unique procedure for the migration energy ε^m_{ia} and the bound energies ε^b_{ia} of the complexes of interstitial *W* atoms with *Re* in *W-Re* alloy to be determined [445, 446].

Essentially, the experiment is as follows: the samples irradiated (needles) in which the interstitials prove in specific situations to be captured by the impurity traps, are annealed during a specified time $\Delta t=t-t_0$ (t_0 is the time of the annealing initiation), the temperature $T=T_0+\beta\Delta t$ being increased linearly, within the interval 0.1÷1.0 grad·s^{-1}, where β is the heating rate. The heating is performed stepwise with a step of ΔT=10-30½C starting from the irradiation temperature. The photographic record of autoionic picture is performed at the end of each stage of heating. The increments of the number of increased-in-brightness points which display the crop out of the interstitials or their complexes at the sample surface are estimated from a successive comparison of the images for each pair of the following-each-other stages. A qualitative dependency of the number of the interstitials croping out at the sample surface is shown in Fig. 9.2 as a function of time, as a linear-law of heating takes place.

Describe the diffusion processes causing the k-type point defect exit to the surface in terms of the concentrations $C_k(t)$ averaged over the volume. Consider, for this purpose, equations (2.11)-(2.14), where we shall ignore the following terms on their right-hand sides; (1) div$\mathbf{J}_q^k(\mathbf{r},t)$, as considered are only the concentrations averaged over the volume; (2) the defect generation rate g, since the annealing of already irradiated samples is investigated; (3) the last terms in the equations (2.12) and (2.14) describing the outgoing of the impurities a and vacancies $n=v$ onto the sinks, because for the system *W-Re* on the temperature interval investigated experimentally in Ref. 445, the vacancies and interstitials are fixed, and the impurity transfer is only realized in the form of *ia*-type complexes. We chose the initial conditions of the problem in the form: $C_i(t_0)=C_{i0}$=const; $C_v(t_0)=C_{v0}$=const≠0; $C_a(t_0)=0$; $C_{ia}(t_0)=C_{ia0}=\hat{C}_a$=const, where $\hat{C}_a$ is the total impurity concentration.

The increase in the interstitial outgoing to the sample surface with increasing the temperature (see Fig. 9.2) fits releasing the interstitials from the traps. At low temperatures, the dissociation must be small relative to the defect outgoing to the surface, i.e. $\chi_{ia}(t)<D_{ia}k^2_{ia}$. At elevated temperatures, on the contrary, we have $\chi_{ia}>D_{ia}k^2_{ia}$. Consequently, at the temperature T_m corresponding to the maximum in the dependency of the number $N_{ia}(t)$ of the complexes outgoing to the surface to the instant t_m (see the Diagram in Fig. 9.2), fulfilled should be the following transitive condition:

$$\chi_{ia}(T_m)\approx D_{ia}(T_m)k^2_{ia}. \qquad (9.29)$$

Consider the case of higher temperatures when the dissociation is large and the process of the complex formation is hampered, i.e. $\alpha_{ia}C_iC_a<\chi_{ia}C_{ia}$. Neglecting with allowance for the above-listed assumptions, the term $\alpha_{ia}C_iC_a$ in the equation (2.12), we obtain an equation for the C_a impurity concentration in the form

$$\frac{dC_a(t)}{dt}=\chi_{ia}C_{ia}.$$

Take into account that the relation $C_{ia}(t)=\hat{C}_a-C_a(t)$ must be fulfilled in our model for any t. Now, instead of (2.12), we have the following equation:

$$\frac{dC_a}{dt}+\chi_{ia}(t)C_a(t)=\chi_{ia}(t)\hat{C}_a,$$

whose solution has the form

$$C_a(t)=\hat{C}_a\exp\left[-\int_{t_0}^{t}\chi_{ia}(t')dt'\right]\int_{t_0}^{t}\chi_{ia}(t')\exp\left(\left[\int_{t_0}^{t'}\chi_{ia}(\tau)d\tau\right]dt',\right. \tag{9.30}$$

where $\chi_{ia}=N_i^{ia}v_0\exp\left\{-\varepsilon_{ia}^b/kT_0\left[1+\frac{\beta}{T_0}(t-t_0)\right]\right\}$. If one evaluates the integral in (9.30) approximately, it can be obtained that

$$C_a(t)\approx\hat{C}_a\left(1-\frac{\varepsilon_{ia}^b\beta}{N_i^{ia}v_0kT_0^2}\right)\left[1-e^{-N_i^{ia}v_0(t-t_0)}\right],$$

and hence,

$$C_{ia}(t)\approx\hat{C}_a\left\{1-\left(1-\frac{\varepsilon_{ia}^b\beta}{N_i^{ia}v_0kT_0^2}\right)\left[1-e^{-N_i^{ia}v_0(t-t_0)}\right]\right\}. \tag{9.31}$$

The expression for the rate of outgoing the complexes *ia* to the sample surface has the form

$$g_{ia}^s(t) = D_{ia}(t)k_{ia}^2C_{ia}(t). \tag{9.32}$$

Now we can write the number of the complexes outgoing to the sample (needle) surface to the time instant t_* as follows:

$$N_{ia}(t_*)=\int_{t_0}^{t_*}g_{ia}^s(t')\frac{V(t')}{2}dt', \tag{9.33}$$

where $V(t')$ is the volume of the top hemisphere of the L-radius needle. Substitute expressions (9.31), (9.32) into (9.33) and evaluate approximately the integrals, then we obtain

$$N_{ia}(t_*)\simeq\frac{2}{3}\pi L^3\hat{C}_a a^2\nu_0 k_{ia}^2(t_*-t_0)\exp\left\{-\frac{\varepsilon_{ia}^m}{kT_0[1+\frac{\beta}{T_0}(t_*-t_0)]}\right\}.$$

So, with a knowledge of the number of the complexes outgoing to the surface in the time interval t_*-t_0, one can determine the value of migration energy in the form

$$\varepsilon_{ia}^m\approx kT_0\left[1+\frac{\beta}{2T_0}(t_*-t_0)\right]\times \tag{9.34}$$

$$\ln\left[\frac{2}{3}\pi L^3\frac{\hat{C}_a}{N_{ia}(t_*)}a^2\nu_0 k_{ia}^2(t_*-t_0)\right].$$

Using the condition (9.29), the bound energy ε_{ia}^b can be evaluated as

$$\varepsilon_{ia}^b\approx\varepsilon_{ia}^m+kT_m\ln\frac{N_i^{ia}}{a^2k_{ia}^2}. \tag{9.35}$$

The numerical values of the energies can be obtained for the following basic parameters entering (9.34) and (9.35): $t_*=60$ s, $N_{ia}(t_*)\simeq10^2$, $\beta=15$ grad$\cdot$s^{-1}, $\nu_0=10^{12}$s^{-1}, $N_{ia}^{ia}\simeq10$, $k_{ia}^2\approx L^{-2}$, $L\simeq50$ nm. Now we have $\varepsilon_{ia}^m\simeq$ 0.9 eV, $\varepsilon_{ia}^b\simeq$ 2 eV.

9.4. Structural Sink Adsorption of Radiation Point Defects

The macroscopic properties of the materials under irradiation are dictated by the radiation point defect interaction with both the structure imperfection of these materials and the defect structure forming under the irradiation. Specifically, it is seen from the expression (8.22) for the radiation swelling rate that this structure essentially depends on the efficiency K_D^j of the j-point defect adsorption by the dislocations and on the rates $I_q^j(R_q)$ of the j-point defect precipitation on different R_q-sized q sinks. To date, the adsorption abilities of the structure imperfections were derived from the experiments on radiation swelling . However, such a method is not correct enough, as it is based on particular physical models of the radiation swelling phenomenon itself, and the values of preferential factors derived from processing the experimental data on alloys, in their turn, must essentially depend on the structure of the material investigated and on the external conditions, i.e. on the temperature. When coupled with simulated computations of the process of transferring the atoms to the surface only whether guess on the macroscopic changes is made, the use of autoionic microscopy permits, with some degree of certainty, to understand such parameters as $I_q^j(R_q)$ and K_D^j, and hence k_j^2, which appear in all the expressions including the defect concentrations, as well as, specifically, in the expressions (9.34) and (9.35) [446-448].

Illustrate the possibility of determining with autoionic microscope the sink strengths from the experimental data on outgoing the interstitial from the needle point irradiated when isothermal annealing is used. Consider, as in Section 9.3, the equation for the average precipitations of j point defects without regard for the recombination in a pure material in the form

$$\frac{dC_j(t)}{dt}=-D_jC_j(t)k_j^2. \tag{9.36}$$

For sinks, we consider the surface (the sink strength S^j_s) and any of the inner sinks q (the sink strength S^j_q). We assume that there can be, in the point, merely one of the inner sinks (which fits the statistics of the autoionic investigations with a layer-by-layer sample evaporation [37,351]. Now the equation (1.26) takes the form

$$k^2_j=S^j_s+S^j_q. \tag{9.37}$$

A solution to the equation (9.36) is given by the following expression:

$$C_j(t)=C^0_j\exp[-D_j(S^j_s+S^j_q)(t-t_0)]. \tag{9.38}$$

with $C_j(t_0)=C^0_j$.

We consider two cases: (1) a point without inner sinks ($k^2_{j1}=\tilde{S}^j_s$) and (2) a point with an inner q sink ($k^2_{j2}=k^2_j$ by (9.37)). As in Section 9.3, $N(t_*)$ designates the number of the j defects outgoing to the point surface to the time instant t_*($N_1(t_*)$ and $N_2(t_*)$ for the cases listed, respectively). For $N_2(t_*)$ (see (9.33)), we have

$$N_2(t_*)=\frac{V}{2}\int_{t_0}^{t_*} dt'\, D_j C_j(t')S^j_s\ . \tag{9.39}$$

Substituting (9.38) into (9.39), we obtain

$$\frac{S^j_s}{S^j_s+S^j_q}\left[1\text{-}\exp\left[-D_j(S^j_s+S^j_q)(t_*-t_0)\right]\right]=\frac{3N_2(t_*)}{2\pi L^3 C^0_j}\ . \tag{9.40}$$

The difference between $N_1(t_*)$ and $N_2(t_*)$ in the adequate samples is in the defect adsorption by the inner sink q. Consequently, one can write

$$N_1(t_*)\text{-}N_2(t_*)=\frac{V}{2}\int_{t_0}^{t_*} dt' D_j C_j(t')S_q^j \ . \qquad (8.41)$$

Substituting (9.38) into (9.41), we have the relation

$$\frac{S_q^j}{S_s^j+S_q^j}\left[1\text{-}\exp\left[\text{-}D_j(S_s^j+S_q^j)(t_*\text{-}t_0)\right]\right]=\frac{3(N_1\text{-}N_2)}{2\pi L^3 C_j^0} \ . \qquad (9.42)$$

Divide (9.42) into (9.40), and obtain

$$\frac{S_q^j}{S_s^j}=\frac{N_1\text{-}N_2}{N_2} \ . \qquad (9.43)$$

Transform (9.40) with allowance for (9.43) to the form

$$\exp\left[\text{-}D_j S_q^j\left[1+\frac{S_s^j}{S_q^j}\right]\right](t_*\text{-}t_0)=1\text{-}\frac{3N_2(t_*)}{2\pi L^3 C_j^0}\left[1+\frac{S_q^j}{S_s^j}\right],$$

which gives

$$S_q^j=\frac{N_1(t_*)\text{-}N_2(t_*)}{N_1(t_*)D_j(t_*\text{-}t_0)}\ln\frac{2\pi L^3 C_j^0}{2\pi L^3 C_j^0\text{-}3N_1(t_*)} \ . \qquad (9.44)$$

Hence, the sink strength q inside the point can be evaluated by (9.44) from two series of measurements for the samples without inner sinks ($N_1(t_*)$) and with such sinks identified by the layer-by-layer sample evaporation after the values $N_2(t_*)$ have been properly measured.

It is to be noted that the strength $\tilde{S}_s^j$ of a point-surface sink can be evaluated from the data for the sample without an inner sink. In this case, we have

$$N_1(t_*)=\frac{V}{2}\int_{t_0}^{t_*} D_j\tilde{C}_j(t')\tilde{S}_s^j dt',$$

where $\tilde{C}_j(t)=C_j^0\exp[-D_j\tilde{S}_s^j(t-t_0)]$. Now, performing the calculation analogously to (9.41) and (9.44), we obtain

$$\tilde{S}_s^j=\frac{1}{D_j(t_*-t_0)}\ln\frac{2\pi L^3C_j^0}{2\pi L^3C_j^0-3N_1(t_*)}.$$

However, in order to calculate the material properties, it is required a knowledge of the rates $I_q^j(R_q)$ and the efficiencies K_D^j of defect precipitation (see, e.g., (8.22)). If the sinks q are distributed in the material in their dimensions R_q with the function $f_q(R_q)$, then their total concentration C_q is expressed in the form (see (7.10))

$$C_q=\int_{R_0}^{R_q} f_q(R'_q)dR'_q . \tag{9.45}$$

The expression (2.19) (see Chapter 2) gives the rate of j-defect outgoing per a unit volume onto the q sinks. Multiply and divide the right-hand side of the relation (2.19) by (9.45), we write

$$D_jC_j(t)S_q^j=C_q\langle I_q^j(t)\rangle, \tag{9.46}$$

where

$$\langle I_q^j(t)\rangle=\frac{\int_{R_{q_0}}^{R_q} I_q^j(R'_q)f_q(R'_q)dR'_q}{\int_{R_{q_0}}^{R_q} f_q(R'_q)dR'_q} \tag{9.47}$$

is the average rate of j defect adsorption by the q sinks. If (9.46) is substituted into (9.41), we obtain

$$\int_{t_0}^{t_*} \langle I_q^j(t')\rangle dt' = \frac{3[N_1(t_*)-N_2(t_*)]}{2\pi L^3 C_q}. \tag{9.48}$$

Differentiate (9.48) with respect to the time, then we have

$$\langle I_q^j(t)\rangle = \frac{3}{2\pi L^3 C_q}\left(\frac{dN_1}{dt}-\frac{dN_2}{dt}\right). \tag{9.49}$$

Therefore, with a knowledge of the functions $N_1(t)$ and $N_2(t)$ from the experiment, one can follow the kinetics of $\langle I_q^j(t)\rangle$ changing.

The analytical calculations [93,381] yield the expressions (6.35) for the rates $I_q^j(R_q)$ and (6.36) for the adsorption efficiencies K_D^j. Then, substituting (6.35) into (9.47), and further into (9.48), we shall have

$$\int_{t_0}^{t} \langle R_q(t')\rangle dt' \approx \frac{3[N_1(t)-N_2(t)]}{2\pi L^3 C_q \alpha_q^j D_j (C_j^+ + \kappa_q^j)},$$

or, analogously to (9.49),

$$\langle R_q(t)\rangle \simeq \frac{3\left(\frac{dN_1}{dt}-\frac{dN_2}{dt}\right)}{2\pi L^3 C_q \alpha_q^j D_j (C_j^+ + \kappa_q^j)}. \tag{9.50}$$

Referring to (9.50), the reliability of the theoretical dependency (6.35) can be tested if, aside from $N_1(t)$ and $N_2(t)$, one measures the values $\langle R_q(t)\rangle$ in various samples investigated with the layer-by-layer evaporation.

If some edge dislocation will prove as an internal sink, then (see

(2.20)), we have

$$D_j C_j(t) S_D^j = \rho_D K_D^j, \tag{9.51}$$

where $C_j(t) \equiv C_j^m(t, k_j^2)$. Further, substituting (6.36) into (9.51) and then into (9.41), we obtain

$$\frac{2\pi L^3}{3} \rho_D \alpha_D^j D_j (C_j^+ + \kappa_D^j)(t - t_0) = N_1(t) - N_2(t),$$

from where we derive

$$\alpha_D^j = \frac{3[N_1(t) - N_2(t)]}{2\pi L^3 \rho_D D_j (C_j^+ + \kappa_D^j)(t - t_0)} \, . \tag{9.52}$$

The expression (9.52) estimates, in an explicit form, the preferential factor α_D^j for the dislocation using the defects j.

Of course, the expression listed here have been derived from simple (9.36)-type equations without regard for the recombination, the impurity reactions, the complex formation and dissociation, and so on. However, they demonstrate some possibilities of such experimental data handling. It can be easily applied to more realistic models of physical processes, since it permits standardized computations.

It was shown by this means that the direct autoionic experiment makes it possible to evaluate the values of sink strengths, the average rates of defect adsorption by the sinks of various nature or the dimensions of such sinks, as well as to reveal the functional dependencies of the adsorption rates on the sink dimensions.

CONCLUSION
PHYSICAL UNITY OF RADIATION DAMAGE PROCESSES IN MATERIALS

Total combination of currently available achievements in radiation damage physics allows, in principle, to go into a description of the interrelations between various aspects of radiation damage in solids, basing on a close integration of the results of analytical and computer calculations [6,15,23, 26, 29-32, 38, 43, 61, 93, 99, 104, 126, 135, 145, 200, 202, 204, 206, 208, 218, 248, 274, 276, 294, 449-452], as well as specialized physical experiments (e.g. Refs. 37, 296, 299, 351, 453-459).

The mechanisms and models of radiation swelling, radiation creep and radiation precipitation are most developed to date. In so far as the macrochanges mentioned are different facets of the same radiation damage of a material, then it is natural that the physical processes leading to such changes prove to be common for those all phenomena (see Diagram in Fig. 2.7). Using as an example the outlined macrochanges in the properties of the materials under irradiation, we have illustrate their intimate physical interplay and the possibilities in the elucidation of general physical regularities in a diagrammatic form (Fig. 1).

Radiation swelling describes changes in the volume of the sample under irradiation and is related to the matter rearrangement. To put it differently, one can say that the radiation swelling can be evaluated from a knowledge of the kinetics of ensemble of radiation and structure defects. When evaluated theoretically, it is necessary to solve a complex system of closed and self-consistent integro-differential equations for the point defect concentrations, their clusters (voids, dislocation loops), the structure defects in the material (dislocations, secondary-phase precipitates, and so on), as well as the equations for the defect sizes, their distribution functions in size, and the sink strengths for the point defects on different sinks.

Radiation creep, i.e. the material constant-load deformation under the action of irradiation, is, in fact, a sufficiently slow transfer of radiation and structure defects. Compared to the calculations made for the radiation swelling, it is here important to implicitly take into account the mobility of such structural defects as dislocations in the field of radiation defect ensemble. It is to note that all evolution peculiarities being taken into account in considering the radiation swelling, contribute to the radiation creep, too.

Radiation hardening, i.e. an increase in the yield strength in materials exposed to radiation, is, above all, closely associated, with the deceleration of the slip dislocations by the radiation and structure defects in the material. It is necessary to consider the interactions of the defects of a different origin. The difficulties arising in the description , such as a random character and multiplicity of the distribution of barrier obstacles on the path of moving dislocations, can be most successfully overcome by the computer simulations.-

The elaboration of the models listed has brought to light a number of difficulties in developing the radiation damage theory. Because the systems of equations which need to be solved sometimes include from ten to thousands of simultaneous equations, it is clear that a purely analytical solution is impossible to find, and qualitative estimates can be only obtained. At the same time the computer simulations still yield only isolated results, and it is possible to gather some statistics only in a few cases. Their main role is reduced to revealing some individual mechanisms of radiation damage, finding one or another of the parameters suitable for a further use, for example as the initial conditions in the analytical models (see Fig. 2).

Nowadays qualitative data can be obtained by a combination of analytical and computer calculations. Some prerequisites for such complex theoretical investigations of radiation damage in solids are now available. It is necessary to concentrate physicist strengths on this important problem tying the developed models, general principles of constructing a unitary theory of radiation damage, as well as special fine physical experiments on the modeling materials. This all permits to develop, in the nearest future, the procedures for the combined evaluations of changing the properties of the modeling materials exposed to radiation for the purpose of constructing a proper physical theory of radiation damage in solids which could predict the behavior of a wide range of the materials under a hard radiation.

REFERENCES

1. Kelly B.T. Irradiation Damage to Solids, Pergamon Press, 1968.
2. Thompson M.W. Defects and Radiation Damage in Metals. Cambridge: University Press, 1969.
3. Lehman Chr. Interaction of Radiation with Solids and Elementary Defect Production, North-Holland Company, 1977.
4. Ma B.M. Nuclear Reaction Materials and Applications, Van Nostrand Reinhold Company Inc., 1983.
5. Ohtsuki Y.H. Charged Beam Interaction with Solids. London-New York: Taylor and Francis Ltd., 1983.
6. Kirsanov V.V., Suvorov A.L., Trushin Yu.V. Processes of Radiation Defect Formation in Metals. M.: Energoatomizdat, 1985 (in Russian).
7. Zelensky V.F., Nekhludov I.M., Chernyaeva T.I. Radiation Point Defectfects and Swelling of Metals. Kiev: Naukova Dumka, 1988 (in Russian).
8. Parshin A.M. Structure and Radiation Damage of Corrosion- Steady Steels and Alloys. Chelyabinsk: Metallurgy, 1988 (in Russian).
9. Konobeevsky S.T. Influence of Irradiation to Materials. M.: Atomizdat, 1967 (in Russian).
10. Sputtering by Particle Bombardment. I. Physical Sputtering of Single-Element Solids/ Ed. R.Behrisch. Berlin-Heidelberg-New York: Springer-Verlag, 1981.
11. Sputtering by Particle Bombardment. II. Sputtering of Alloys and

Compounds, Electron and Neutron Sputtering, Surface Topography/ Ed R Behrisch. Berlin-Neidelberg- New York-Tokyo: Springer-Verlag, 1983.

12. Gerasimov V.V., Monakhov A.S. Materials of Nuclear Engineering. M.: Energoizdat, 1982 (in Russian).
13. Abramovich M.F., Votinov S.I., Poltuchovsky A.G. Radiation Material Science from APP. M.: Energoatomizdat, 1984 (in Russian).
14. Ibragimov Sh.Sh., Kirsanov V.V., Pyatiletov Yu.S. Radiation Damage of Metals and Alloys. M.: Energoatomizdat, 1985 (in Russian).
15. Akhiezer I.A., Davydov L.N. Introduction to the Theoretical Physics of Metals and Alloys. Kiev: Naukova Dumka, 1985 (in Russian).
16. Shalaev A.M. Radiation-Induced Processes in Metals. M.: Energoatomizdat, 1988 (in Russian).
17. Proceedings of the School on the Theory of Defects in Crystals and Radiation Damage (Telavi, 1965). V. 1-3. // Tbilisi: Institute of Physics of the GSSR Academy of Sciences, 1966 (in Russian).
18. Radiation and Other Defects in Solids (Tbilisi, 1973)/ Tbilisi: Institute of the GSSR Academy of Sciences, 1974 (in Russian).
19. Radiation Physics of Crystals. Collected Papers/ Sverdlovsk: Proseedings of the Metal Physics Institute, Ural Research Center of the USSR Academy of Sciences, 1977 (in Russian).
20. Modern Problems in Condenced Matter Sciences/ Eds. V.M.Agronovish, A.A.Maradudin. V. 13// In: Physics of Radiation Effects in Crystals/ Eds. R.A. Johnson and A.N. Orlov, North-Holland, 1986.
21. Theory and Somputerized Stimulating of Defect Structures in Crystals.

Collected Papers/ Eds.A.N. Orlov, B.A. Grinberg. Sverdlovsk: Ural Research Center of the USSR Academy of Sciences 1986 (in Russian).

22. Questions of the Theory of Defects in Crystals. Collected Papers/ Eds. S.V. Vonsovsky, M.A. Krivoglaz. L.: Nauka, 1987 (in Russian).
23. Computerized Modeling in Investigation of Materials/ Ed. D.B. Pozdneev. M.: Mir, 1974 (in Russian).
24. Kosevich A.M., Saralidze Z.K., Slesov V.V.// *JETP*. V. 52. P. 1073-1080. 1967 (in Russian).
25. Bystrov l.N., Ivanov l.i., Platov Yu.M.// *DAN of the USSR*. 1969. V. 185. P. 309-312 (in Russian).
26. Katz J.L., Wiedersich M.// *J. Chem. Phys.*. 1971. V. 55. P. 1414-1475.
27. Russel K.C.// *Scripta Met.* 1972. V. 6. P. 209-214.
28. Konobeev Yu.V., Subbotin A.V., Golubov S.I.// *Rad. Effects*. 1973. V. 20. P. 265-271.
29. Heald P.T., Speght M.V.// *Phil. Mag*. 1974. V. 30. P. 869- 875.
30. Bullough R., Hayns M.R.// *J. Nucl. Mater*. V. 57. P. 348- 352.
31. Heald P.T., Bullough R.// *Vacancies. Proc. Int. Conf. Point Defects Behaviour and Diffusion Processes/* London. 1977. P. 134-138.
32. Saralidze Z.K.// *Sov. Phys. Sol.* 1978. V. 20. p. 378-384.
33. Saralidze Z.K.// *Sov. Phys. Sol.* 1978. V. 20. P. 2716- 2720.
34. Brailsford A.D., Bullough R.// *Phil. Mag*. 1979. V. 27. P. 49-64.
35. Wolfer W.G.// *Phil. Mag. A*. 1981. V. 43. P. 61-70.
36. Ehrlich K.J.// *J. Nucl. Mater*. 1981. V. 100. P. 149-166.
37. Suvorov A.L. Autoionic Microscopy of Radiation Defects in Metals. M.:

Energoatomizdat. 1982 (in Russian).

38. Kirsanov V.V., Orlov A.N.// *Uspekhi Fiz. Nauk.* 1984. V. 142. P. 219-264 (in Russian).
39. Ehrlich K.J.// *J. Nucl. Mater.* 1985. V. 133/134. P. 119- 126.
40. Abov Yu.G., Ivanov L.I., Suvorov A.L., et al.// *Preprint of ITEF/* M. 1985 (in Russian).
41. Weber H.W.// *Advances in Cryogenic Engineering Materials.* 1986. V. 32 P. 853-864.
42. Schneider W., Wassilev C., Ehrlich K.Z.// *Metallkunde.* 1986. Bd. 77. S. 611-619.
43. Slesov V.V., Sagaglovich V.V.// *Uspekhi Fiz. Nauk.* 1987. V. 151. P. 67-104 (in Russian).
44. Umesawa A., Crabtree G.W., Weber H.W., et al.// *Phys. Rev. B.* 1987. V 36. P. 7151-7154.
45. Wassilev C., Ehrhich K., Bergmann H.-J.// *Proc. 13th Int. Sympos. Influence of Radiation on Material Properties/* Eds. F.A. Garner, C.H. Henager, N. Igata. Philadelphia. 1987. P. 33-53.
46. Bagley K.Q., Little E.A., Ehrlich K., et al.// *Materials for Nuclear Reactor for Application.* London: BHES. 1987. P. 37-46.
47. Kirk M.A., Baker M.C., Weber H.W., et al.// *Mat. Res. Sympos. Proc.* 1988. V. 99. P. 209-214.
48. Weber H.W., Khier W., Wacenovsky M.// *Advances in Cryogenic Engineering Materials.* 1988. V. 34. P. 1033- 1039.
49. Khavanchak K., Senesh D., Shchgolev V.A.// *Phys. Chem. Processing of*

Materials. 1989. No. 2. P. 5-10 (in Russian).

50. Orlov V.V., Altovsky I.V.// *Questions of Atomic Science and Technics. Ser. Radiation Damage Physics and Radiation Material Science*. 1981. No. 1(15). P. 9-16.
51. Seeger A.J.// *J. Nucl. Mater*. 1989. V. 165. P. 7-17.
52. Seeger A., Kronmller H.// *J. Nucl. Mater*. 1989. V. 166. P. 18-22.
53. Frank W., Seeger A.// *J. Nucl. Mater*. 1989. V. 166. P. 74-80.
54. Frank W., Weller M., Seeger A.// *J. Nucl Mater*. 1989. V. 166. P. 82-87.
55. Veprek S., Mattenberger F., Heintze M.// *J. Vac. Sci. Technol. A*. 1989. V. 7. P. 69-76.
56. Chernikov V.N., Zakharov A.P.// *J. Nucl. Mater*. 1989. V. 165. P. 89-100.
57. Mansur L.K., Farrell K. Research by the Division of Materials Sciences. U.S. Dapartment of Energy. 1989.
58. Mansur L.K.// *Canadian J. Phys*. 1990. V. (Proc. Conf. on Kinetics of Nonhomogeneous Processes, Canada, Oct., 1989).
59. Cocke D.L., Jucik-Rajman M., Veprek S.// *J. Electrochem. Soc*. 1989. V. 136. No. 12. P. 3655-3662.
60. Veprek S., Sarrott F.-A., Rambert S., Taglauer E.// *J. Vac. Sci. Technol*. 1989. V. A7(4). P. 2614-2629.
61. Abromeit C.// *Int. J. Mod. Phys. B*. 1989. V. 13. No. 9. P. 1301-1342.
62. Naundorf V., Abromeit C.// *Nucl. Instrum. and Methods in Physics Research*. 1989. V. B43. P. 513-519.

63. Abromeit C.// *Defect and Diffusion Forum*. 1990. V. 66-69. P. 1153-1168.

64. Estrin Yu., Abromeit C., Aifantis A.E.// *Phys. Stat. Sol. (b)*. 1990. V. 157. P. 117.

65. Abromeit C., Trinkaus H., Wollenberger H., Can J.// *Physics*. 1990.

66. Vell B., Wollenberger H.// *J. Nucl. Mater*. 1989. V. 169. P. 126-130.

67. Wollenberger H.// *Nucl. Instrum. and Methods on Physics Research*. 1990. V. B48. P. 493-498.

68. 6th Symposium Thermodynamics of Nuclear Materials. Canada. 1985.// *J Nucl. Mater*. 1985. V. 130.

69. Proc. of the First Int. Conf. on Fusion Reactor Materials.// *J. Nucl. Mater*. 1985. V. 133/134.

70. 2th Int. Conf. Fusion Reactor Materials (ICFRM-2). Chicago. 1986.// *J. Nucl. Mater*. 1986. V. 141-143.

71. The Relation between Mechanical Properties and Microstructure under Fusion Irradiation Conditions. Denmark. 1985.// *Rad. Effects*. 1987. V. 101. No. 1/4.

72. Proc. Int. Conf. on Fundamental Mechanisms of Radiation-Induced Creep and Growth, Canada. 1987// *J. Nucl. Mater*. 1988. V. 159.

73. Proc. 3rd Int. Conf. on Fusion Reactor Materials (TCFRM-3). Karlsruhe, RG, 1987// *J. Nucl. Mater*. 1988. V. 155-157.

74. Proc. Int. Symposium on the Utilization of Multipurpose Research Reactors and Related International Cooperation. Grenoble, 1987. Vienna: IAEA. 1988.

75. Proc. Int. Conf. "Energy Pulse and Particle Beam Modification of Materials", Dresden, 1987. Berlin: Academie-Verlag. 1988.

76. 8th Conf. on Plasma-Surface Interactions in Controlled Fusion Devices, Jlich, 1988// *J. Nucl. Mater.* 1989. V. 162-164.

77. Radiation Materials Science// Proc. Int. Conf. on Radiation Materials Science, Alushte, May 22-25, 1990/ Kharkov. 1990.

78. Bystrov L.N., Ivanov L.I., Platov Yu.M.// In: Atom Ordering and Its Influence to the Alloy Properties. Kiev: Naukova Dumka. 1968. P. 250-254 (in Russian).

79. Bystrov A.N., Ivanov L.I., Platov Yu. M.// In: Structural Materials for the Thermonuclear Fusion Reactors/ M.: Nauka. 1983. P. 5-18 (in Russian).

80. Balandin Yu.F., Gorynin I.V., Zvezdin Yu.I., et al. NPS Structural Materials/ M: Energoatomizdat. 1984 (in Russian).

81. Platov Yu.M., Simakov S.V.// *Sov. Phys. Met. and Met. Sci.* 1986. V. 61. P. 213-217 (in Russian).

82. Kirsanov V.V., Kislitsin S.V.// *Izvestia of the Kaz. SSR Academy of Sciences.* 1984. V. 6. P. 5-7 (in Russian).

83. Bystrov L.N., Ivanov L.I., Platov Yu.M.// *Phys. Stat. Sol. (a).* 1971. V. 8. P. 375-381.

84. Ivanov L.I., Lazorenko V.M., Platov Yu.M., et al.// Doklady of *SSSR* Ac. Sci. 1981. V. 257. P. 1175-1178 (in Russian).

85. Kirsanov V.V., Kislitsin S.V.// *Izvestia of the Kaz. SSR Academy of Sciences.* 1983. V. 6. P. 1-7 (in Russian).

86. Bakaj A.S., Platov M.P.// *Sov. Phys. Met. and Met. Sci.* 1987. V. 64. P. 458-463 (in Russian).

87. Bakaj A.S., Fateev M.P.// *Letters to JETP*. 1983. V. 9. P. 1287-1291 (in Russian).

88. Ivanov L.I., Platov Yu.M., Pletnev M.N.// *Phys. Chem. Mat. Processing*. 1975. No. 1. P. 72-76 (in Russian).

89. Orlov A.N., Trushin Yu.V.// In: Computerized Modeling of Crystal Defects/ L.: A.F. Ioffe Physical-Technical Institute of the USSR Academy of Sciences. 1979. P. 74-102 (in Russian).

90. Orlov A.N., Trushin Yu.V.// In: Computers and Modeling Defects in Crystals/ L.: A.F. Ioffe Physical-Technical Institute of the USSR Academy of Sciences. 1982. P. 51-67 (in Russian).

91. Trushin Yu.V., Orlov A.N., Samsonidze G.G.// *Questions of Atomic Science and Technics. Ser. Radiation Damage Physics and Radiation Material Science.* 1983. No. 3(26). P. 14-20 (in Russian).

92. Orlov A.N., Parshin A.M., Trushin Yu.V.// *Sov. Phys. Techn. Phys.* 1983. V. 28(12). P. 1455-1458.

93. Trushin Yu.V., Orlov A.N.// *Sov. Phys. Techn. Phys.* 1986. V. 31(7). P. 763-767.

94. Trushin Yu.V.// Questions of Crystal Defect Theory/ L.: Nauka. 1987. P. 133-144 (in Russian).

95. Orlov A.N., Parshin A.M., Trushin Yu.V.// *Phys. Chem. Mat. Processing*. 1987. No. 3. P. 22-27 (in Russian).

96. Trushin Yu.V.// In: Computerized Modeling of Defects in Metals/ L.:

Nauka. 1990. P. 119-145 (in Russian).

97. Trushin Yu.V.// *Superpure Substance*. 1991. No. 3. P. 50-56 (in Russian).

98. Trushin Yu.V.// *Sov. Phys. Techn. Phys*. 1991. V. 36(1). P. 42-45.

99. Trushin Yu.V.// *J. Nucl. Mater*. 1991. V. 185. P. 279-285.

100. Trushin Yu.V.// Proc. Int. Conf. on Physics of Irradiation Effects in Metals, May 20-24, 1991/ SiÂfok, Hungary: T-19.

101. Wigner E.F.// *J. Appl. Phys*. 1946. V 17. P. 857-870.

102. Orlov A.N., Trushin Yu.V.// *Questions of Atomic Science and Technics. Ser. Radiation Damage Physics and Radiation Material Science*. 1979. No. 1(9). P. 1-9.

103. Trushin Yu.V.// *Ibid*. 1980. No. 1(12). P. 3-13.

104. Orlov A.N., Trushin Yu.V.// *Rad. Effects*. 1981. V. 56. P. 193-204.

105. Beavan L.A., Scanlan P.M., Seidman D.N.// *Acta Met*. 1971. V. 19. P. 1339-1350.

106. Kirsanov V.V.// In: Reactor Materials/ M.: Central Institute Atominform. 1978. V.1. P. 340-353 (in Russian).

107. Wei C.-Y., Seidman D.N.// *Phil. Mag*. 1981. V. 43A. P. 1419-1439.

108. Kirsanov V.V.// In: Radiation Pzysics of Solids and Reactor Material Science/ M.: Atomizdat. 1970. P. 27-34 (in Russian).

109. Leibfried G.// *J. Appl. Phys*. 1959. V. 30. P. 1388-1396.

110. Gibson J.B., Goland A.N., Vineyard G.H., et. al.// *Phys. Rev*. 1960. V. 120. P. 1229-1237.

111. Frere R.// *Phys. Stat. Sol*. 1962. V. 3. P. 1252-1259.

112. Oen O.S., Robinson M.T.// *Appl. Phys. Lett.* 1963. V. 2. P. 83-85.

113. Garber P.I., Fedorenko A.I.// *Uspekhi Fiz. Nauk.* 1964. V. 83. P. 385-432 (in Russian).

114. Kirsanov V.V.// *Sov. Phys. Sol.* 1977. V. 19. P. 1184-1189 (in Russian).

115. Zabolotny V.T., Ivanov L.I., Suvorov A.L., et al.// *Atomic Energy.* 1980. V. 48. P. 326-327 (in Russian).

116. Kalashnikov N.P., Remizovich V.S., Rezanov M.I.// Fast Particle Collisions in Solids/ M.: Atomizdat. 1980 (in Russian).

117. Kumakhov M.A., Shrimer G.// Atomic Collisions in Crystals/ M.: Atomizdat. 1980 (in Russian).

118. Kirsanov V.V.// *Phil. Mag. A.* 1981. V. 43. P. 1441-1445.

119. Suvorov A.L.// In: Radiation Defecis in Metals/ Alma-Ata: Nauka. 1981. P. 23-32 (in Russian).

120. Kinchin G.N., Pease R.S.// *Rep. Progr. Phys.* 1955. V. 12. P. 1-51.

121. Sneider W.S., Neufeld J.// *Phys. Pev.* 1955. V. 97. P. 1636-1643.

122. Robinson M.T.// *Phil. Mag.* 1965. V. 12. P. 741-765; V. 17. P. 639-643.

123. Sigmund P.// *Appl. Phys. Lett.* 1974. V. 25. P. 169-171.

124. Littmark U., Gras-Marti A.// *Appl. Phys.* 1978. V. 16. P. 147-253.

125. Winterbon K.B., Sanders J.B.// *Rad. Effects.* 1979. V. 39. P. 39-44.

126. Littmark U., Ziegler J.F.// *Phys. Rev. A.* 1981. V. 23. P. 64-72.

127. Fedder S., Littmark U.// *J. Appl. Phys.* 1981. V. 52. P. 4259-4265.

128. Littmark U., Fedder S.// *Nucl. Instr. Meth.* 1982. V. 194. P. 607-610.

129. Kirk M.A., Robertson J.M., Jenkins M.L., et al.// *J. Nucl. Mater.*

1987. V. 149. P. 21-28.

130. Baroody B.M.// *Phys. Rev.* 1958. v. 112. P. 1571-1576; V. 116. P. 1418-1423.

131. Kostin M.D.// *J.* Appl. Phys. 1966. V. 37. P. 3801-3809.

132. Chadderton L.T., Morgan D.V., Torrens I.// *Solid State Comm.* 1966. V. 4. P. 391-394.

133. Torrens I., Chadderton L.I.// *Phys. Rev.* 1967. V. 159. P. 671-682.

134. Felder R.M.// *J. Phys. Chem. Sol.* 1967. V. 28. P. 1383- 1387.

135. Winterbon K.B., Sigmund P., Sanders J.B.// *Kgl. dansk. selsk. vid. Mat.-fys. Medd.* 1990. V. 37. P. 1-73.

136. Jackson R.O., Leighley N.P., Edwards D.R.// *Phil. Mag.* 1972. V. 25. P. 1169-1193.

137. Andersen H.U., Sigmund P., Kong D.// *Kgl. danske. vid. selsk. Mat.-fis. Medd.* 1974. V. 39. No. 3.

138. Coutler C., Parkin D.M.// *J. Nucl. Mater.* 1980. V. 88. P. 249-260.

139. Mercle K.L.// Proc. Sympos. Nature Small Defects Clusters/ England, Harwell: AERE-R5269. 1966. P. 8-21.

140. Mercle K.L., Averback R.S.// Proc. Int. Conf. on Fundamental Aspects of Radiation Damage in Metals, Gatlinburg, USA, 1975/ Washington. 1976. V. 1. P. 127-133.

141. Kirsanov V.V.// *Rad. Effects.* 1980. V. 46. P. 167-174.

142. Kirsanov V.V.// In: Defects in Crystals and Their Computerized Modeling/ L.: A.F. Ioffe Physical-Technical Institute. 1980. P. 134-135 (in Russian).

143. Pletnev M.I., Platov Yu.M.// *Sov. Met. Phys. and Met.-Sci.* 1975. V. 40. P. 304-310 (in Russian).

144. Dworschak F., Birtcher R.C., Averback R.S.// *Rad. Effects.* 1987. V. 100. P. 313-322.

145. Kirsanov V.V. Computerized Experiment in Atomic Material Science. 1990. M.:Energoatomizdat. 304 P (in Russian).

146. Indenbom V.L.// *Sov. Tech. Phys. Lett.* 1979. V. 5. P. 489-492 (in Russian).

147. Zhetbaeva M.P., Indenbom V.L., Kirsanov V.V., et al.// *Sov. Tech. Phys. Lett.* 1979. V. 5. P. 1157-1161 (in Russian).

148. Orlov A.N., Trushin Yu.V.// *Questions of Atomic Science and Technics. Ser. Radiation Damage Physics and Radiation Material Science.* 1985. No. 2(35). P. 14-26.

149. Oen O.S., Robinson M.T.// *J. Appl. Phys.* 1963. V. 34. P. 302-312.

150. Dederichs P.H.// *Phys. Stat. Sol.* 1065. V.10. P. 303-318.

151. Sanders J.B.// *Physica.* 1966. V. 32. P. 2147-2160.

152. Trushin Yu.V.// *Sov. Phys. Sol. State.* 1974. V. 16. P. 3435-3436 (in Russian).

153. Sanders J.B., Winterbon K.B.// *Rad. Effects.* 1974. V. 22. P. 109-115.

154. Trushin Yu.V.// *Preprint of PTI-488/ L.: A.F. Ioffe PTI of USSR Academy of Sciences.* 1975 (in Russian).

155. Trushin Yu.V.// *Sov. Phys. Met. and Met. Sci.* 1975. V. 40. P. 15-20.

156. Sanders J.B., Roosendaal H.E.// *Rad. Effects.* 1975. V. 24. P. 161-172.

157. Akhiezer A.I., Akhiezer I.A.// *Questions of Atomic Science and*

Technics. Ser. Radiation Damage Physics and Radiation Material Science. 1975. No. 1(12). P. 3-7.

158. Akhiezer I.A., Ginzburg A.Z.// *Ukr. Phys. J.* 1977. V. 22. P. 1233-1237, 1450-1454 (in Russian).
159. Trushin Yu.V.// In: Radiation Physics/ Sverdlovsk: Institute of Metal Physics of Ural Research Center of the USSR Academy of Sciences. 1977. P. 66-80 (in Russian).
160. Williams M.M.R.// *Rad. Effects*. 1978. V. 37. P. 131-145.
161. Sanders J.B., Roosendaal H.E., Vitalis R.// *Rad. Effects*. 1978. V. 38. P. 201-210.
162. Williams M.M.R.// *J. Phys. D*. 1978. V. 11. P. 801-821.
163. Winterbon K.B.// *Rad. Effects*. 1979. Vol. 39. P. 31-38.
164. Weinberg A.M., Wigner E.P. The Physical Theory of Neutron Chain Reactors. The University of Chicago Press, 1959.
165. Lindhard J., Nielsen V., Scharff M. et al.// *Kgl. dansk. vid. selskab. Mat.-fys. medd.* 1963. Vol. 33. No 10.
166. Trushin Yu.V., Platovskikh Yu.A., Popkov K.K. et al.// *Izv. Akad. Nauk, Bel.SSR, Ser. of Phys. and Energ. Nauk*. 1973. No 3. P. 5-15 (in Russian).
167. Lindhard J., Scharff M, Schiott H.E.// *Kgl. dansk. vid. selskab. Mat.-fys. medd.* 1963. Vol. 33. No 14.
168. Agranovich V.M., Kirsanov V.V.// *Uspekhi Fiz. Nauk*. 1976. Vol. 118. P. 3-51 (in Russian).

169. Beeler J.R. Radiation Effects Computer Experiments. North-Holland, Amsterdam, 1983.

170. Gann V.V., Volobuev A.V., Yamnitsky V.A.// *Questions of Atomic Scier and Technics. Ser. Radiation Damage Physics and Radiation Material Science.* 1980. No 1 (12). P. 49-53 (in Russian).

171. Guinan M.W., Kinney J.H.// *J. Nucl. Mater.* 1981. Vol. 103/104. P. 1319-1325.

172. Kirsanov V.V.// *Questions of Atomic Science and Technics. Ser. Radiation Damage Physics and Radiation Material Science.* 1980. No 1 (12). P. 28-40 (in Russian).

173. Beeler J.R.// Proc. Int. Conf. on Fundamental Aspects of Radiation Damage in Metals. Gatlinburg, USA, 1975/ Washington. 1976. Vol. 1. P. 127-133.

174. Gann V.V., Rozhkov V.V.// *Questions of Atomic Science and Technics. Ser. Radiation Damage Physics and Radiation Material Science.* 1981. N 2 (16). P. 37-38 (in Russian).

175. Crystal Defects and Their Computer Modelling. Ed. Yu.A. Osipjan. L.: Nauka, 1980 (in Russian).

176. Computer Simulating of Defects in Metals. Ed. Yu.A. Osipjan. L.: Nauka, 1990 (in Russian).

177. Sokursky Yu.N// *Questions of Atomic Science and Technics. Ser. Radiation Damage Physics and Radiation Material Science.* 1978. No 2 (7). P. 49-65 (in Russian).

178. Seidman D.N.// *Surface Science.* 1978. Vol. 70. P. 532- 565.

179. Modern Electron Microscopy in Matter Study. M.:Nauka. 1982 (in Russian).

180. Seeger A.// *Comments Solid State Phys*. 1972. Vol. 4. P. 79-83.

181. Orlov A.N., Trushin Yu.V. Point Defect Energies in Metals. M.: Energoatomizdat, 1983 (in Russian).

182. Dederichs P.H., Leibfried G.// *Z. Physik*. 1962. Bd. 170. S. 320-335.

183. Christian J.W. The Theory of Transformation in Metals and Alloys. Part 1, Pergamon Press, 1975.

184. Lyubov B.Ya. Kinetic Theory of Phase Transformations. M.:Metallurgy, 1969 (in Russian).

185. Lyubov B.Ya. Diffusion Processes in Inhomogeneous Solids. M.: Nauka, 1981 (in Russian).

186. Lifshitzs I.M., Slyozov V.V.// *JETP*. 1958. Vol. 35. P. 479-492 (in Russian).

187. Lifshitz I.M., Slyozov V.V.// *J. Phys. Chem. Solids*. 1961. Vol. 19. P. 35-50.

188. Wagner C.// *Z. Elektrochem*. 1961. Vol. 65. P. 581-591.

189. Glowinsky L.D.// *J. Nucl. Mater*. 1976. Vol. 61. P. 8-21.

190. Norris D.I.R.// *Radiation Effects*. 1972. Vol. 14. P. 1- 37.

191. Little E.A., Bullough R., Wood M.H.// *Proc. Roy. Soc. London*. 1980. Vol. A 372. P. 565-579.

192. Becker R., Dring W.// *Ann. d. Physik*. 1935. Vol. 24. P. 719-752.

193. Bondarenko A.I., Konobeev Yu. V.// *Phys. Stat. Sol*. 1976. Vol. 31. P. 195-205.

194. Russell H.C.// *Acta Met.* 1978. Vol. 26. P. 1615-1630.-

195. Pethenkin V.A., Strokova A.M.// *Sov. Phys. Tech. Phys.* 1982. Vol. 27(9). P. 1051-1056.

196. Turchin S.I.// *Preprint of the Atomic Energy Institute*-3687/ M., 1982 (in Russian).

197. Altovsky I.V., Turchin S.I.// *Preprint of the Atomic Energy Institute*-3083/ M., 1979 (in Russian).

198. Koptelov E.A.// *Rad. Effects.* 1980. Vol. 45. P. 163-168.

199. Ryazanov A.I., Maximov L.A.// *Preprint of the Atomic Energy Institute*-2648/ M., 1976 (in Russian).

200. Akhiezer I.A., Davydov L.N.// *Ukr. Phys. J.* 1982. Vol. 27. P. 961-972 (in Russian).

201. Wilkes P.// *J. Nucl. Mater.* 1979. Vol. 83. P. 166-175.

202. Mansur L.K.// *J. Nucl. Mater.* 1979. Vol. 83. P. 109-127.

203. Wolfer W.G., Glasgov V.W.// *Acta Met.* 1985. Vol. 33. P. 1997-2004.

204. Margvelashvili I.G., Saralidze Z.K.// *Sov. Phys. Sol. State.* 1973. Vol. 15. P. 2665-2668 (in Rusian).

205. Pethenkin V.A.// *Sov. Phys. Tech. Phys.* 1980. Vol.25(3). P. 349-354.

206. Brailsford A.D., Bullough R., Hayns M.R.// *J. Nucl. Mater.* 1976. Vol. 60. P. 246-256.

207. Brailsford A.D., Bullough R.// *J. Nucl. Mater.* 1978. Vol. 69/70. P. 434-450.

208. Mansur L.K.// *Nucl. Technol.* 1978. Vol. 40. P. 5-34.

209. Krishan K.// *Radiation Effects.* 1980. Vol. 45. P. 169- 184.

210. Mansur L.K.// *Acta Met.* 1981. Vol. 29. P. 575-381. -

211. Heald P.T., Speight H.V.// *Acta Met.* 1975. Vol. 23. P. 1389-1399.

212. Galimov R.R., Goryachev S.B.// *Phys. Stat. Sol. (b).* 1989. Vol. 153. P. 443-454.

213. Galimov R.R., Goryachev S.B.// *Phys. Stat. Sol. (b).* 1989. Vol. 154. P. 43-54.

214. Wolfer W.G., Ashkin M.// *J. Appl. Phys.* 1976. Vol. 47. P. 791-800.

215. Heald P.T.// *Phil Mag.* 1975. Vol. 31. P. 551-558.

216. Ryazanov A.I., Borodin V.A.// *Rad. Effects.* 1981. Vol. 56. P. 179-185.

217. Subbotin A.V.// *Atomic Energy.* 1983. Vol. 54. P. 342-346.

218. Bullough R., Quigley T.M.// *J. Nucl. Mater.* 1981. Vol. 104. P. 1397-1402.

219. Bullough R., Quigley T.M.// *J. Nucl. Mater.* 1983. Vol. 113. P. 179-191.

220. Rauh H., Bullough R.// *Phil Mag. A.* 1985. Vol. 52. P. 333-356.

221. Ham F.S.// *J. Appl.Phys.* 1959. Vol. 30. P. 915-926.

222. Friedel J. Dislocations. Pergamon Press. 1964.

223. Hirth J.P., Lothe J. Theory of Dislocations. McGraw-Hill Book Company, New York-London-St.Luis. 1969.

224. Arfken G. Mathematical Methods in Physics. M.: Atomizdat, 1970 (in Russian).

225. Turchin S.I.// *Preprint of the Atomic Energy Institute-* 3326/ M. 1980 (in Russian).

226. Klyavin O.V., Likhodedov N.P., Orlov A.N.// *Sov. Phys. Sol. State.*

1986. Vol. 28(1). P. 84-87.

227. Masuda K., Sugano Y., Sato A.// *Phil. Mag. B.* 1981. Vol. 43. P. 869-879.

228. Kosevich A.M., Saralidze Z.K., Slyozov V.V.// *Sov. Phys. Sol. State.* 1964. Vol. 6. P. 3383-3391 (in Russian).

229. Saralidze Z.K., Slyozov. V.V..// *Sov. Phys. Sol. State.* 1965. Vol. 7. P. 904-911 (in Russian).

230. Brown L.M., Kelly A., Mayer R.M.// *Phil. Mag.* 1969. Vol. 19. P. 721-741.

231. Ryazanov A.I.// *Preprint of the Atomic Energy Institute-* 2621/ M., 1976 (in Russian).

232. Kroupa E.// *Proc. Summer School Held in Hrazany.* 1964/ Prague: Academia, 1966. P. 275-316.

233. Dokhner R.D.// *Sov. Phys. Sol. State.* 1969. Vol. 11. P. 1124-1131 (in Russian).

234. Plishkhin Yu.M., Podchinyonov I.E.// *Sov. Phys.Met. and Met. Sci.* 1971. Vol. 32. P. 254-258 (in Russian).

235. Bullough R., Newmann R.// *Rep. Progr. Phys.* 1970. Vol. 33. P. 101-148

236. Eshelby G. Continuous Dislocation Theory. M.:Izd. Inostr. Liter. 1963 (in Russian).

237. Teodosiu C. Elastic Models of Crystal Defects. Berlin, Heidelberg, New York, 1982, Springer Verlag.

238. Mikhailova Yu.V., Maximov L.A.// *JETP.* 1970. Vol. 59. P. 1368-1377 (Russian).

239. Harkness S.D., Che-Yu Li// *Metal. Trans.* 1971. Vol. 2. P. 1457-1470.

240. Brailsford A.D., Bullough R.// *J. Nucl. Mater.* 1972. Vol. 44. P. 121-135.

241. Wiedersich H.// *Rad. Effects.* 1972. Vol. 12. P. 111-125.

242. Roger W., Powell R.W., Russell K.C.// *Rad. Effects.* 1972. Vol. 12. P. 127-131.

243. Bullough R., Hayns M.R.// *J. Nucl. Mater.* 1975. Vol. 55. P. 237-245.

244. Mansur L.K., Wolfer W.G.// *J. Nucl. Mater.* 1978. Vol. 69/ 70. P. 825-829.

245. Gorbatov G.Z., Roytburd A.L.. Tyomkin D.K.// *Sov. Phys. Met. and Met. Sci.* 1981. Vol. 52. P. 790-799 (in Russian).

246. Koehler J.S.// *J. Appl. Phys.* 1975. Vol. 46. P. 2423- 2428.

247. Brailsford A.D.// *J. Nucl. Mater.* 1981. Vol. 102. P. 77- 86.

248. Brailsford A.D., Bullough R.// *Phil. Trans. Roy. Soc. London.* 1981. Vol. 302. P. 87-137.

249. Bullough R., Murphy S.M., Wood M.H.// *J. Nucl. Mater.* 1984. Vol. 122. P. 489-494.

250. Bullough R., Chomien N.M.// *J. Nucl. Mater.* 1985. Vol. 127. P.47-55.

251. Brailsford A.D.// *J. Nucl. Mater.* 1986. Vol. 139. P. 163- 178.

252. Saralidze Z.K., Slyozov V.V.// *Sov. Phys. Sol. State.* 1965. Vol. 7. P. 1605-1611 (in Russian).

253. Slyozov V.V.// *Sov. Phys. Sol. State.* 1967. Vol. 9. P. 3448-3455 (in Russian).

254. Slyozov V.V., Sagalovich V.V.// *Preprint of the Kharkov Physical and*

Technical Institute N 82-41. 1982 (in Russian).

255. Schulson E.M.// *J. Nucl. Mater.* 1979. Vol. 83. P. 239–264.
256. Boquet J.-L., Martin G.// *J. Nucl. Mater.* 1979. Vol. 83. P. 186-194.
257. Cauvin R., Martin G.// *Phys. Rev.B.* 1981. Vol. 23. P. 3322-3332, 3333-3348.
258. Maydet S.I., Russell K.C.// *J. Nucl.Mater.* 1977. Vol. 64. P. 101-114.
259. Cauvin R., Martin G.// *J. Nucl. Mater.* 1979. Vol. 83. P. 67-78.
260. Zeldovich Ya.B.// *JETP*. 1942. Vol. 12. P. 525-538 (in Russian).
261. Bakaj A.S., Turkin A.A.// *Sov. Tech. Phys. Lett.* 1987. Vol. 13(11). P. 535-536.
262. Bakaj A.S., Turkin A.A.// *J. Nucl. Mater.* 1988. Vol. 152. P. 331-333.
263. Ibragimov Sh.Sh., Kirsanov V.V., Melikhov V.D., et al.// *Questions of Atomic Science and Technics. Ser. Radiation Damage Physics and Radiation Material Science.* 1983. No 3(26). P. 54-58 (in Russian).
264. Ibragimov Sh.Sh., Kirsanov V.V., Melikhov V.D., et al.// *Rad. Effects.* 1985. Vol. 84. P. 281-287.
265. Bakaj A.S.// *Questions of Atomic Science and Technics. Ser. Radiation Damage Physics and Radiation Material Science.* 1983. No 3(26). P. 89-91 (in Russian).
266. Bakaj A.S., Turkin A.A.// *Questions of Atomic Science and Technics. Ser. Radiation Damage Physics and Radiation Material Science.* 1987. No 1(39). P. 12-24 (in Russian).
267. Bakaj A.S., Fateev M.P.// *Sov. Phys. Met. and Met. Sci.* 1988. Vol. 66. P. 239-246 (in Russian).

268. Okamoto P.P., Rehn L.E.// *J. Nucl. Mater.* 1979. Vol.-83. P. 2-33.

269. Erck R.A., Potter D.I., Wiedersich H.// *J. Nucl. Mater.* 1979. Vol. 80. P. 120-125.

270. Sklad P.S., Mitchell T.E.// *Scripta Met.* 1974. Vol. 8. P. 1113-1118.

271. Wagner A., Seidman D.N.// *J. Nucl. Mater.* 1979. Vol. 83. P. 48-56.

272. Takayama T., Ohnuki S., Takahashi K.// *Scripta Met.* 1980. Vol. 14. P. 1105-1110.

273. Davydov L.I., Kiryukhin N.M.// *Questions of Atomic Science and Technics. Ser. Radiation Damage Physics and Radiation Material Science.* 1980. No 2(13). P. 10-12 (in Russian).

274. Wiedersich H.// In: Physics of Radiation Effects in Crystals/ Eds. Johnson R.A., Orlov A.N. North-Holland, 1986. P. 226-280.

275. Wiedersich H., Okamoto P.R., Lam N.Q.// *J. Nucl. Mater.* 1979. Vol. 83. P. 98-108.

276. Johnson R.A., Lam N.Q.// *Phys. Rev.* 1976. Vol. 13. P. 4364-4375.

277. Lam N.Q., Okamoto P.R., Wiedersich H., et al.// *Met. Trans.* 1978. Vol. 9A. P. 1707-1714.

278. Johnson R.A., Lam N.Q.// *J. Nucl. Mater.* 1978. Vol. 69. P. 424-432.

279. Cahn J.W.// *Acta Metall.* 1956. Vol. 4. P. 449-459.

280. Cahn J.W.// *Acta Metall.* 1957. Vol. 5. P. 169-172.

281. Gomez-Ramirez R., Pound G.// *Metall. Trans.* 1973. Vol. 4. P. 1563-1570.

282. Militzer M., Trushin Yu.V.// *Sov. Phus. Techn. Phys.* 1989. Vol. 34(12). P. 1374-1376.

283. Grinchuk P.P., Kirsanov V.V.// *Sov. Phys. Met. and Met. Sci.* 1974. Vol. 38. P. 756-765 (in Russian).

284. Frost H.J., Russell K.C.// *Acta Metall.* 1982. Vol. 30. P. 953-960.

285. Orlov A.N., Samsonidze G.G., Trushin Yu.V.// *Rad. Effects.* 1986. Vol. 97. P. 45-66.

286. Orlov A.N., Samsonidze G.G., Trushin Yu.V.// In: Theory and Computer Modelling of Defect Structures in Crystals/ Sverdlovsk, 1986. P. 33-43 (in Russian).

287. Averback R.S., Rehn L.E., Wagner W. et al.// *Phys. Rev. B.* 1983. Vol. 28. P. 3100-3110.

288. Wagner W., Rehn L.E., Wiedersich H., et al.// *Phys. Rev. B.* 1983. Vol. 28. P. 6780-6794.

289. Girifalco L.A., Herman H.A.// *Acta Metall.* 1965 Vol. 13 P. 583 590.

290. Kahn H., Girifalco L.A.// *Acta Metall.* 1966. Vol. 14. P. 749-753.

291. Garkusha I.P., Lyubov B.Ya.// *Sov. Phys. Met. and Met. Sci.* 1970. Vol. 29. P. 449-457 (in Russian).

292. Vershok V.A., Gorbatov G.Z.// *Sov. Phys. Sol.* 1979. Vol. 21. P. 508-514 (in Russian).

293. Akhiezer I.A., Davydov L.N.// *Metal Physics.* 1981. Vol. 3. No 1. P. 3-21; No 3. P. 3-15 (in Russian).

294. Arsenin V.Ya. Methods of Mathematical Physics and Special Functions. M.: Nauka, 1974 (in Russian).

295. Parshin A.M.// *Questions of Atomic Science and Technics. Ser. Radiation Damage Physics and Radiation Material Science.* 1978. No 3

(8). P. 34-38 (in Russian).

296. Zelensky V.F., Neklyudov I.M., Matvienko V.V.// In: *Reactor Material Science*/ M., Central Research Institute "Atominform", 1978. Vol. 2. P. 21-43 (in Russian).

297. Zelensky V.F., Parshin A.M., Neklyudov I.M.// *Questions of Atomic Science and Technics. Ser. Radiation Damage Physics and Radiation Material Science*. 1980. No 2 (13). P. 18-22 (in Russian).

298. Parshin A.M.// Ibid. 1980. No 3 (14). P. 20-29 (in Russian).

299. Gorynin I.V., Parshin A.M.// *Atomic Energy*. 1981. Vol. 50. P. 319-324 (in Russian).

300. Gorynin I.V., Zelensky V.F., Parshin A.M. et al.// In: Radiation Defects in Metals/ Alma-Ata: Nauka. 1981. P. 265-272 (in Russian).

301. Woo C.H., Frank W.// *J. Nucl. Mater*. 1986. Vol. 140. P. 214-227.

302. Woo C.H., Frank W.// *J. Nucl. Mater*. 1987. Vol. 148. P. 121-135.

303. Carpenter P.W., Yoo M.H.// *Metall. Trans*. 1987. Vol. 9A. P. 1739-1748.

304. Baron M.A.// *J. Nucl. Mater*. 1979. Vol. 83. P. 128-138.

305. Williams T.M., Titchmarsh J.M., Arkell D.R.// *J. Nucl. Mater*. 1982. Vol. 107. P. 222-224.

306. Mansur L.K., Yoo M.H.// *J. Nucl. Mater*. 1978. Vol. 74. P. 228-241.

307. Brailsford A.D.// *J. Nucl. Mater*. 1975. Vol. 56. P. 7-17.

308. Nichols F.A.// *J. Nucl. Mater*. 1980. Vol. 90. P. 29-43.

309. Venker H., Ehrlich K.// *J. Nucl. Mater*. 1976. Vol. 60. P. 347-349.

310. Gittus J.H., Watkin J.S.// *J. Nucl. Mater*. 1974. Vol. 64. P. 300-302.

311. Bykov V.N., Vakhtin A.P., Dmitriev V.D., et al.// *Atomic*

Energy. 1973. Vol. 34. P. 247-250 (in Russian).

312. Parshin A.M. Structure, Strenght and Plasticity of Stainless and High-Temperature Steels and Alloys Used in Shipbuilding. L.: Korablestroenie. 1972 (in Russian).

313. Smolman R.,Ashbi K. Modern Metallograghy. M.: Atomizdat. 1970 (in Russian).

314. Utevsky L.M. Diffraction Electron Microscopy in Metal Science. M.: Metallurgy. 1973 (in Russian).

315. Rakhman V.M., Madorsky A.Ya., Obukhovsky V.V. Metal Science. L.: Sudpromgiz. 1960. P. 235-240 (in Russian).

316. Parshin A.M., Ushkov S.S., Yarmolovich I.I.// In: Light Alloy Technology./ M.: All-Union Institute of Light Alloys, 1974. II. P. 53-58 (in Russian).

317. Zelensky V.F., Neklyudov I.M., et al.// In: Reactor Material Science. Vol.2. M.: Central Research Institute "Atominform". 1978. P. 3-19 (in Russian).

318. Johnson W.G., Rosolowsky J.H., Turkalo A.M. et al.// *J. Nucl. Mater.* Vol. 54. P. 24-40.

319. Parshin A.M., Zelensky V.F., Kursevich I.P., et al.// *Questions of Atomic Science and Technics. Ser. Radiation Damage Physics and Radiation Material Science*. 1980. No 2 (13). P. 13-17 (in Russian).

320. Zelensky V.F., Parshin A.M., Neklyudov I.M., et al.// *Ibid*. 1981. No 2 (16). P. 57-61 (in Rissian).

321. Parshin A.M., Kolosov I.E., Korshunova T.E., et al.// In: Radiation

Effects in Metals and Alloys. Alma-Ata: Nauka. 1985.-P. 203-209 (in Russian).

322. Guseva M.I., Gordeeva G.V., Baranova E.K., et al. *Atomic Energy*. 1986. Vol. 60. P. 406-408 (in Russian).

323. Neklyudov I.M., Petrusenko Yu.T., Sleptsov A.N., et al.// *Preprint of Kharkov Physical and Technical Institute* 89-63, Kharkov, 1989 (in Russian).

324. Khachaturyuan A.G. Theory of Phase Transformations and Structure of Solid Solutions. M.:Nauka. 1974 (in Russian).

325. Physical Metallurgy. Eds. R.W. Chan, P. Haasen, North- Holland Physics Publishing, 1983, Amsterdam-Oxford-New York- Tokyo.

326. English C.A., Eyre B.L., Bartlett A.F., et al.// Proc. Int. Conf. on Fundamental Aspects of Radiation Damage in Metals, Gatlinburg, USA, 1975/ Washington. 1976. V. 2. P. 924-928.

327. English C.A.// Proc. Int. Conf. "Appl. Ion Beams Materials", 1975. London-Bristol: Univ. Warwick. 1976. P. 257-261.

328. Averback R.S., Mercle K.L.// *J. Nucl. Mater*. 1983. V. 118. P. 83-90.

329. Averback R.S., Thompson L.J., Mogle J., et al.// *J. Appl. Phys.* 1982. V. 53. P. 1342-1349.

330. Babaev V.P., Zabolotny V.T., Komissarov A.P., et al.// *Phys. Chem. Mat. Processing*. 1987. No. 2. P. 3-7 (in Russian).

331. Collins A.G., Pcate J.M., Borders J.A.// *Appl. Phys. Lett.* 1976. V. 28. P. 314-316.

332. Zabolotny V.T., Ivanov L.I., Platov Yu.M., et al.// *Questions of*

Atomic Science and Technics. Ser. Radiation Damage Physics and Radiation Material Science. 1985. No. 4(37). P. 17-21 (in Russian).

333. Wang Z.L., Westendorp J.R., Saris F.W.// *Nucl. Instr. Meth.* 1983. V. 209/210. P. 115-124.

334. Herschitz R., Seidman D.N.// *Acta Met.* 1984. V. 32. P. 1141-1154.

335. Jonson R.A. // In: Diffusion in BCC-Metals/ M. 1969. P. 212-241.

336. Leamy H.J.// *Acta Met.* 1967. V. 15. P. 1839-1851.

337. Swanson M.L., Home L.W., Matsunami N.// In: Proc. Int. Conf. on Solids/ M.: Moscow State Univ. 1977. P. 238-240.

338. Vaisfeld A.M., Gann V.V., Pavlenko V.I., et al.// *Questions of Atomic Science and Technics. Ser. Radiation Damage Physics and Radiation Material Science.* 1984. No. 4(32). P. 3-9 (in Russian).

339. Gann V.V., Yudin O.V.// *Ibid.* 1985. No. 2(35). p. 84-92 (in Russian).

340. Yudin O.V.// *Ibid.* 1987. No. 1(39). P. 25-29 (in Russian).

341. Gann V.V., Vaisfeld A.M., Yamnitsky V.A.// *Ibid.* 1983. No. 2(25). P. 20-23 (in Russian).

342. Gann V.V., Marchenko I.G.// *Ibid.* 1987. No. 2(40). P. 61-62 (in Russian).

343. Gann V.V., Yudin O.V.// *Ibid.* 1987. No. 3(41). P. 73-75 (in Russian).

344. Stathopoulos A.Y., English C.A., Eyre P.L., et al.// *Phil. Mag.* 1981. V. 44. P. 309-332.

345. Trushin Yu.V., Orlov A.N., Evlampiev I.K.// In: Radiation Physics of Metals and Alloys// Tbilisi: Institute of Physics of Georgian Academy of Science. 1969. P. 3-6 (in Russian).

346. Evlampiev I.K., Trushin Yu.V.// In: Radiation Defects in Metals/ Alma-Ata: Nauka. 1981. P. 69-75 (in Russian).

347. Evlampiev I.K., Orlov A.N., Trushin Yu.V.// In: Radiation Physics of Semiconductors and Related Materials/ Tbilisi: Tbilisi State University. 1980. P. 472-475.

348. Mayer H., Boning K., Dimitrov C.// *Phys. Stat. Sol.* 1973. V. 15. P. 91-93.

349. English C.A., Eyre P.L., Wadley H., et al.// Proc. Int. Conf. on Fundamental Aspects of Radiation Damage in Metals, Gatlinburg, USA, 1975/ Washington. 1976. V. 2. P. 918-924.

350. Suvorov A.L., Kukavadze G.M., Bobkov A.F.// In: Reactor Material Science/ M.: Atominform Central Research Institute. V. 3. 1978. P. 143-161 (in Russian).

351. Suvorov A.L. Surface Structure and Properties of Metal Atomic Layers. M.: Energoatomizdat. 1989 (in Russian).

352. Kirsanov V.V., Osipova Z.Ya.// *Preprint of RIAR,* P-147. Dmitrovgrad. 1972.

353. Kirsanov V.V.// *Questions of Atomic Science and Technics. Ser. Radiation Damage Physics and Radiation Material Science.* 1974. No. 1(1). P. 3-7 (in Russian).

354. Nelson R.S., Thompson M.W., Montgomery H.// *Phil. Mag.* 1962. V. 2. P. 1385-1389.

355. Pugacheva T.S.// *Sov. Phys. Sol. State.* 1967. V. 9. P. 102-195 (in Russian).

356. Makarov A.A., Demkin N.A., Lyashchenko V.G.// *Questions of Atomic Science and Technics. Ser. Radiation Damage Physics and Radiation Material Science.* 1981. No. 2(16). P. 53-56 (in Russian).

357. Kapinos V.G., Kevorkyan Yu.R.// *Ibid.* 1983. No. 2(25). P. 3-10 (in Russian).

358. Balarin M.// Kernenergie. 1964. Bd. 7. S. 434-443.

359. Gann V.V., Marchenko I.G.// *Questions of Atomic Science and Technics. Ser. Radiation Damage Physics and Radiation Material Science.* 1983. No. 5(28). P. 99, 100 (in Russian).

360. Kirsanov V.V.// *Ibid.* 1984. No. 1(29), 2(30). P. 35-45 (in Russian).

361. Dokhner R.D.// In: Questions of Theory of Defects in Crystals. Col. Papers/Ed.: S.V. Vonsovsky, M.A. Krivoglaz, L.: Nauka, 1987. P. 145-159 (in Russian).

362. Kirsanov V.V.// Ibid. P. 160-173.

363. Demkin M.A., Makarov A.A.// *Questions of Atomic Science and Technics. Ser. Radiation Damage Physics and Radiation Material Science.* 1984. No. 4(32). P. 20-22 (in Russian).

364. Babaev V.P., Zabolotny V.T., Suvorov A.L.// *Ibid.* 1985. No. 4(37). P. 7-9 (in Russian).

365. Dzhaparidze S.K., Orlov A.N., Trushin Yu.V.// *Ibid.* 1981. No. 1(15). P. 31, 32 (in Russian).

366. Dzhaparidze S.K., Orlov A.N., Trushin Yu.V.// *Proc. Int. Conf. "Defects in Insulating Crystals"*, Riga. 1981. P. 177-178.

367. Dzhaparidze S.K., Orlov A.N., Trushin Yu.V.// *Sov. Phys. Tech. Phys.*

1982. V. 27(6). P. 651-654.

368. Dzhaparidze S.K., Orlov A.N., Trushin Yu.V.// *Sov. Phys. Tech. Phys.* 1982. V. 27(6). P. 654-657.

369. Dzhaparidze S.K., Orlov A.N., Trushin Yu.V.// *Sov. Phys. Tech. Phys.* 1985. V. 30(6). P. 718-719.

370. Dzhaparidze S.K., Orlov A.N., Trushin Yu.V.// *Sov. Phys. Tech. Phys.* 1985. V. 30(9). P. 1066-1067.

371. Abrahamson A.A.// *Phys. Rev.* 1969. V. 178. P. 76-79.

372. Gaydenko V.T., Nikitin V.K.// *Chem. Phys. Lett.* 1970. V. 7. P. 360-362.

373. Dzhaparidze S.K.//*Preprint of Institute of Physics, Georgian Academy of Science,* TT-1, Tbilisi, 1981 (in Russian).

374. Kontorova T.A., Frenkel Ya.I.// *JETP*. 1933. V. 8. P. 89-95 (in Russian).

375. Lin H.C., Mitchell T.E.// *Acta Metall.* 1983. V. 36. P. 863-872.

376. Banerjee S., Urban K.// *Phys. Stat. Sol.* 1984. V. 81. P. 145-162.

377. Orlov A.N., Trushin Yu.V.// In: Radiation Defects in Metal Crystals. Col. Papers/ Alma-Ata: Nauka, 1978. P. 30-39 (in Russian).

378. Orlov A.N., Pompe W., Trushin Yu.V.// *Proc. Int. Conf. "Energy Pulse Modification Semicond. and Related Materials"*/ Berlin-Verlag, 1984. P. 635-639.

379. Trushin Yu.V., Pompe W.// *Sov. Tech. Phys. Lett.* 1985. V. 11(4). P. 162-163.

380. Pompe W., Bobeth M., Trushin Yu.V.// *Proc. Int. Conf. on Energy Pulse*

and Particle Beam Modification of Materials/ Akademie Verlag, Berlin, 1988. P. 375-378.

381. Trushin Yu.V.// *Sov. Phys. Tech. Phys.* 1987. V 32(2). P. 136-138.
382. Kamke E. Differentialgleichungen, Leipzig, 1959.
383. Samsonidze G.G., Trushin Yu.V.// *Sov. Phys. Tech. Phys.* 1988. V. 33(1). P. 24-29.
384. Samsonidze G.G., Trushin Yu.V.// *Preprint of Physical-Technical Institute* No. 1089, L., 1986 (in Russian).
385. Orlov A.N., Samsonidze G.G., Trushin Yu.V.// *Sov. Phys. Tech. Phys.* 1986. V. 31(7). P. 768-772.
386. Ditkin V.A., Prudnikov A.P. Integral Transformations and Operator Calsulus. M.: Fizmatgiz, 1961 (in Russian).
387. Ham F.S.// *J. Phys. Chem. Sol.* 1958. V. 6. P. 335-351.
388. Ham F.S.// *J. Appl. Phys.* 1959. V. 30. P. 1518-1525.
389. Girifalco L.A., Behrendt D.R.// *Phys. Rev.* 1961. V. 124. No. 2. P. 420-427.
390. Kahn H., Girifalco L.A.// *J. Phys. Chem. Sol.* 1975. V. 36. P. 919-925.
391. Kudinov G.M., Lyubov V..Ya., Shmyakov V.A.// *Izv. of Academy of Sciences, Series: Metals.* 1979. No. 2. P. 176-180 (in Russian).
392. Odintsov D.D.// *Questions of Atomic Science and Technics. Ser. Radiation Damage Physics and Radiation Material Science.* 1983. No. 3(26). P. 21-25 (in Russian).
393. Orlov A.N., Samsonidze G.G., Trushin Yu.V.// *Preprint of Physical-Technical Institute* No. 1018, L., 1986 (in Russian).

394. Trushin Yu.V.// *Proc. 3rd Int. Conf. "Energy Pulse and Particle Beam Modification of Materials"* (EPM-89) 4-8 Sept., 1989, Drezden, Akademie Verlag, Berlin, 1990. P. 552-554.

395. Trushin Yu.V.// *Sov. Met. Phys. Met. Sci.* 1991. No.. 7. P. 101-104 (in Russian).

396. Karlslaw G., Eger D. Heat Conductivity of Solids. M.: Nauka, 1964 (in Russian).

397. Rubinstein L.I. *Stephan's Problem.* Riga: Zveigzne. 1967.

398. Tien R.H.// *Metal Trans.* 1981. V. 12A. P. 1548-1551.

399. Simmich O.// *Phys. Stat. Sol.* 1986. V. 93. P. 105-112.

400. Samsonidze G.G., Orlov A.N., Trushin Yu.V.// *Sov. Met. Phys. Met. Sci.* 1983. V. 55. P. 676-684 (in Russian).

401. Samsonidze G.G., Orlov A.N., Trushin Yu.V.// *Sov. Tech. Phys. Lett.* 1983. V. 9(5). P. 263-264.

402. Golubov S.I.// *Sov. Met. Phys. Met. Sci.* 1987. V. 64. P. 879-885.

403. Dubinko V.I., Ostapchuk P.N., Slesov V.V.// *J. Nucl. Mater.* 1989. V. 161. P. 239-260.

404. Asimov P.M.// *Acta metall.* 1966. V. 14. P. 1005-1007.

405. Nakai K., Kinoshita C., Mitchell T.E., et al.// *J. Nucl. Mater.* 1985. V. 133/134. P. 694-697.

406. Gudlatt H.-J., Naundorf V., Macht M.-P., et al.//*J. Nucl. Mater.* 1983. V. 118. P. 73-77.

407. Wahi R.P., Koch R., Abromett C., et al.// *J. Nucl. Mater.* 1985. V. 127. P. 175-186.

408. Knoll R.W., Kulcinski C.L.// *J. Nucl. Mater.* 1985. V.-131. P. 172-196.

409. Balluffi R.W.// *J.* Nucl. Mater.// 1978. V. 69/70. P. 240-263.

410. Parshin A.M., Trushin Yu.V.// *Sov. Tech. Phys. Lett.* 1983. V. 9(5). P. 243-244.

411. Orlov A.N., Trushin Yu.V.// In: Radiation Defects in Metals. Col. Papers/ Alma-Ata: Nauka, 1988. P. 84-89. (in Russian).

412. Demin N.A., Konobeev Yu.V., Tolstikova O.V.// *Questions of Atomic Science and Technics. Ser. Radiation Damage Physics and Radiation Material Science.* 1982. No. 3(22). P. 13-19 (in Russian).

413. Bakaj A.S., Zelensky V.F., Kolosov I.E., et al.// *Ibid.* 1983. No. 5(28). P. 3-11 (in Russian).

414. Orlov A.N., Parshin A.M., Trushin Yu.V.//In: Radiation Effects in Metals and Alloys. Col. Papers/ Alma-Ata: Nauka, 1985. P. 178-182 (in Russian).

415. Trushin Yu.V.//*Sov. Tech. Phys. Lett.* 1991. V. 17(3). P. 175-177.

416. Trushin Yu.V.// *J. Nucl. Mater.* 1991. V. 185. P. 268-272.

417. Trushin Yu.V.//*Sov. Phys. Tech. Phys.* 1992. V. 62. P. 1081-1094 (in Russian).

418. Trushin Yu.V.// *Sov. Phys. Tech. Phys.* 1992. V. 62. P. 1095-1103 (in Russian).

419. Turchin S.I., Altovsky I.V. *Questions of Atomic Science and Technics. Ser. Radiation Damage Physics and Radiation Material Science.* 1983. No. 4(27). P. 18-23 (in Russian).

420. Bakaj A.S.// *Sov. Tech. Phys. Lett.* 1983. V. 9(11). P. 552-553.

421. Orlov A.N., Trushin Yu.V.// *Sov. Tech. Phys. Lett.* 1988. V. 14(8). P. 595-596.

422. Trushin Yu.V., Eldishev Yu.N.// *Proc. Int. Conf. on Ion Implantation and Ion Beam Equipment*/ Elenite, Bulgaria, 1990. P. 203-208.

423. Trushin Yu.V.// *Sov. Tech. Phys. Lett.* 1991. V. 17(3). P. 177-178.

424. Kirsanov V.V.// In: Computerized Modeling of Defects in Metals. Col. Papers/ L.: Nauka, 1990. P. 52-64 (in Russian).

425. Takahashi H., Takegama T., Nakahigeshi S., et al.// *Nucl. Sac. Trans.* 1980. V. 35. P. 230-231.

426. Bramman J.I., Brown C., Watkins J.S., et al.// *Proc. Int. Conf. on Radiation Effects in Breeder Reaction Structural Materials,* USA, Arizona, 1977/ Eds. M.L. Bleiberg, J.W. Bennett. P. 479-482.

427. Shober H.R., Ingle K.W.// *J. Phys. F.* 1980. V. 10. P. 575-581.

428. Kim S.J., Nicolet M.-A., Averback R.S., et al.// *Appl. Phys. Lett.* 1985. V. 46. No. 2.P. 154-156.

429. Kim S.J., Nicolet M.-A., Averback R.S., et al.// *Phys. Rev. B.* 1988. V. 37. No 1. P. 38-49.

430. Lifshits I.M., Kaganov M.I., Tanatarov L.V.// *Atomic Energy.* 1959. V. 6. No. 9. P. 391-402 (in Russian).

431. Vineyard G.H.// *Rad Effects.* 1976. V. 29. No. 4. P. 245- 248.

432. Guinan M.W., Kinney J.H.// *J. Nucl. Mater.* 1981. V. 104. P. 1319-1324.

433. Rubia T.D., de la, Averback R.S., Benedek R., et al.// *Phys. Rev. Lett.* 1987. V. 59. No. 17. P. 1930-1933.

434. Vatnik M.P., Trushin Yu.V.// *Sov. Tech. Phys. Lett.* 1990. V. 16(4). P. 260-262.

435. Katin V.V., Martynenko Yu.V., Yavlinsky Yu.N.// *Preprint of Institute of Atomic Energy,* No. 4661/6. M. 1988 (in Russian).

436. Kalbitzer S., Eumann H.// In: Ionic Implantation in Semiconductors and Other Materials M.: Mir. 1980. P. 65-91 (in Russian).

437. Ryssel H., Ruge J. Ionenimplantation. B.G. Teubner, Stuttgart, 1978.

438. Ryssel H., Ruge J.// In: Methods of Surface Analysis. M.: Mir, 1979. P. 18-59 (in Russian).

439. Holloway P.H.// *Appl. Surf. Sci.* 1986. V. 26. No. 4. P. 550-560.

440. Nicholson R.B.// *Phase Transformations.* Ohio: ASM, 1970. P. 269-312.

441. Schulze G.E.R. Metallphysik, Akademie-Verlag, Berlin, 1967.

442. Christian J.W. The Theory of Transformations in Metals and Alloys. Part I. Equilibrium and General Kinetic Theory. Pergamon Press.

443. Maslenkov S.B., Braslavskaya G.S.// *Izv. of USSR Academy of Sciences. Metals.* 1984. No. 1. P. 203-206 (in Russian).

444. Orlov A.N., Perevezentsev V.N., Rybin V.V. Grain Boundaries in Metals. M.: Metallurgia. 1980 (in Russian).

445. Guseva M.I., Suvorov A.L., Trushin Yu.V., et al.// *Sov. Phys. Tech. Phys.* 1991. V. 36(10). P. 1126-1129.

446. Trushin Yu.V.// *Sov. Phys. Met. and Met. Sci.* 1992. No. 4. P. 53-69 (in Russian).

447. Trushin Yu.V., Suvorov A.L., Dolin D.E., et al.// *Sov. Tech. Phys. Lett.* 1990. V. 16(9). P. 676-678.

448. Suvorov A.L., Trushin Yu.V., Dolin D.E., et al.// *Proc._Int. Conf. on Physics of Irradiation Effects in Metals.* SiÂfok, Hungary, May 20-24, 1991. T-16.

449. Orlov A.N., Trushin Yu.V.// *Sov. Met. Phys. and Met. Sci.* 1976. V. 41. P. 925-932 (in Russian).

450. Kalnin Yu.N., Pirogov F.V.// *Phys. Stat. Sol.(b).* 1977. V. 84. P. 521-527.

451. Pirogov F.V.// In: Computerized Modeling of Defects in Crystals. L.: Nauka. 1990. P. 71-84 (in Russian).

452. Kirsanov V.V., Trushin Yu.V.// In: Computerized Modeling of Defect Crystal Structure. L.: A.F. Ioffe Physical-Technical Institute. 1987. P. 28-33 (in Russian).

453. Okamoto P.R., Wiedersich H.// *J. Nucl. Mater.* 1974. V. 53. P. 336-345.

454. English C.A., Eyre B.L., Shoaib K., et al.// *J. Nucl. Mater.* 1975. V. 58. P. 220-226.

455. Rizk R., Vajda P., Maury F., et al.// *J. Appl. Phys.* 1976. V. 47. P. 45-47.

456. Dimitrov C., Dimitrov O., Dworschak F.// *J. Phys. F.* 1978. V. 8. P. 1031-1052.

457. Rehn L.E., Robrock K.H., Jaques H.// *J. Phys. F.* 1978. V. 8. P. 1835-1844.

458. Ehrlich K.// *J. Nucl. Mater.* 1981. V. 100. P. 149-166.

459. Ehrlich K.// *Zeitschrift fr Metallkunde.* 1986. Bd. 77. S. 611-619.

Subject Index

Two-Atom Crystals, Crowdion-Free Path in, 141

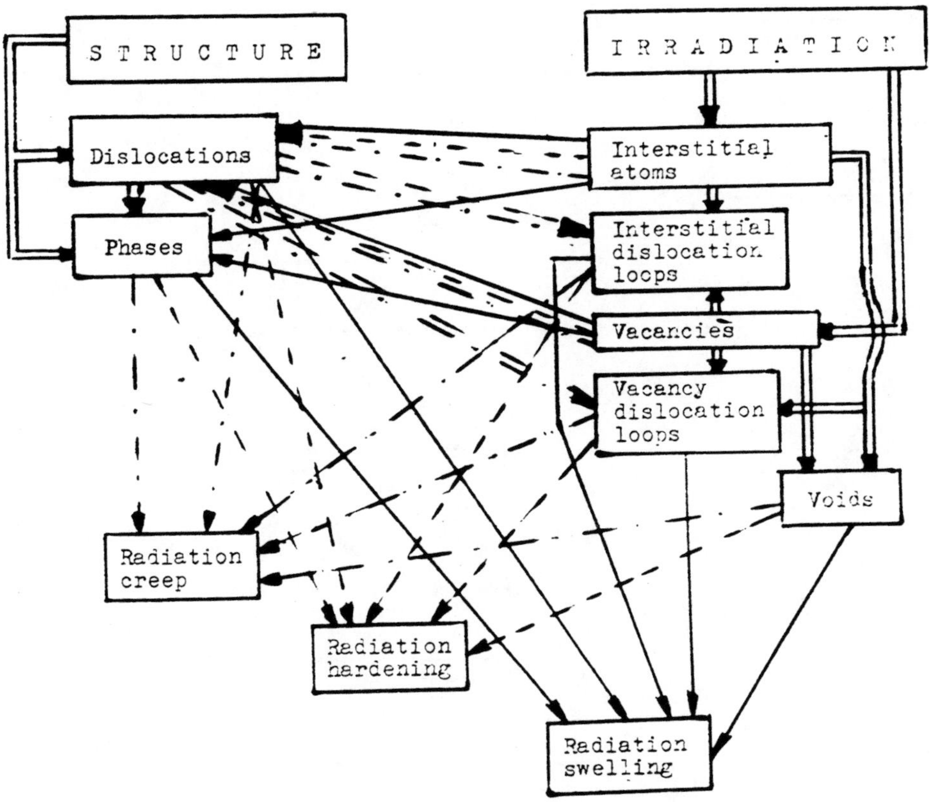

[Fig. 1. Diagram of the interrelations of physical processes of radiation swelling, radiation creep and radiation precipitation.

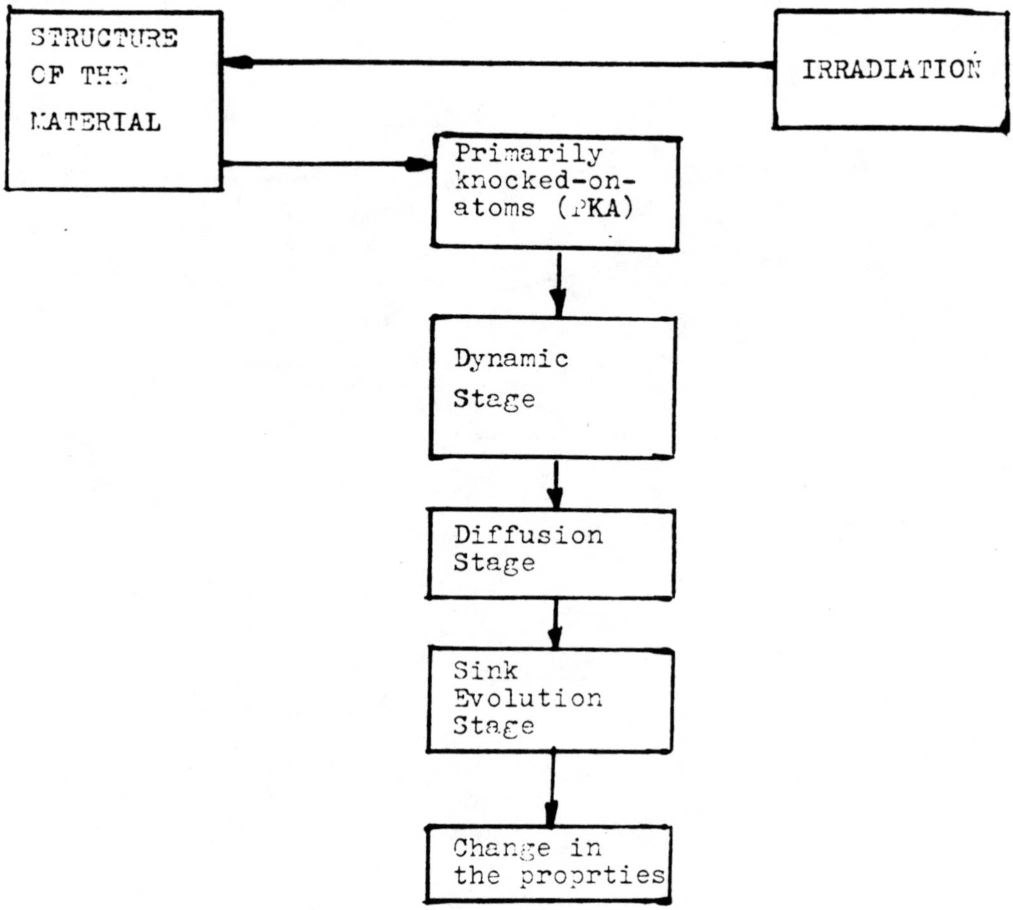

[Fig. 1.1. Diagram of essential stages of radiation damage in materials: [I - dynamic stage; [II - diffusion stage; [III - sink evolution stage.

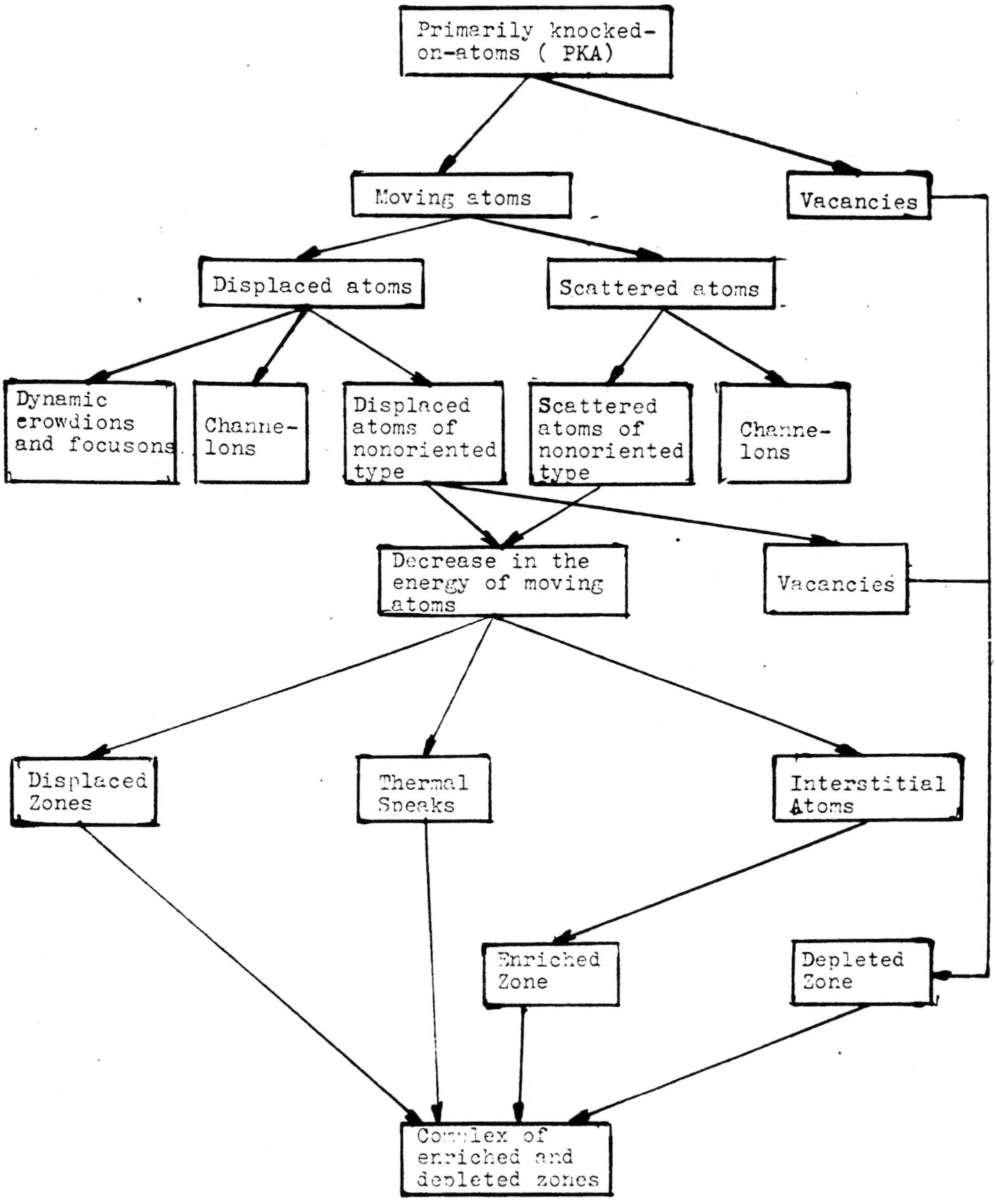

[Fig. 1.2. Diagram of developing the radiation cascade in a crystal.

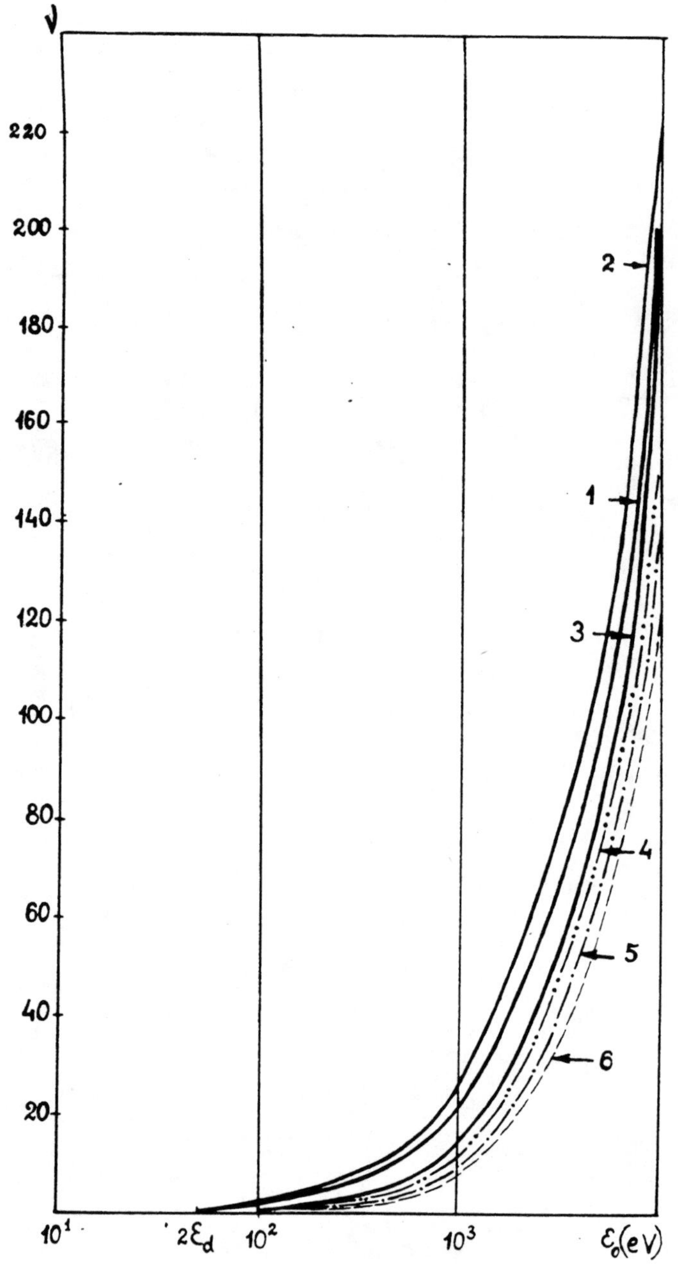

[Fig. 1.3. PKA-energy dependence of the copper cascade functions calculated with account for various factors: 1 - Kinchin-Pease (1.8), 2 - Sneider-Neufeld (1.9), 3 - for an isotropic material (1.12), 4 - with account for dynamic crowdions (1.13), 5 - with account for a heavy lead impurity (3.10), 6 - with account for a heavy impurity and dynamic crowdions (3.11) [6,94].

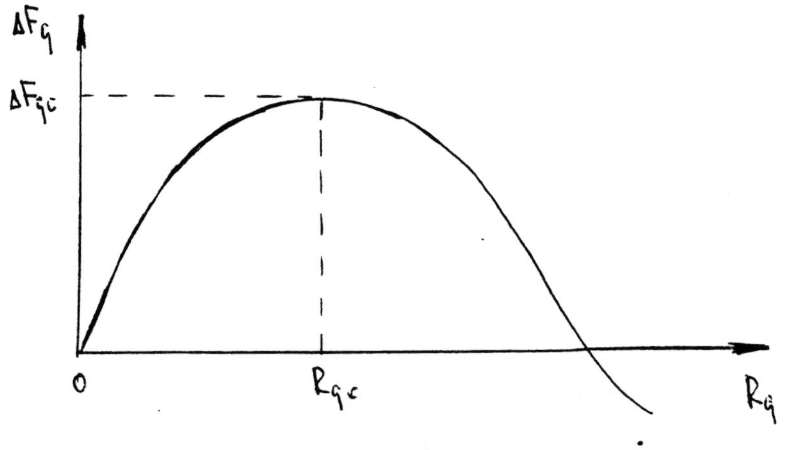

[Fig. 1.4. Change in the free energy of a point defect cluster in dependence on its dimensions Rq

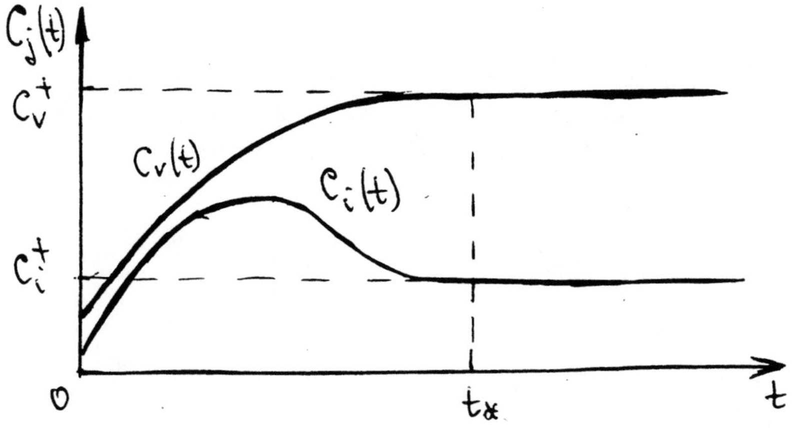

[Fig. 1.5. Change in the concentrations of intrinsic point defects under irradiation with time (diagram).

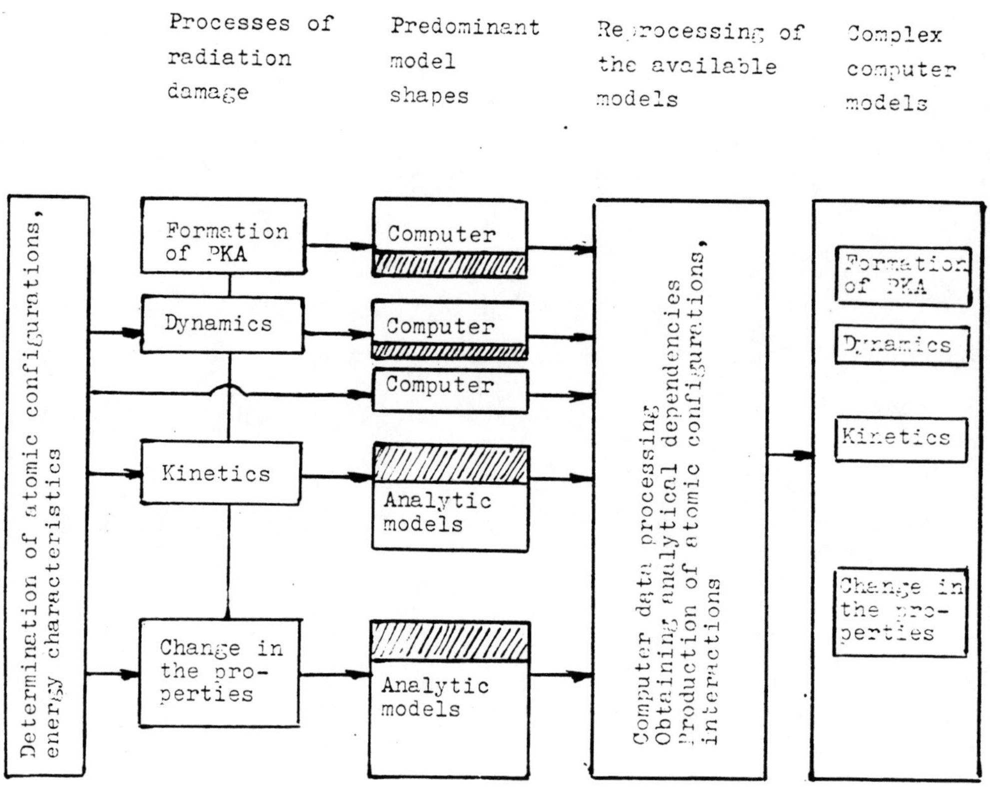

[Fig. 2. Diagram of developing the models of radiation damage of materials. The dashed areas are proportional to a less part of the analytical or computer methods.

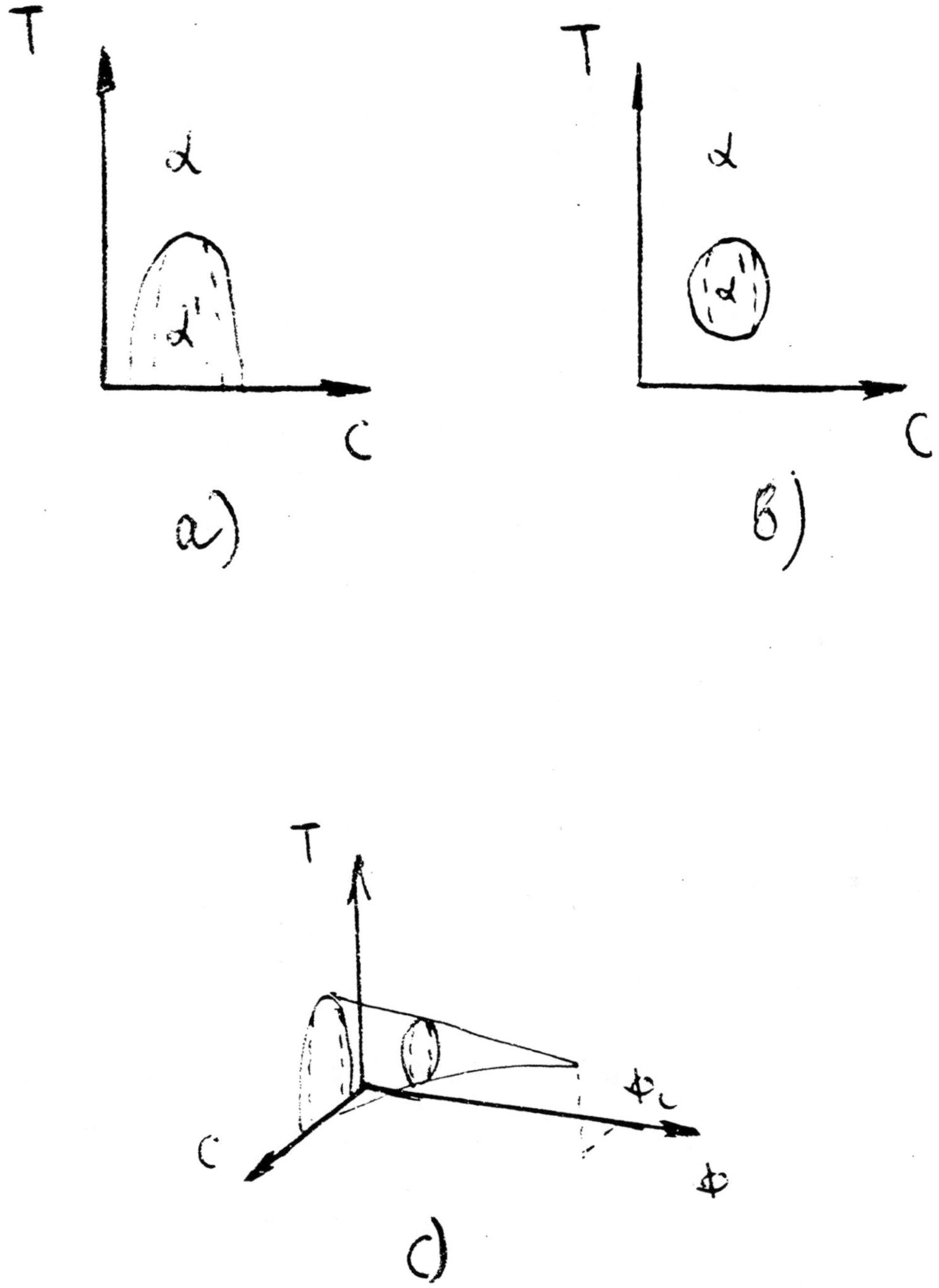

[Fig. 2.1. Phase diagram of an alloy ordering under irradiation (a and a' phases):
[(a) without irradiation; (b) under irradiation; (c) under irradiation with the flux density more than critical one (diagram).

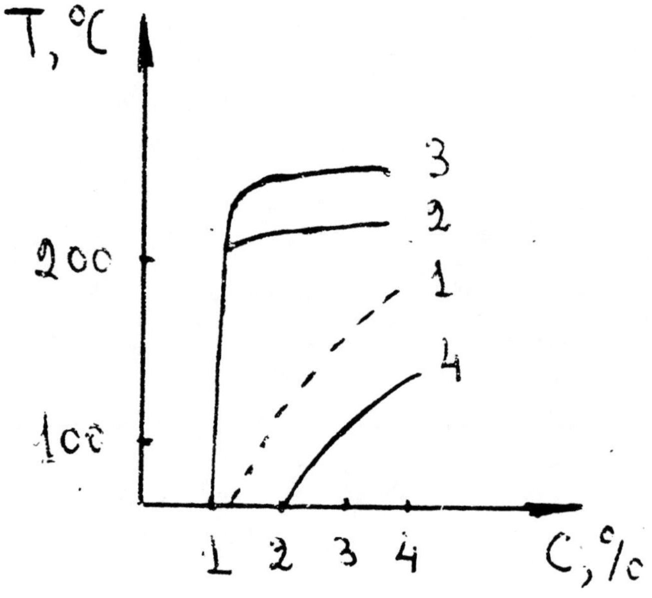

[Fig. 2.2. Solubility boundary of Zn in Al under the irradiation by fast electrons

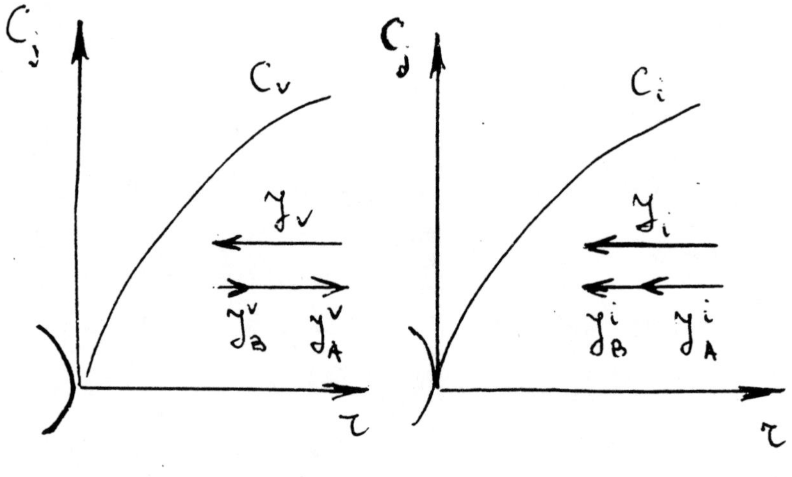

[Fig. 2.3. Reverse Kirkendall effect induced by the vacancy flux (on the left) and the interstitial flux (diagram).

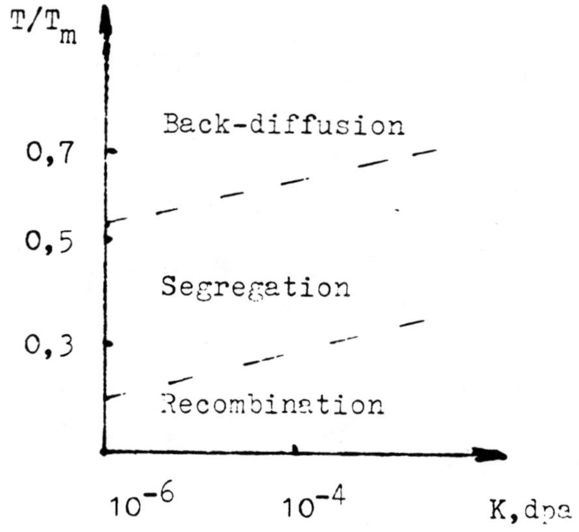

[Fig. 2.4. Temperature interval of the segregation under irradiation.

Diagram 1

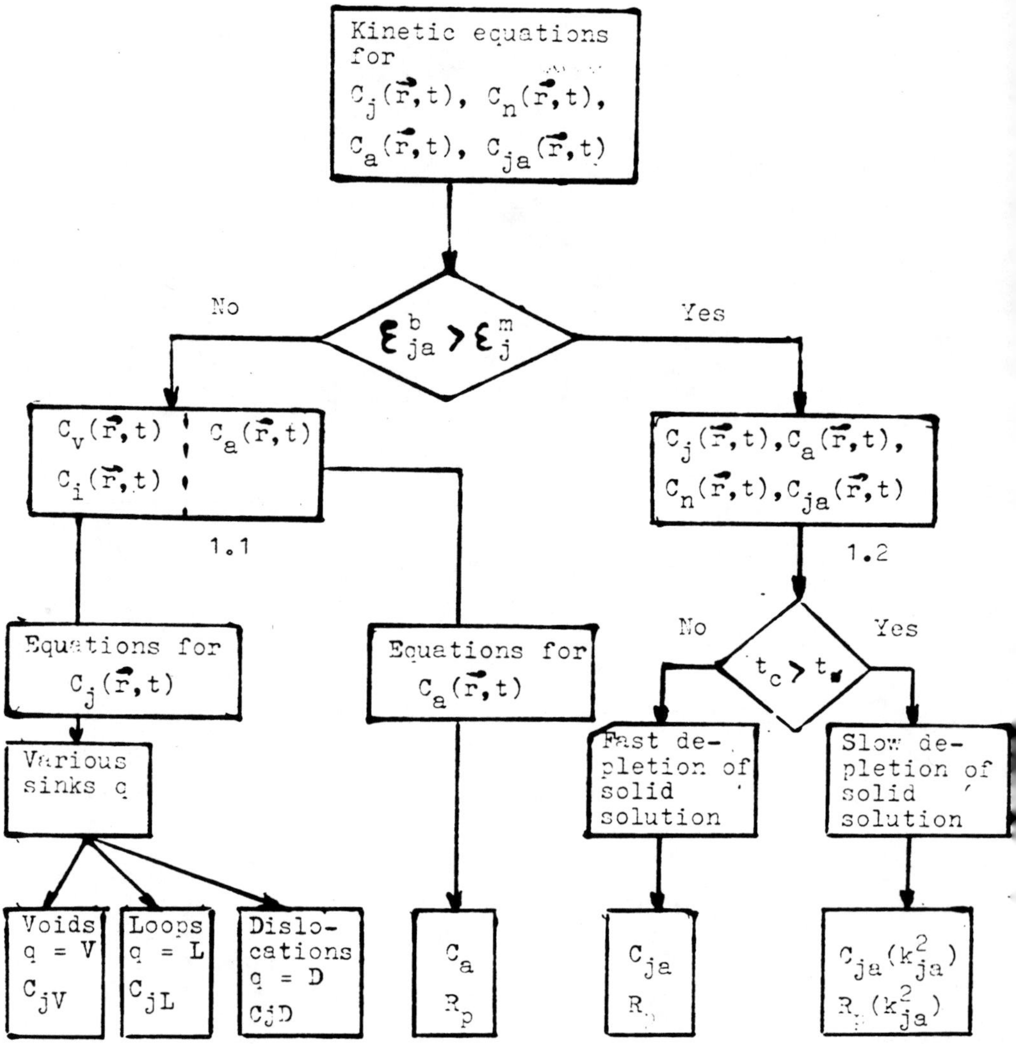

[Fig. 2.5. The diagram of balance equations describing the mobile-point-defect kinetics in crystals (Diagram 1).

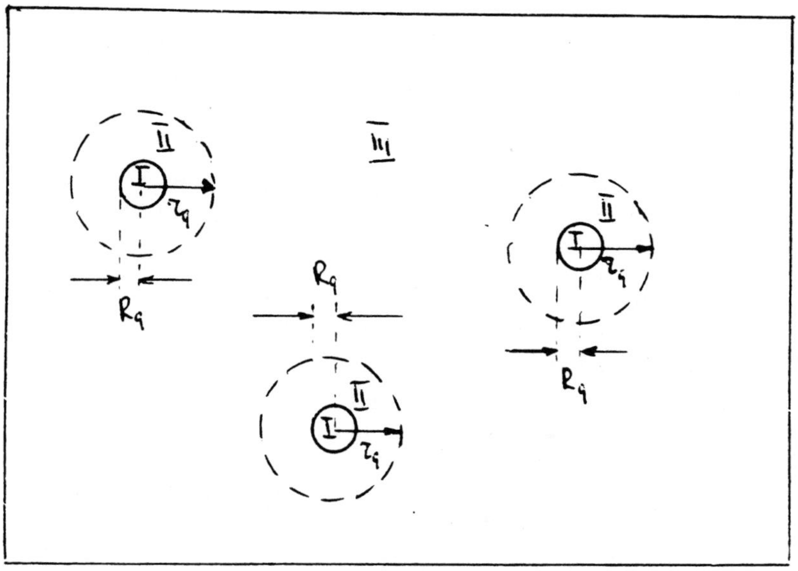

[Fig. 2.6. Diagram of the effective-medium method:[(I) sinks; (II) adsorbing medium near a q -sink; (III) adsorbing medium with the volume-spreading sinks of all types; R is the q - sink radius; r is the dimension of region II.

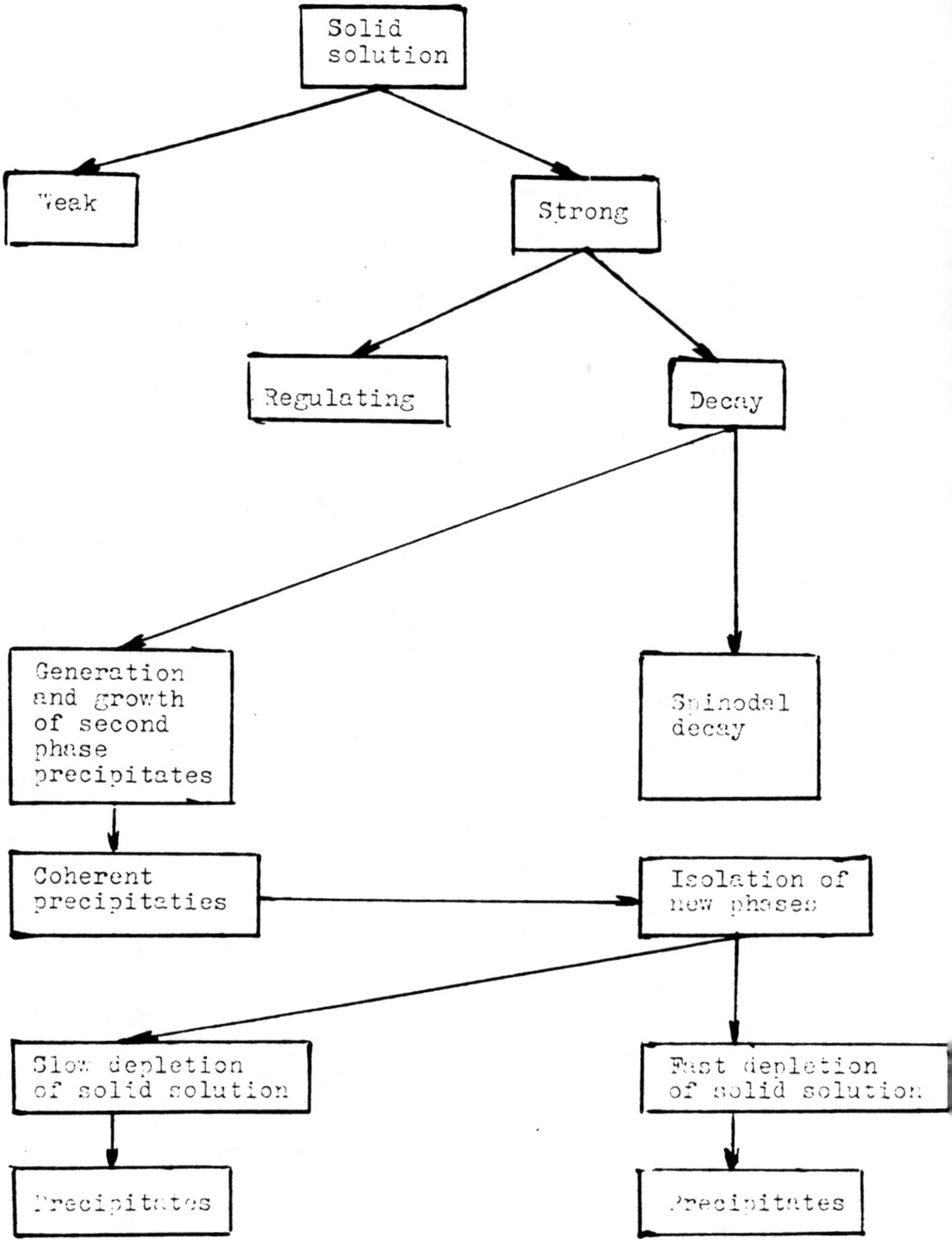

[Fig. 2.8. Diagram of the decomposition stages for solid solutions under irradiation.

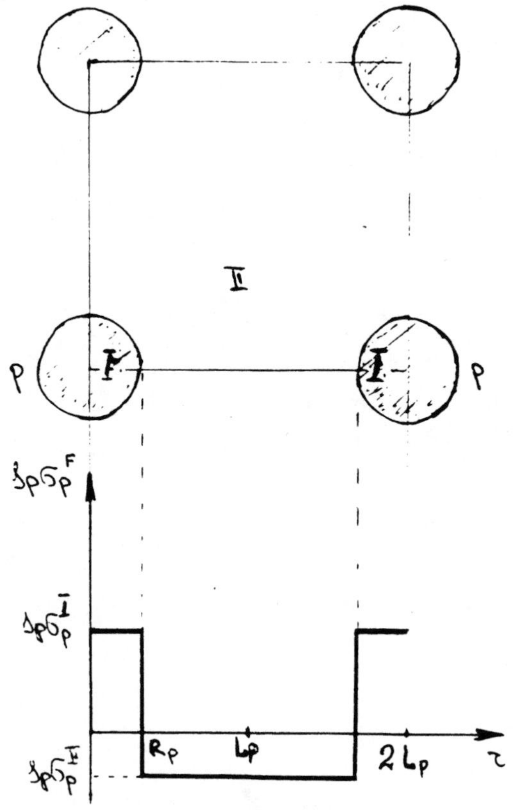

Fig. 2.9. Diagram of r -dependence of Sps for spherical coherent secondary-phase precipitates of R -radius. The precipitates are separated by 2 L between each other. Regions: (I) precipitate; (II) solid solution.

a

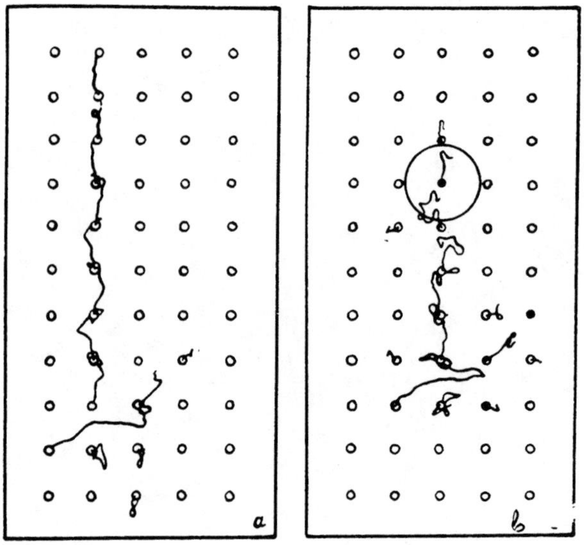

b

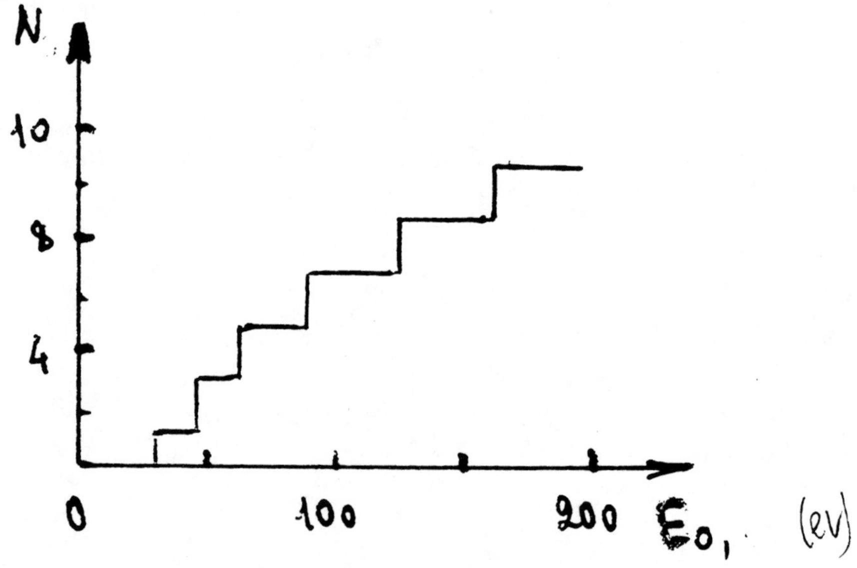

[Fig. 3.1. The interaction between a collision cascade in the bcc-crystal of 2Fe with impurities. Shown are the paths of atoms in a cascade from the PKA with the initial energy 100 eV and the direction of motion making the angle y = 31 with <100> in the plane (001). Impu rity - dark circles, (a) pure case, (b) impurity (m /n =7). The formation of a mixed dumbbell interstitial is emphasized [142].

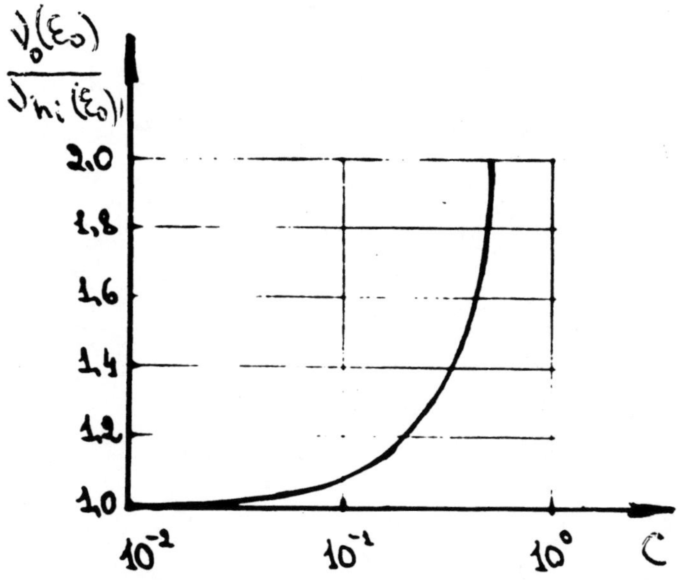

[Fig. 3.2. The dependence of the cascade ratio n (e)/ n (e) on the relative concentration of lead impurity in copper [345].

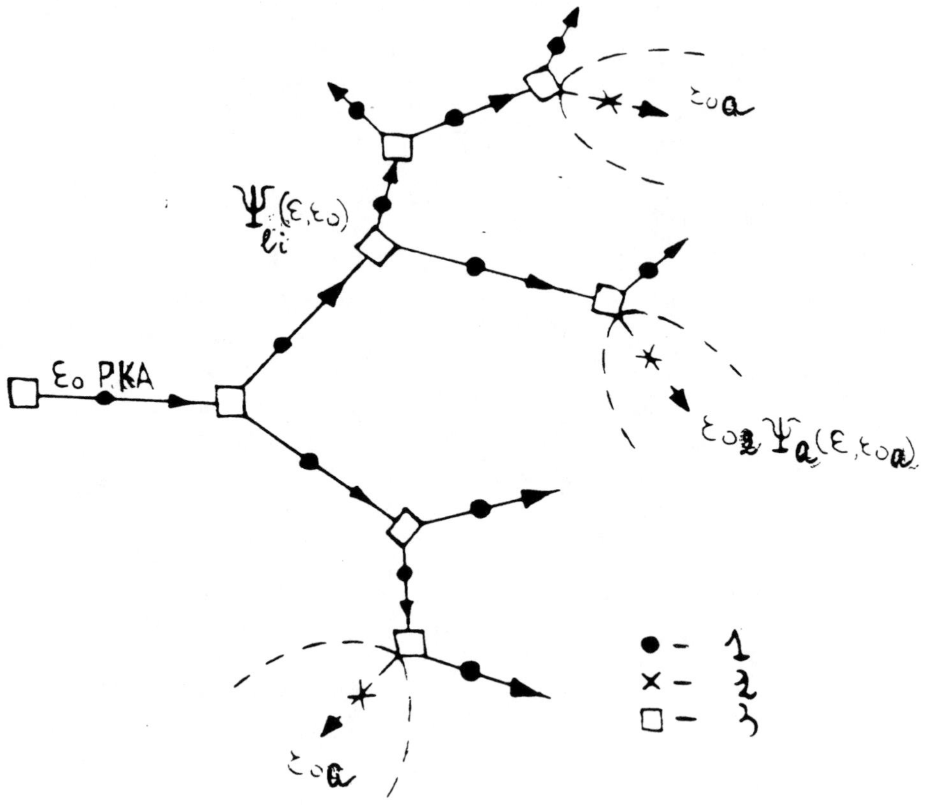

[Fig. 3.3. Diagram of the cascade of moving atoms in a crystal with light impurities: I - matrix atom (m); 2 - impurity atom (m); 3 - vacancy.

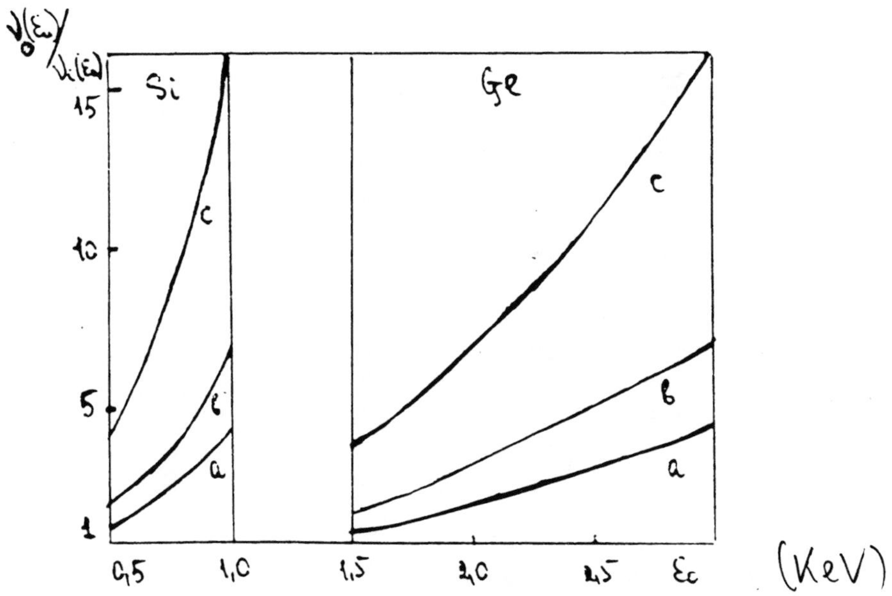

[Fig. 3.4. The dependence of the cascade function ratio on the PKA energy without account and with account for the channeling n (e)/ n (e) in silicon and germanium for the directions <110> (e = 0.5 keV) and various channeling probabilities P ; P ; P [347]: (a) P = 0.25; (b) P = 0.30; (c) P = 0.35.

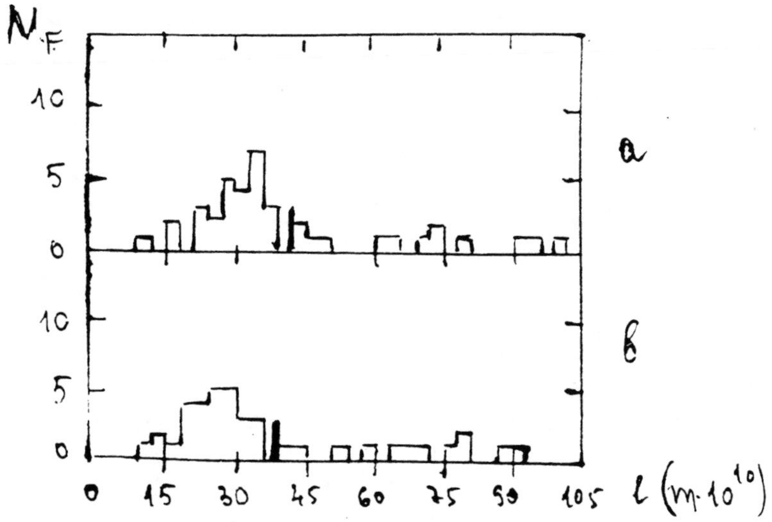

[Fig. 3.5. The distribution of focused substituting collision chains of N atoms in the path length l in the technical-purity tungsten irradiated by the deuterons of the energy 5.8 and 12 MeV at T = 330 K: (a) along the directions <110>, (b) along the directions <111> [37, 115].

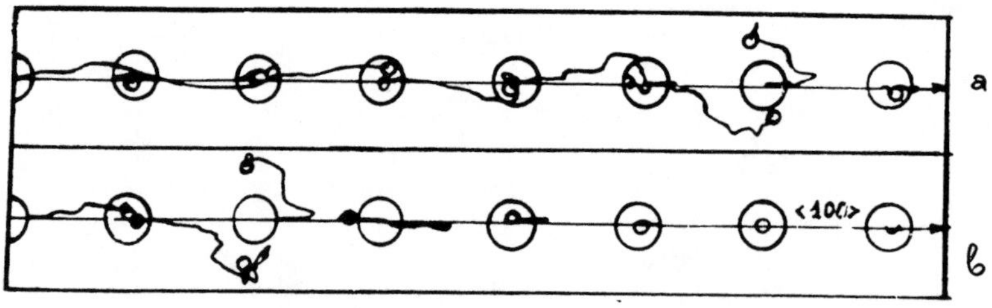

[Fig. 3.6. Chopping in the propagation of a heavy impurity dynamic crowdion on a heavy impurity: (b, the impurity is emphasized by the dark circle) in the plane (110) of a - Fe . The environment atoms are not shown, (a) is a pure case [114].

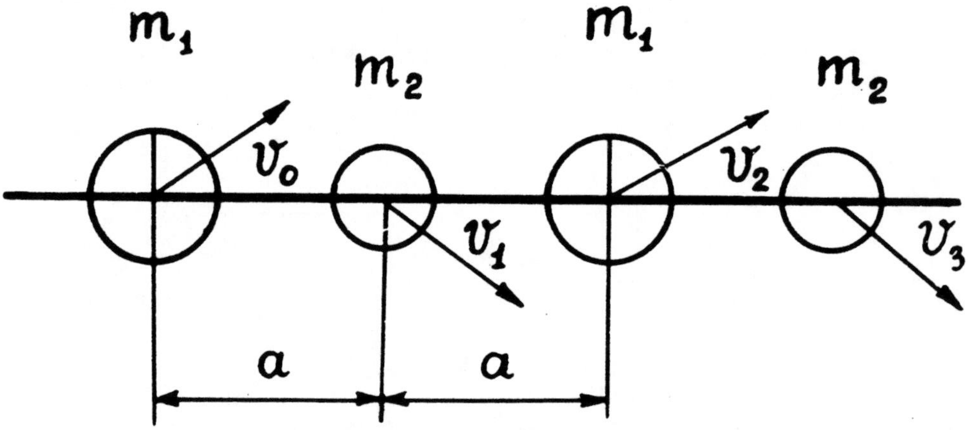

[Fig. 3.7. History of collision for the atoms in a two-atom chain.

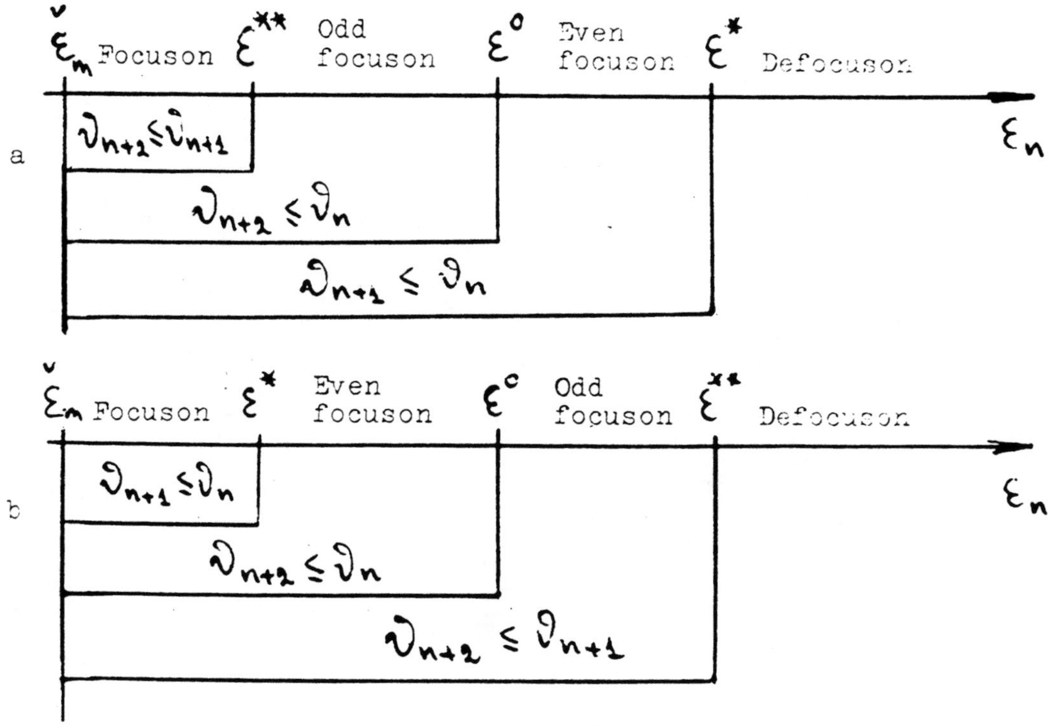

[Fig. 3.8. Focuson and defocuson types at small angles y for l >1 (a) and l <1 (b), is the minimum energy defined by condition (3.35).

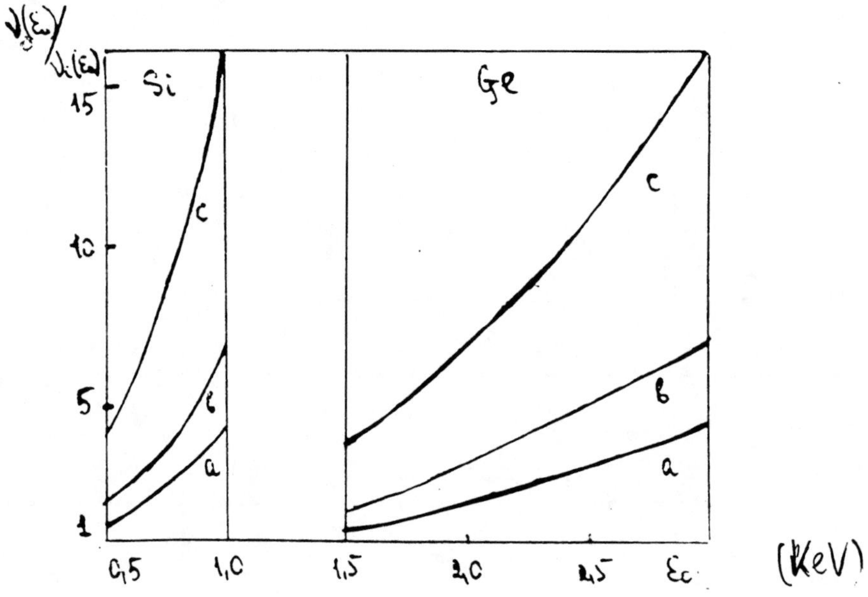

Fig. 3.9

a

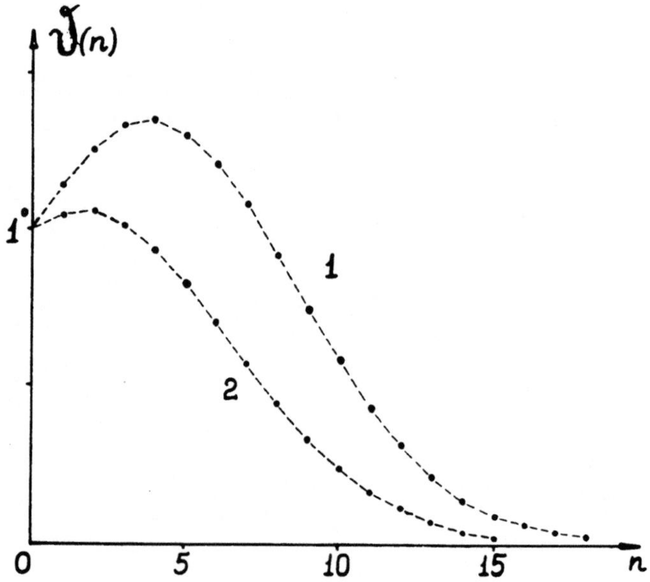

b

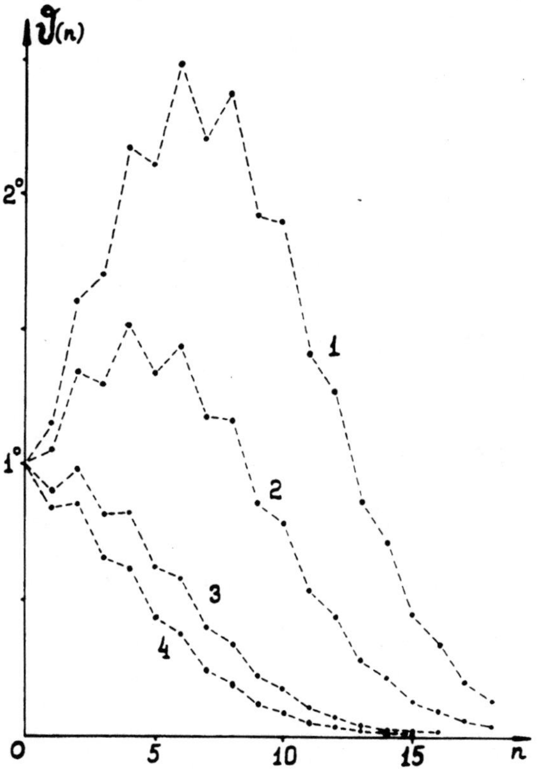

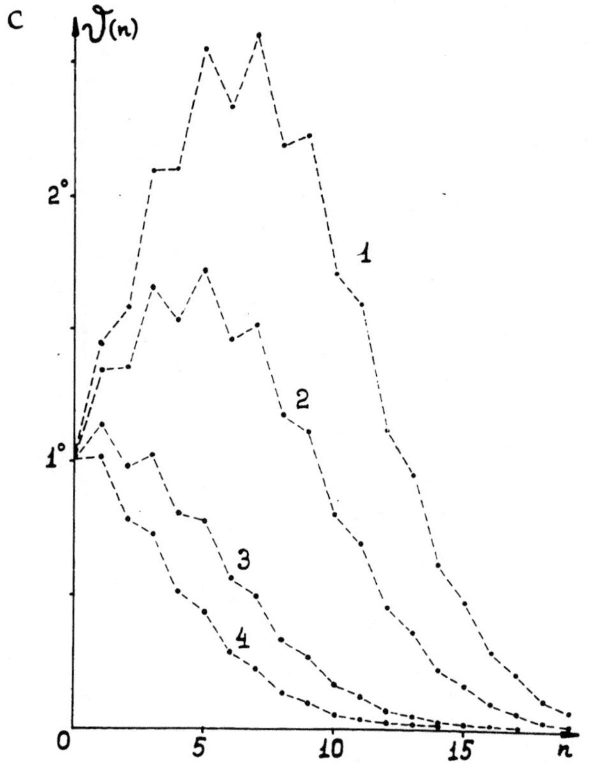

[Fig. 3.9. The dependence of the angle y on the collision number for the system Fe-Al when l = 1 (a), 2.07 (b), 0.48 (c), and e = 25 (1), 31 (2), 32(3), 43 (4), 50 (5), 65 (6,7) eV.

a

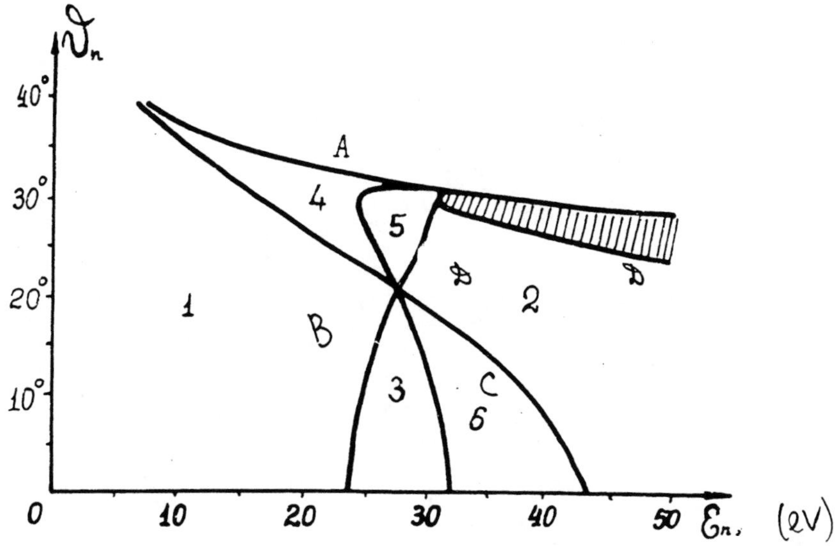

b

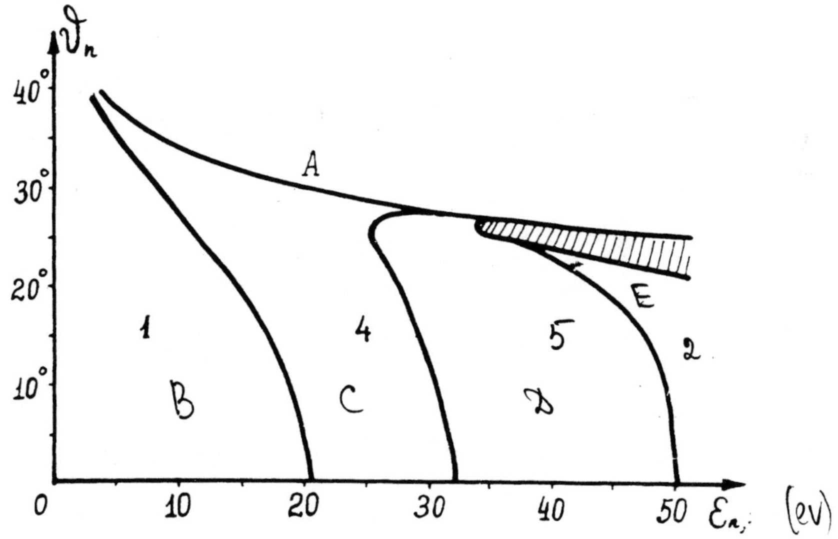

[Fig. 3.10. Diagram of conditions of initiating various types of focusons and defocusons for the system Fe-Al : [(a) I = m / m =2.07 ; Fe Al [(b) I = m / m = 0.48 . Al Fe [Arabic figures correspond to the types of pulse propagation over the cell, indicated in (3.42). The areas of forbidden y and values are dashed.

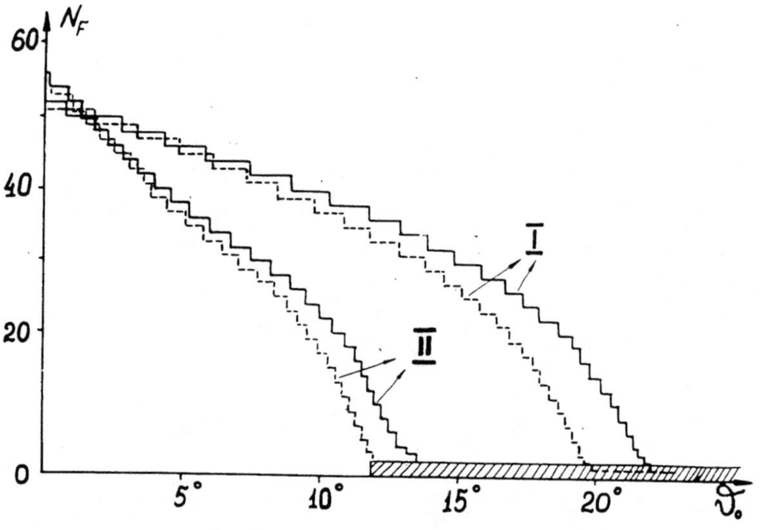

[Fig. 3.11. The dependence of focuson path length N on the initial angle y : I - at e = 60 , II - at e = 100 eV. A solid line corresponds to I = 2.07, a dashed one to I = 0.48, a dashed area is the case of violation of condition (3.47).

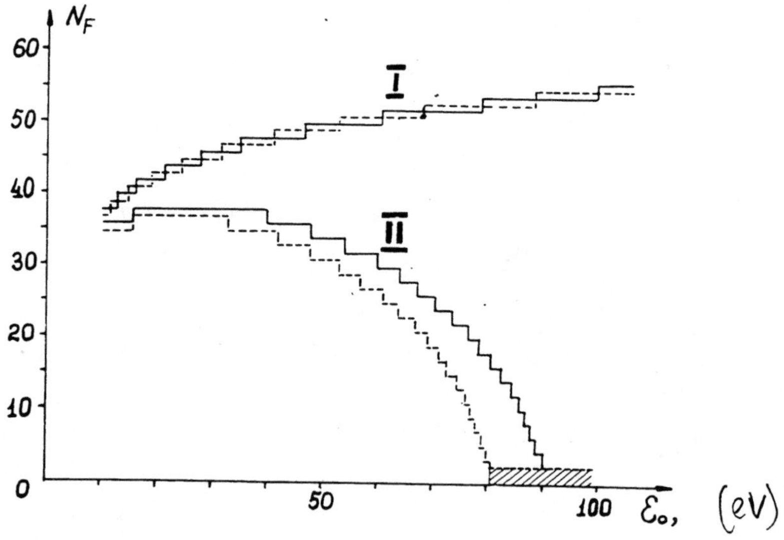

[Fig. 3.12. The dependence of focus on path length N on the initial energy e : (I) at y = 0 , (II) at y = 15 . A solid line means I = 2.07, a dashed one I = 0.48.

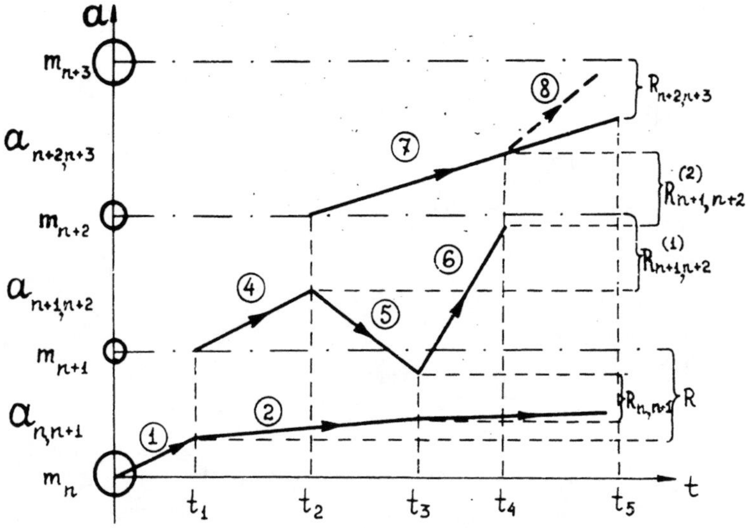

[Fig. 3.13. Diagram of atomic motion in a chain. Arrows indicate the directions of atomic motion. Figures at the line segments correspond to the motion with the energy: (1) e ; (2) e' ; (3) e\9" ; (4) e ; (5) e' ; (6) e\9" ; (7) e ; (8) e' ; R are the radii of collisions between the atoms of the masses m and m ; s =1 , 2 indicate a primary and iterated collision.

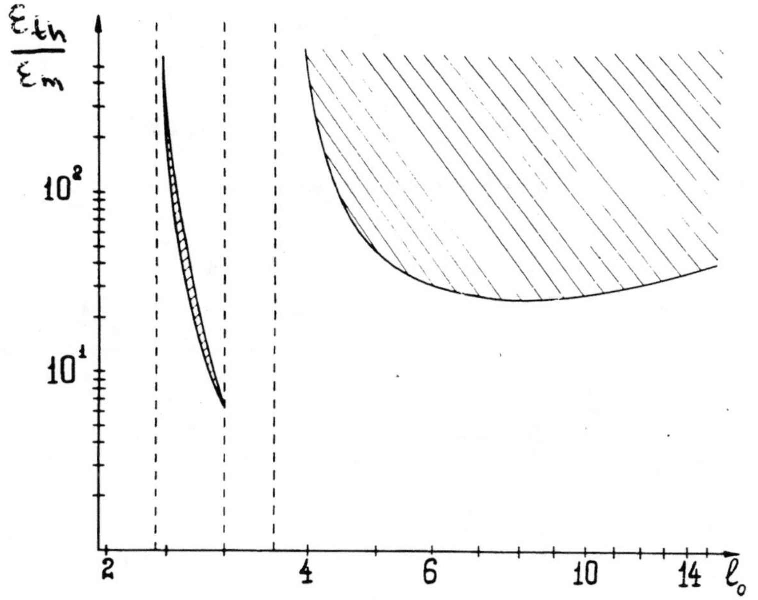

[Fig. 3.14. The dependence of e on the mass ratio of the interacting atoms. The dashed areas indicate the values at which condition (3.62) is obeyed; is the minimum energy necessary for the first collision to take place. The lower curve in the first subregion is specified by the condition of existing all collisions involved.

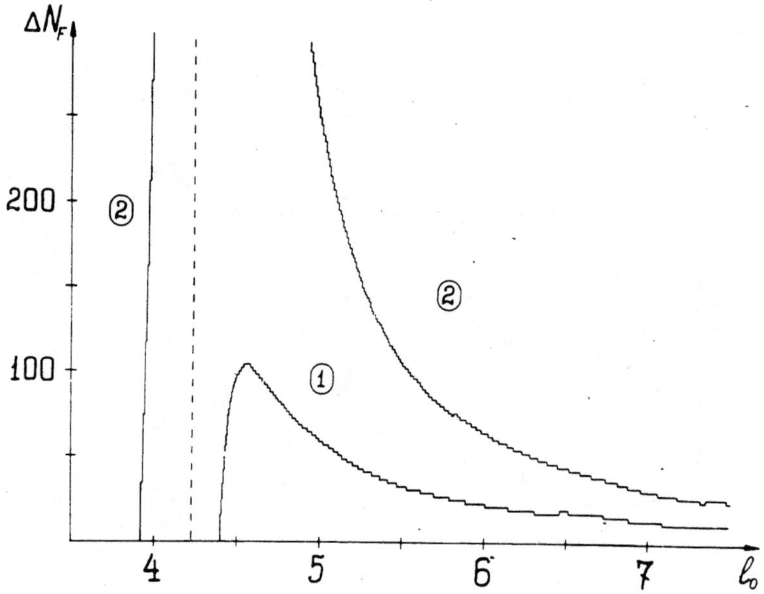

[Fig. 3.16. The arrangement of Fe (1) and Al (2) atoms along the direction <100> in the Fe\ Al crystal .

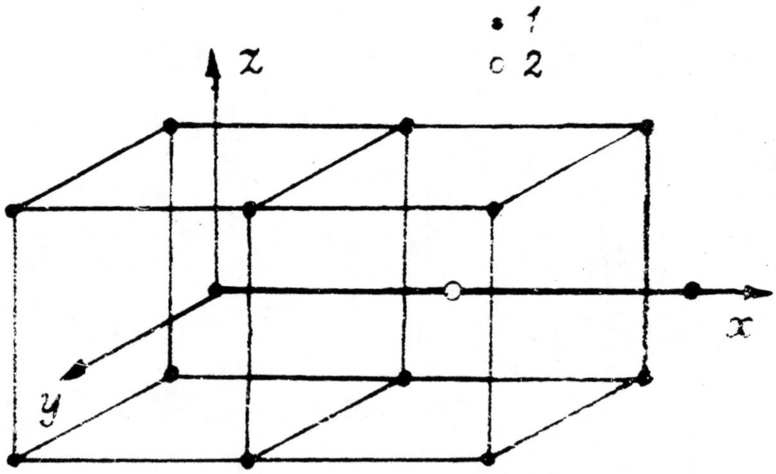

[Fig. 3.17. The dependence of the substituting collision length N along the direction <100> in the crystal FeAl on the energy of the incoming Fe atom.

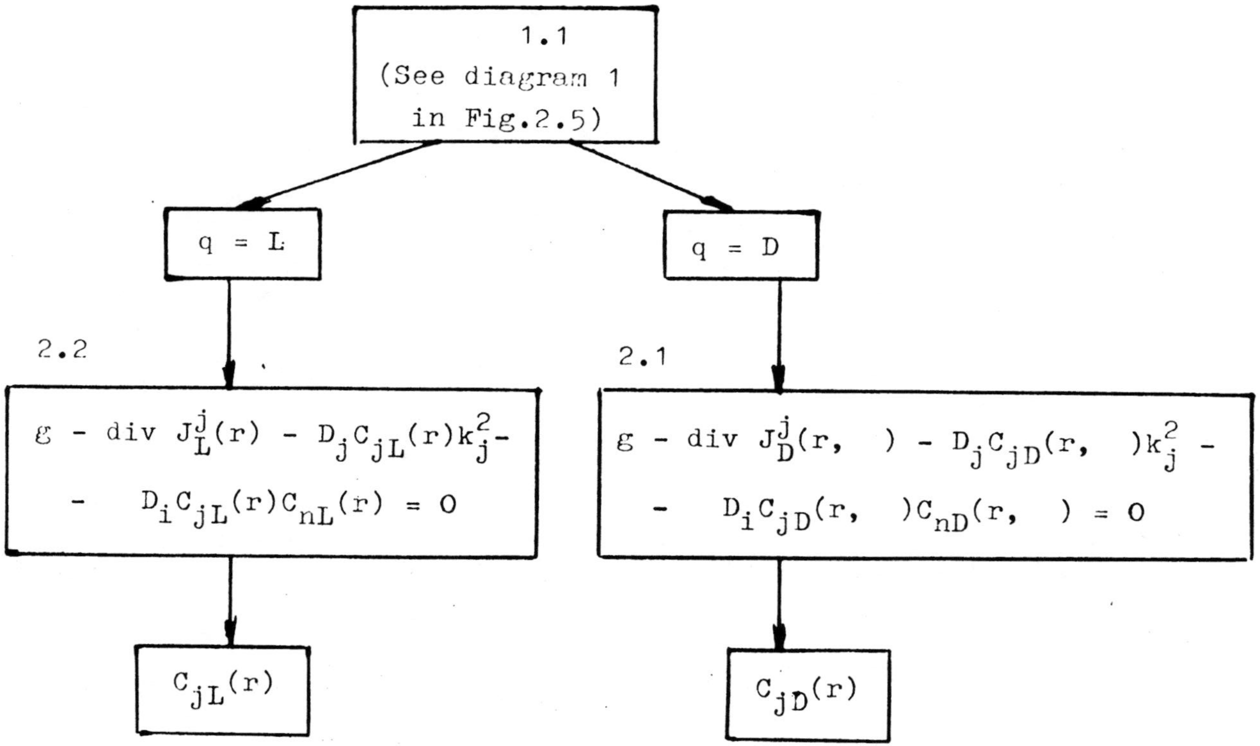

[Fig. 4.1. Diagram of solving the balance equations nearby the edge dislocations (q = D) and the small-sized interstitial loops (q = L) (Diagram 2) (a continued Diagram I-1.1 in Fig. 2.15).

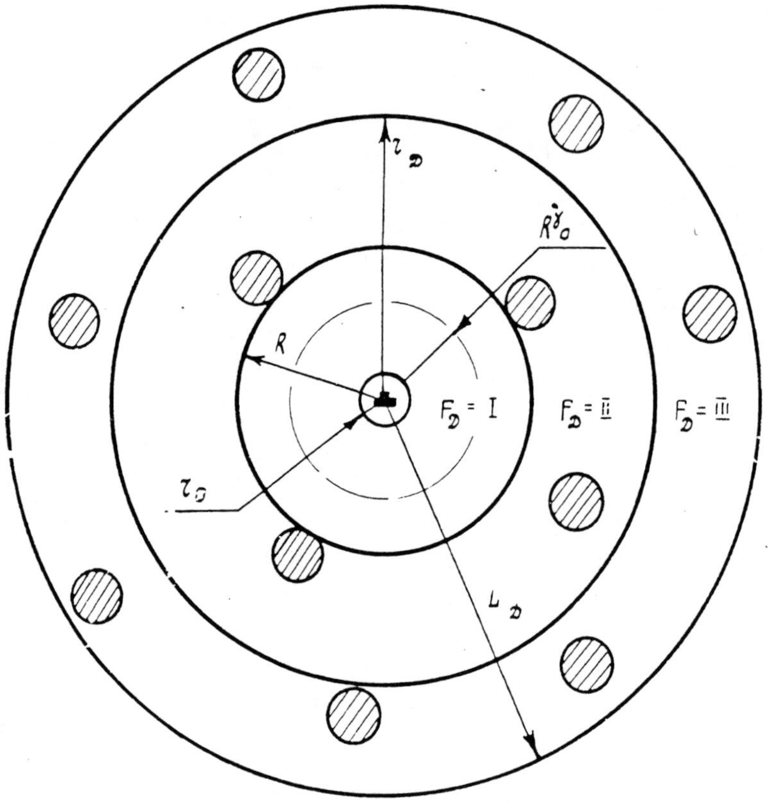

[Fig. 4.2. Diagram of dividing the solid solution volume nearby an edge dislocation into coaxially cylindrical regions F = I,II,III: D [- coherent precipitates of the secondary phase; [r - dislocation core radius, R   is defined in (1.35), R , r, L   are the dimensions of regions I, II, III, DD respectively.

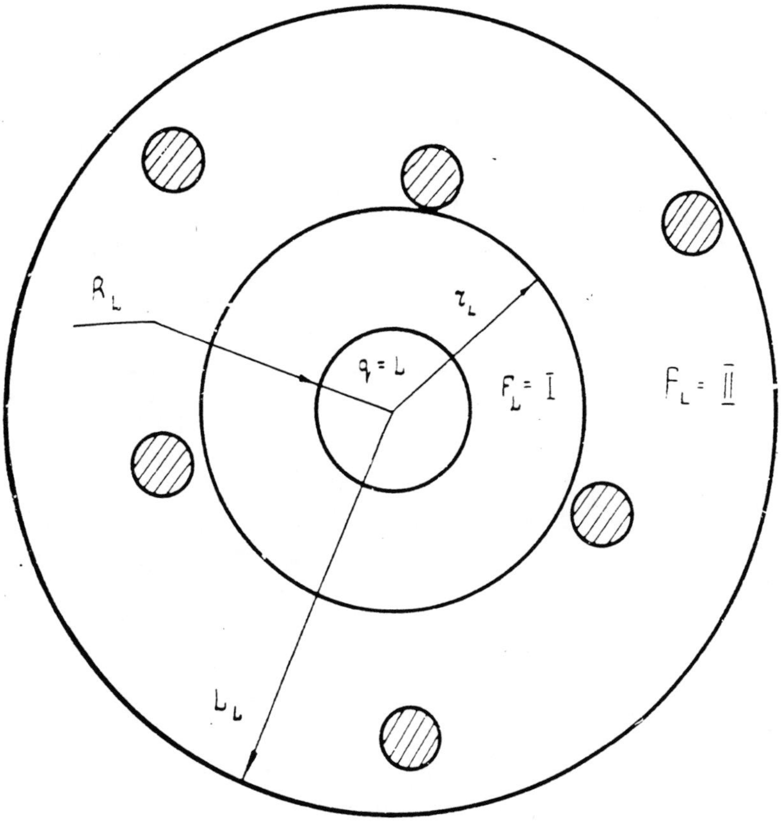

[Fig. 4.3. Diagram of dividing the solid solution volume bordered on the dislocation small-range loop of a small radius R into the regions F = I, II:
[8 - spherical sinks; [r , L - the radii of regions I and II, respectively.

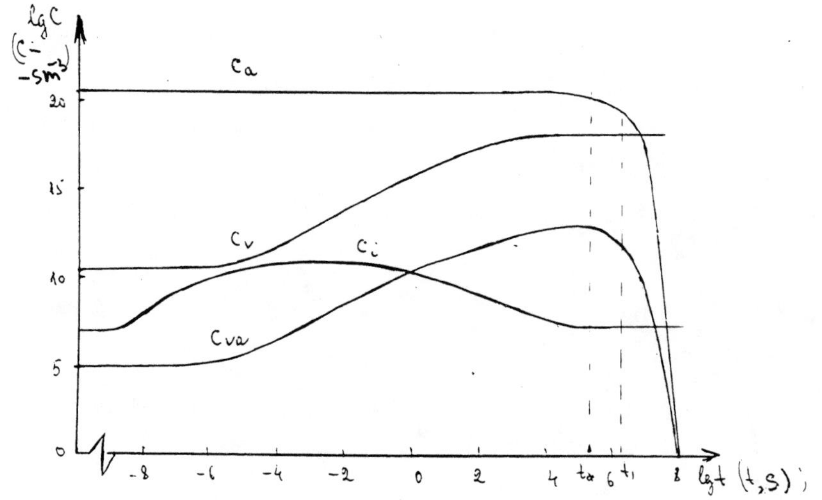

[Fig. 5.2. The time-dependence of the point defect concentrations for a modeling Ni - based alloy (see Table 5.1):is that of impurity atoms, Ca that of interstitials, Cv that of vacancies, Ci that of the vacancy-impurity complexes.

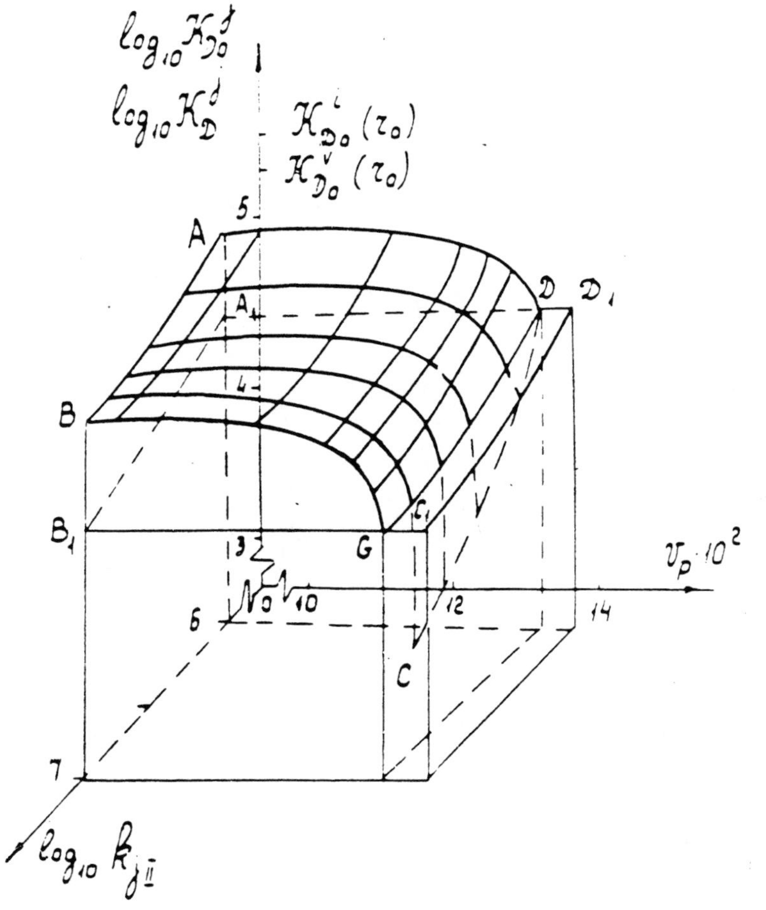

[Fig. 6.1. The dependencies of the efficiencies of point-defect adsorption by an edge dislocation K (r) as functions of the volume precipitate fraction v and the sink strength sum k , which have been calculated by (6.19) (e = - e = 0.05, II I T = 600 K). K (r) is related to the ABCD surface, K (r) to the A B C D surface. The values K (r) have been calculated by `1 1 1 *(6.20) for a pure crystal (E = 0) in the absence of sinks (k = 0) in the regions between the dislocations (L = r = R).

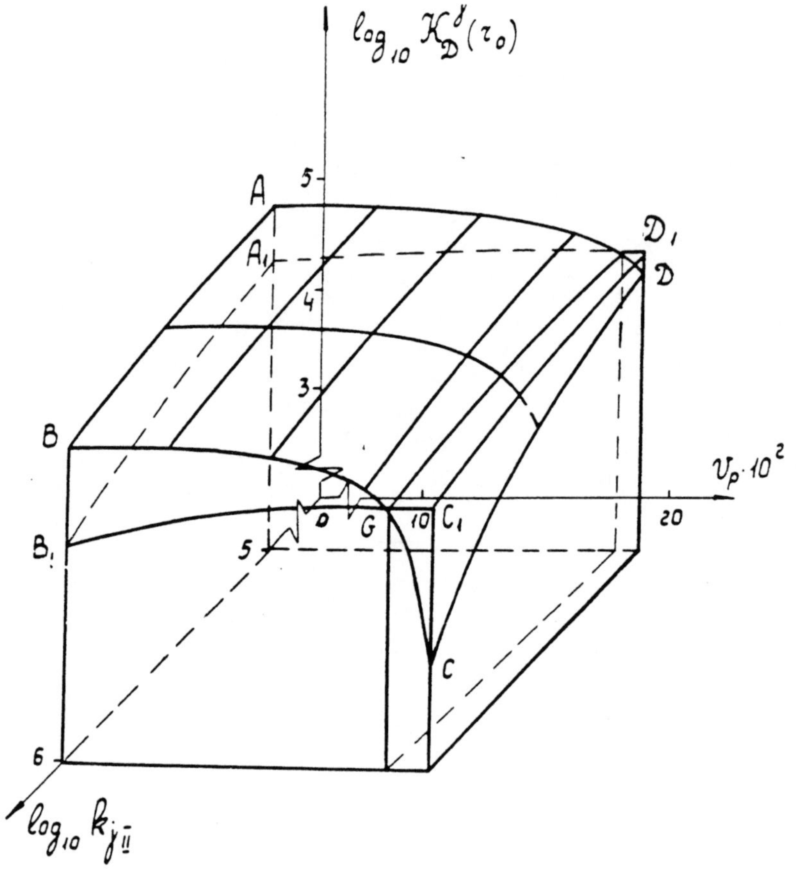

[Fig. 6.2. The dependencies of the efficiencies of point- B defect adsorption by an edge dislocation K (r) as functions of the volume precipitate fraction v and the sink strength sum k , which have been calculated by (6.19) at e = - e = - 0.01 II I and T = 600 K: [K (r) is related to the ABCD surface, [K (r) to the ABCD surface.

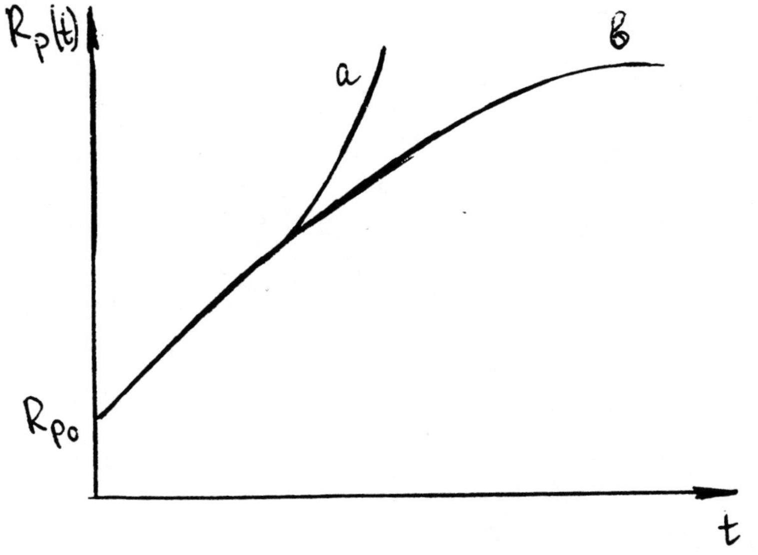

[Fig. 7.1. The dependence of the precipitate radius $R(t)$ on the time t [183] for: [(a) isolated spherically symmetric precipitates, [(b) nonisolated ones.

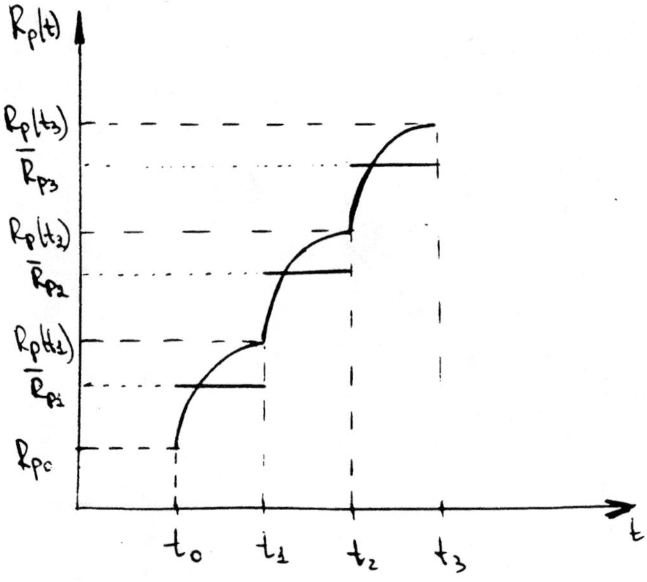

[Fig. 7.2. A qualitative dependence of changing the average precipitate radius R on the time t .

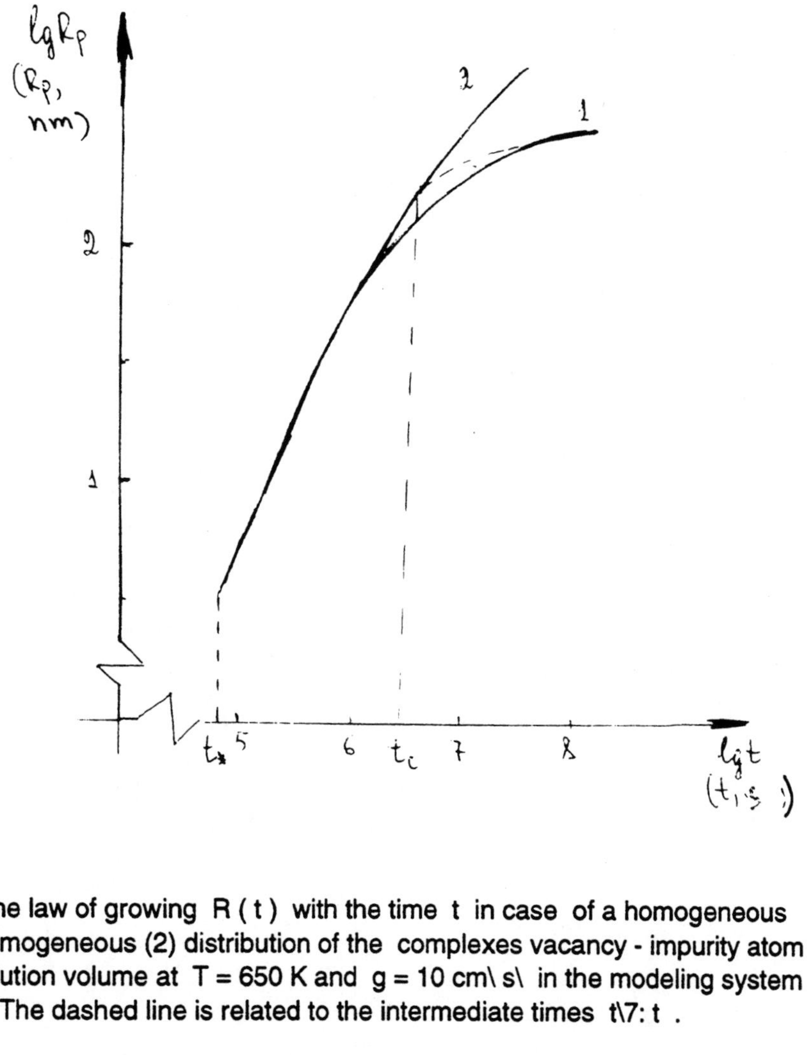

[Fig. 7.3. The law of growing R (t) with the time t in case of a homogeneous (1) and inhomogeneous (2) distribution of the complexes vacancy - impurity atom in the solid-solution volume at T = 650 K and g = 10 cm\ s\ in the modeling system (see Table 5.1). The dashed line is related to the intermediate times t\7: t .

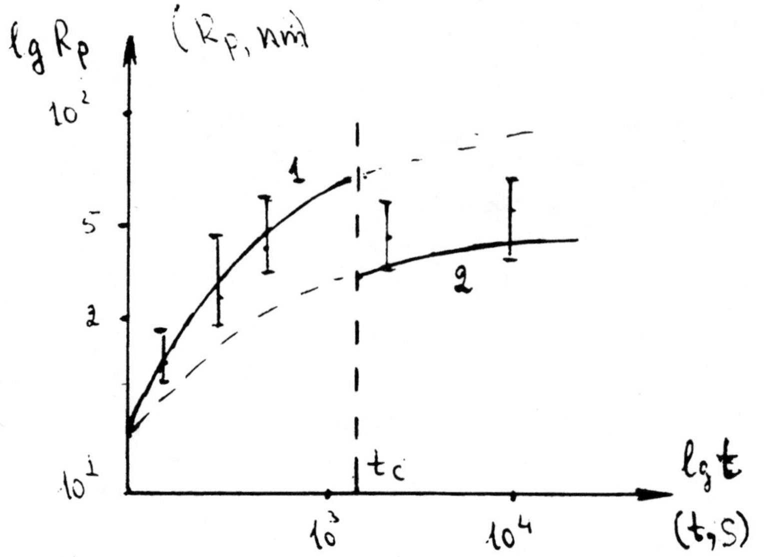

[Fig. 7.4. The dependence of the precipitate radius R (t) on the time t in the maximum depletion of the solid solution Cu-Be at T = 583 K and g ~ 4 V 10\ cm\ s\ (F = 2.9 V 10\ el.cm\ s\ ). The solid lines: (1) calculated by (7.16); (2) calculated by (7.13). The points correspond to the experimental data [405].

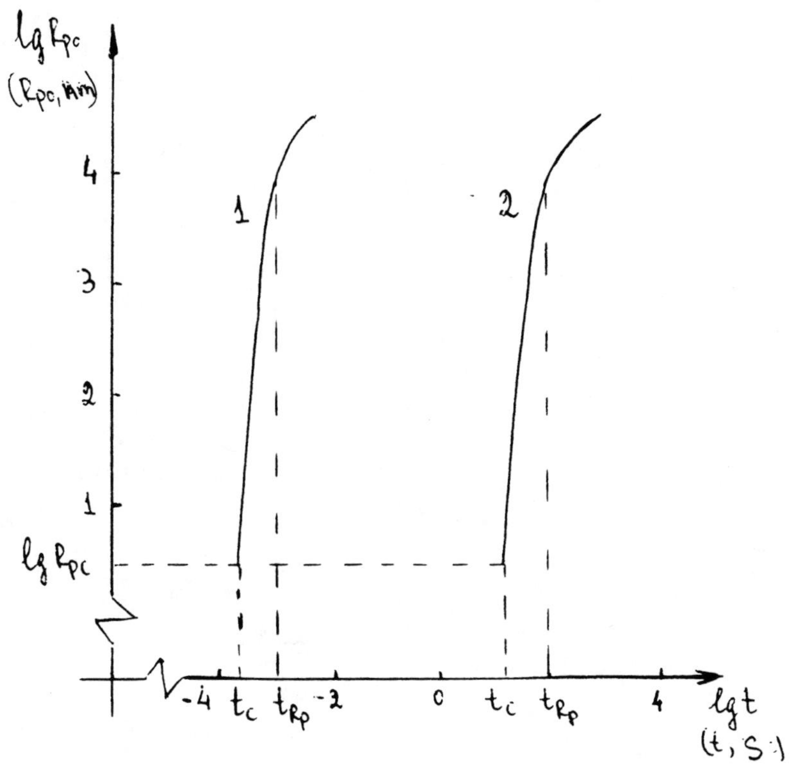

[Fig. 7.5. The dependence of the precipitate radius R (t) on the time t in the solid solution Al-Zn. Curve 1 is related to T =600 K; curve 2 to T =300 K.

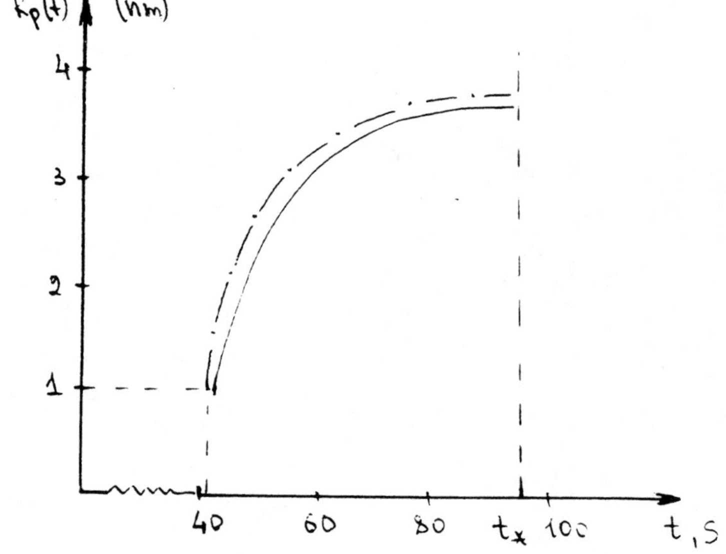

[Fig. 7.6. Comparison of the dependence R (t) in the fast depletion of the solid solution Zn-Al, which has been calculated by (7.19) - V)-) with the experiment [289]

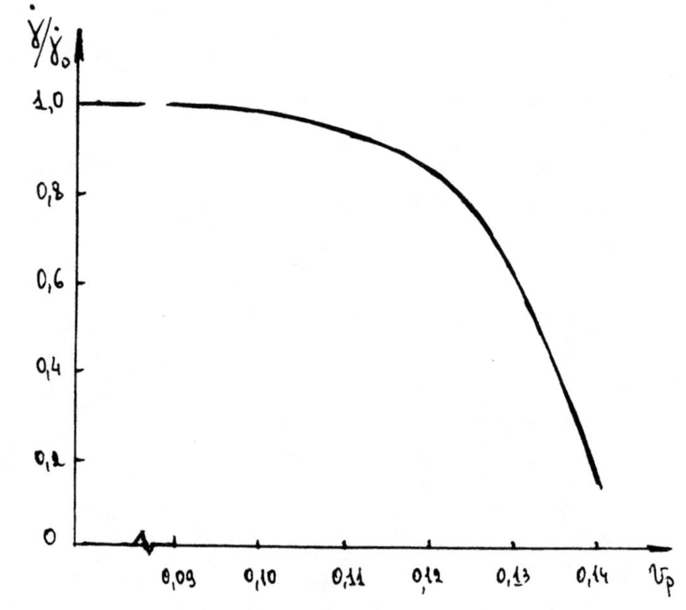

[Fig. 8.1. The dependence of the ratio between the radiation swelling rates according to (8.52) on a volume fraction of coherent precipit ates for the Ni-based material (see Chapter 5, Table 5.1).

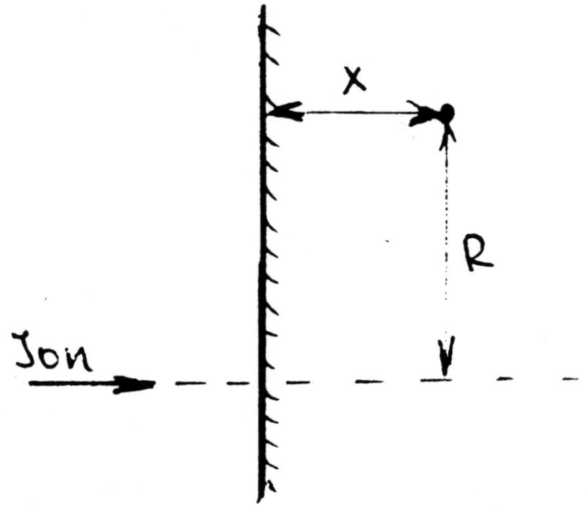

[Fig. 9.1. Diagram of positioning the diffusate in ionic etching.

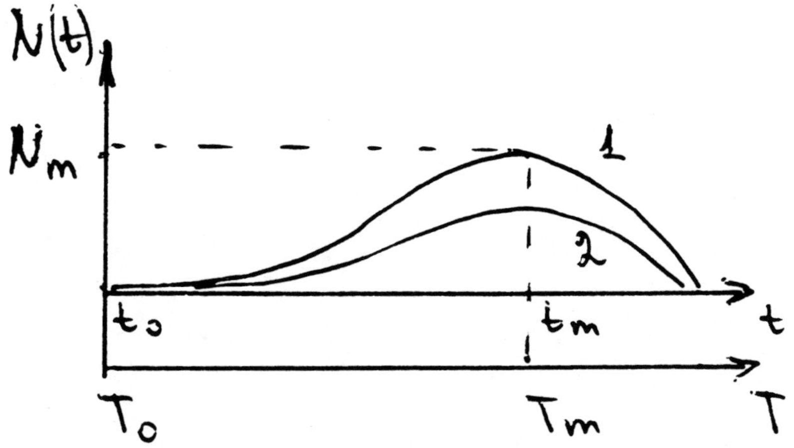

[Fig. 9.2. A qualitative dependence of the number of the interstitials outgoing to the sample surface in linear heating (according to the number of points of elevated brightness on the autoionic image): (1) without internal sinks, (2) with internal sinks.